EXS 71

Toward a Molecular Basis of Alcohol Use and Abuse

Edited by B. Jansson
H. Jörnvall
U. Rydberg
L. Terenius
B.L. Vallee

Birkhäuser Verlag
Basel · Boston · Berlin

Editors

Prof. B. Jansson
Department of Psychiatry
Karolinska Institutet
Huddinge University Hospital
S-141 86 Huddinge
Sweden

Prof. U. Rydberg
Prof. L. Terenius
Department of Clinical Neuroscience
Karolinska Institutet
S-171 76 Stockholm
Sweden

Prof. H. Jörnvall
Department of Medical Biochemistry
and Biophysics
Karolinska Institutet
S-171 77 Stockholm
Sweden

Prof. B.L. Vallee
Center for Biochemical and
Biophysical Sciences and Medicine
Harvard Medical School
Seeley G. Mudd Building
250 Longwood Avenue
Boston, MA 02115
USA

Library of Congress Cataloging-in-Publication Data

Deutsche Bibliothek Cataloging-in-Publication Data
Toward a molecular basic of alcohol use and abuse / ed. by
B. Jansson ... – Basel ; Boston ; Berlin : Birkhäuser, 1994
   (Experientia : Supplementum ; 71)
   ISBN 3-7643-2940-8 (Basel ...)
   ISBN 0-8176-2940-8 (Boston)
NE: Jansson, Bengt [Hrsg.]

© 1994 Birkhäuser Verlag, PO Box 133, CH-4010 Basel, Switzerland
Camera-ready copy prepared by the authors
Printed on acid-free paper produced from chlorine-free pulp
Printed in Germany

ISBN 3-7643-2940-8
ISBN 0-8176-2940-8

9 8 7 6 5 4 3 2 1

Contents

VI

Section 2: Biological markers and special clinical features

Section 3: Enzymatic aspects

Preface

Since the dawn of Western civilization, ethanol consumption has profoundly affected human affairs. In the course of history, the consequences of ethanol use and abuse have focused attention variously on its nutritional, psychological, medical, sociological, toxicological, philosophical, legal and religious implications. In this century, public attention has been directed progressively to the chronic medical consequences including morbidity and mortality, and the resultant economic burdens. There have often been hopes of decisive preventive or therapeutic action. However, these hopes have been thwarted by the meager knowledge of the biology and pathology including their underlying mechanisms of chronic ethanol abuse. This frustration has been intensified by the increasing realization that ethanol abuse now constitutes one of the major lethal human afflictions but not enough is known about the mechanisms involved. Ethanol related problems are multifactorial in character and this poses both investigative and intellectual challenges. Interdisciplinary barriers of language and knowledge have to be removed or overcome and dialogue established. The best approach, perhaps, is to develop, share and integrate pertinent information from the fields of molecular and cell biology. Advances in these areas have been quite remarkable in recent years and yet they still await translation into pharmacological progress. All the evidence suggests that we are on the verge of uncovering rewarding possiblities.

Such considerations led us to convene a meeting of scientists working at the frontiers of molecular and clinical sciences pertinent to the ethanol field. With the support of the Nobel Foundation, a symposium was held at the Nobel-Forum Conference Center of the Karolinska Institute in Stockholm in the fall of 1993. The participants were asked to emphasize conceptual features rather than experimental details of their work in order to stimulate future collaboration spanning conventional boundaries. Four such major sections now define the organization of the 39 chapters of this volume. Section 1 presents neuropharmacological aspects of ethanol consumption that include receptors, adaptive responses, signal transduction and the resultant reward systems. Section 2 deals with biological markers as well as with molecular genetics and clinical features in studies of families, twins, adoptions, and high risk groups. Section 3 cites

X

advances in the knowledge of alcohol metabolism, including the structure, function and polymorphism of relevant enzymes. The final section, devoted to clinical aspects and pharmacological approaches, features both generalized aspects of clinical treatments and inhibitors of unconventional approaches. The influences and interactions of cultural, social and other environmental factors with those of personality and behavior are not covered explicitly. It is hoped, nevertheless, that the present description of the actions of ethanol will be of interest to those concerned with the making of government policy and with the sociological aspects of ethanol use.

We are grateful to the Nobel Foundation for its support in providing the impetus to discern correlations between molecular events and clinical disease states. In our view, these pages provide both an introduction to current trends and an indication of more intensive studies yet to come on the molecular basis of the use and abuse of ethanol.

Finally, we gratefully acknowledge the skilful secretarial assistance and preparation of the manuscripts by Ann-Margreth Gustavsson.

Bengt Jansson, Hans Jörnvall, Ulf Rydberg, Lars Terenius and Bert Vallee

Karolinska Institutet and Harvard Medical School
Stockholm and Boston, January 1994

Toward a Molecular Basis of Alcohol Use and Abuse
ed. by B. Jansson, H. Jörnvall, U. Rydberg, L. Terenius & B. L. Vallee
© 1994 Birkhäuser Verlag Basel/Switzerland

Alcohol in human history

Bert L. Vallee

Center for Biochemical and Biophysical Sciences and Medicine, Harvard Medical School, 250 Longwood Avenue, Boston, Massachusetts 02115 USA

Keynote: *"Man's supposed love affair with alcohol turns out to have been less of a passionate infatuation than a marriage of convenience and necessity. "*

Summary

The role of ethanol in the history of human development is here summarized under seven topics: I. Alcohol: the substitute for water as the major human beverage; II. Alcohol as a component of the diet and source of calories; III. Alcohol, concentration by distillation; IV. The Reformation, Temperance and Prohibition; V. Potable non-alcoholic beverages: Boiled water (coffee, tea); VI. Purification and sanitation of water; VII. The present and future.

I.   Alcohol as the substitute for water as the major human beverage

Until the 19th century human beings in Western society considered water unsuitable for consumption. The very earliest historic societies, whether Egyptian, Babylonian, Hebrew, Assyrian, Greek or Roman, unanimously rejected water as a beverage. Through the ages water was known to cause acute and chronic but deadly illnesses and to be poisonous and therefore was avoided, particularly when brackish. The Old and New Testaments are virtually devoid of references to water as a common beverage as is the Greek literature, excepting some positive statements regarding the quality of water from mountain springs (Marcuse, 1899; Glen W. Bowersock, personal communication). Bacteriological insight identifying waterborne infections would not occur until the 19th century. Water purification methods were unknown, and in the West - in contrast to the East - the benefits of boiling water (thereby destroying bacteria and other parasites) were neither appreciated nor attributed to that process.

For nearly 10,000 years of known Western history, beer and wine, *not* water, were the major daily thirst quenchers, consumed by all ages. This simple statement does not make an impact commensurate with the consequences of this reality. Seafaring nations and explorers, e.g., knew that drinking water became putrid and fetid quickly. The consumption of milk was not common and was generally considered to be "barbaric". As a component of beer and wine, alcohol was consumed in moderate amounts with few if any adverse individual or social consequences. Most likely the alcohol content of those beverages was then so low that it produced few ill effects and those effects became the center of serious attention and concern only much later. The results of excessive consumption were well known, of course, and were the subject of much debate among philosophers, e.g., Plato, Socrates and Xenophon (O'Brien, 1980). But their primary objective was to balance the inspirational, socializing and tranquilizing effects of alcohol with the lack of judgment, anger and even violence resulting from its abuse. The consequences of excessive consumption and consequent intoxication by beer and wine were largely ignored and considered insignificant relative to their resulting benefits. It must not be forgotten that the "side" effects of beer and wine actually generated *most people's* "normal" state of mind, brought about by a constant low level of alcohol intake, i.e., a mild sense of well-being, counteracting both the prevalent state of fatigue and boredom, and alleviating wide-spread pains of all kinds for which remedies were unknown.

Without doubt, in both beer and wine the alcohol was accompanied by relatively large amounts of acetic and other organic acids. This acidity most likely contributed to the sterilizing effects on the water that was used to dilute beer and wine, often on a two to one basis.

II.   Alcohol as a dietary component and source of calories

Many historians have overlooked the fact that alcohol was a major constituent of the human diet essential to survival not only to replace and maintain the balance of liquids (see above), but also to provide calories. The common diet of ancient civilizations was cereal-based and, hence, composed of carbohydrates providing 4 kcal/gm directly and 7 kcal/gm carbohydrate indirectly when fermented to alcohol, which accounted for a significant percentage of the daily energy intake. In addition to alcohol and carbohydrate, beer also provided essential food additives such as vitamins and minerals. The caloric value and the liquid volume of the diet of the ancients guaranteed survival. The nutritional value of beer (or wine) was on par with the essentiality of bread and indispensable to the support of life. The ancients' expression "Bread and Beer" may have been akin to the present-day reference to "Bread and Butter" (Darby *et al.*, 1976).

III. Alcohol: Purification and concentration by distillation

The discovery of distillation brought about the first major change in the mode and extent of human alcohol consumption in the 9000 years since brewing and viticulture were introduced initially. Most importantly perhaps it marked the transition from the consumption of beer and wine as nutrients to the consumption of alcohol in amounts sufficient to be harmful and calling attention to the downside of the use of alcohol and its abuse. Thus, distillation provided the true benchmark of alcohol consumption and its negative effects.

At the end of the Middle Ages the return of cities as centers of civilization, the expansion of trade and economies and the reform of religious and cultural life, political and legal regulations generated new social systems, including those due to the expansion of European hegemony in the New World, Africa and Asia. The expansion of distillation and consequent condensation and concentration of alcohol into much more potent spirits of low volume coined novel social patterns and subsequent new alliances between peasants, merchants, clergy, aristocrats and other members of the ruling classes.

Distillation, concentrated and isolated ethanol, resulting in the preparation of beverages with known alcohol content and potency and standardizing both dosage and effects. Alcohol was praised as the best medium for the preparation of pharmaceutical distillates and considered a panacea in itself. However, wider interest in the process did not take hold until the mid-13th century.

In assessing the role of alcohol and alcoholic beverages during the Dark Ages the recurrences of Pestilence and the Black Death or Plague must not be forgotten. It was *this* set of medical events, including war, that overwhelmed and overshadowed all other considerations during the Middle Ages. No remedies of any kind were known, and those that were tried proved ineffective against the events that decimated the populations of Europe by as much as 2/3 in one generation, causing devastation, grief and hardship of almost unimaginable magnitude. The extension of drinking spirits seems to have followed very closely in the wake of the Black Death of 1348-1349. Since aqua vitae gave a temporary feeling of warmth and well-being and was believed to have "magic" powers, it was prescribed by physicians in cases of Plague, much as this was to no avail (Forbes, 1970; Wilson, 1973; Claudian, 1970). There ensued a period of virtual medical euphoria which declared alcohol to be a marvelous medicament for serious illness, "an emanation of the divinity possibly a key to everlasting life, an element newly revealed to man but hidden from antiquity because the human race was then too young" (Forbes, 1970). Hieronymous Brunschwygk (Brunschwygk, 1512) so popularized this viewpoint in his "Big Book of Distillation" that it was translated into many languages and became one of the most cited publications of the Middle Ages (Forbes, 1970).

IV.   The Reformation, temperance and prohibition

In spite of the remarkable political and religious upheavals of nearly 3000 years in the Western world, there was no substantial change in the consumption and reaction to alcohol during the rise and fall of the Roman Empire, the migrations of the Germanic tribes and the succeeding Dark Ages with its accompanying religious and political realignments.

Beginning in about 1450 the economic recovery increased urbanization, made available goods and generated new standards of luxury contrasting with the previous abysmal poverty and class conflicts. It initiated an age of ostentation, gluttony and inebriation. Public attitudes from then to the beginning of the 18th century positively supported drinking, much as the negative effects of drunkenness were acknowledged and governments tried to apply force and discipline to impose restrictions on the masses. Subsequent to the Dark Ages intoxication was recognized to interfere with the "new order of rationality" reflecting loss of physical and psychic control at the expense of efficiency, time and order. No doubt, the availability of distilled spirits contributed greatly to excess consumption of alcohol.

Contrary to general perceptions, it is at least of interest that the Church took no militant stand against alcohol. The advocacy of abstinence coincided with the advent of the Reformation (Hirsch, 1953). Much as this might be inferred to imply a correlation, in point of fact, there was neither any advocacy of the Church nor of leaders of the Reformation or for that matter any break between Protestantism and Catholicism on that basis.

Protestant leaders like Luther (1520) and Calvin (1540) viewed wine as one of God's creations meant for man's enjoyment and benefit. In fact, in the course of the Reformation abstinence seems to have been more characteristic of Catholics than of Protestants. Protestantism emphasized the separation of the "secular from the sacred" but relied on civil rather than church authority to affect social behavior based on moral sentiments instead of customs. The austere moral codes of the Huttites, Quakers and Anabaptists exemplified the extremes of views which posterity - not contemporaries - emphasized.

The origins of modern prohibitionism can be traced to a combination of the Anabaptists' moral code and the Calvinist reliance on civil authority (Bainton, 1945). Later on, German Pietism heavily influenced English Methodism and these together with the views of the 17th century Quakers jointly, turned out to be the real pioneers of modern temperance and prohibitionism.

At this juncture it is appropriate to emphasize that calls for temperance and even prohibition go back to Hebrew, Greek and Roman times and were reiterated throughout history. The reasons given were as varied as the political systems, social structures and religious motivations under which they occurred. Ordinances, regulations, laws, decrees and various forms and threats of punishment were as varied as the governmental or religious systems which originated them. It takes little effort to identify edicts, regulations and laws that endeavored to control the intake of alcohol since biblical, Greek and Egyptian times.

V.   Advent of boiled water, coffee and tea: potable non-alcoholic beverages

Boiling of water apparently destroyed its "poisonous" factors, a circumstance that was appreciated at least 5000 years ago in Eastern (Chinese) (Lu Yu, ~ 800 AD) but not in Western Society. This recognition was delayed in the West until the 17th century when brewing coffee and tea led to the boiling of water on a broad scale. As the consumption of coffee and tea increased, that of alcoholic beverages fell dramatically. Following the introduction of coffee in the 17th century, coffee houses and coffee drinking spread quickly throughout Britain. From 1680 to 1730, more coffee was consumed in London than in any other city in the world, and thus by the last years of the century drunkenness among the upper classes there diminished perceptibly (Jacob, 1935; Lillywhite, 1963). This development of alternative beverages began to make alcohol consumption unnecessary as a primary method of maintaining liquid balance. And, by now, it had become even less useful as a method of appropriate caloric intake.

VI.   Sanitation and purification of water

Between 1801 and 1850, the population of Europe and particularly the urban areas increased significantly. Industrialization concentrated ever increasing numbers of people in decreasing amounts of space. Thus, the population of Great Britain more than doubled and in Glasgow, Scotland it even quadrupled (Derry and Williams, 1961). The increasing density of the population and concomitant hygiene considerations brought to the fore the extremely serious problem of purifying water and particularly separating sewage from wastewater. Prior to 1900 sewage was allowed to join the general water runoff into the rivers and lakes, and since these were the source of drinking water, cholera and typhoid fever became ever more widespread.

The advent of bacteriology, beginning with Pasteur and continuing with Ebert, who in 1880 isolated the typhoid bacillus, and Koch, who in 1883 isolated that of cholera (Carr, 1966), proved that the sewage was the direct cause of epidemics due to these organisms and therefore had to be separated from the water supply. Even though John Snow had shown in 1854 that cholera was linked to the water supply in London (Carr, 1966), the greatest problem with sewage was still thought to be the smell, which pervaded the cities.

It finally remained only for the development of sanitation and purification of drinking water in Western industrialized countries before the end of the 19th century to eliminate alcohol entirely as an essential component of the human diet in those countries, whatever other uses were made of it.

VII.   The present and future

In the past, Far Eastern and Western societies differed greatly in their consumption of liquids. The former drank tea prepared from boiling water while the latter drank wine and beer. All of that has, of course, since changed. Yet, these past differences raise important new questions regarding human alcohol metabolism.

In approximately half of the Chinese and Japanese population, the metabolism of alcohol is known to differ significantly from that of virtually all Occidentals, due to genetic differences. This genetic difference gives rise to a non-functional mitochondrial aldehyde dehydrogenase, which results in the "flushing syndrome". Orientals lacking this enzyme are intolerant to ethanol consumption which makes them acutely ill. Therapeutic measures such as the use of Antabuse, which simulates the consequences of this difference, result in the avoidance of alcohol all right, but through a complex of signs and symptoms which signify that this can be dangerous and is toxic. It is not the primary function of Antabuse - and similar agents - to control the amount of alcohol ingested.

This failure to express the mitochondrial aldehyde dehydrogenase is the only specific genetic alteration known to result in the avoidance of alcohol. While the results of attempts to mimic this condition through drugs have not been encouraging, they have focused attention on possible alternative measures that have been employed in the Orient, particularly in China, to deal with problematic alcohol abuse, which only occurs in that half of the population in which there is no genetic alteration.

A role of "appetite" in alcohol consumption has not been explicitly addressed in attempting to control alcohol consumption. At this point the biochemistry of *appetite* is largely unknown both in regard to food and liquids in particular and to pleasure in general. This molecular biology must be understood if a rational therapy is to be designed. An approach to the therapy of alcohol abuse by modification of the appetite for it remains to be defined, though this aim might be achieved empirically. Such approaches, used in China for millennia, indeed account for the recognition of *Radix puerariae* in just this manner. This herb may be the first to provide a class of compounds affecting the appetite for alcohol (Keung and Vallee, 1993, 1994).

The last 500 years have witnessed major experimentation with the forms in which alcohol can be consumed in different combinations, which depend on geography, sociology, economics and medicine and on all other such factors mentioned above. Medicine, pharmacology, psychiatry and nutrition have continued to make major impacts in the last 100 years.

Medical and moral as well as religious arguments have all been advanced and tested to modify excessive consumption of alcohol but none of them, including psychological or psychiatric propositions, have made lasting and impressive impacts. In the case of addiction legal, religious and medical appeals to reason have proved ineffective in altering or regulating human conduct and morality. The history of such efforts and their lack of effectiveness discourages optimism in this regard; there is little evidence that such endeavors have been effective. The Anglo-Saxon and American experiences do not constitute exceptions. The relevant literature abounds with claims of success of such measures through the ages, although they have actually failed and have even confused the issues, not resolved them. The history highlighted here may add further insight and help design means of management of the problem based perhaps on better biological, physiological or pharmacological understanding which is being gained. However, the advocates of such rational approaches should remain mindful of Lord D'Abernon's maxim in 1918: "Those who would give any attention to scientific work on alcohol do so less to gain knowledge than to find arms and arguments to support their preconceived opinion".

Newly available biological and genetic facts could generate suitable public health measures

8

which would illuminate both the origins of and control of the use and abuse of alcohol. Clearly, this historical perspective cannot identify possible alternatives. However, it is the philosophy of this meeting to highlight both new experimental and intellectual avenues that might give novel directions to the solution of remaining unsolved problems.

References

Bainton, R. (1945) The Churches and Alcohol. *Quart. J. Stud. Alc.* 6: 45-58.

Brunschwygk, H. (1512) *Liber de arte Distillandi de Compositis* (Big Book of Distillation), Strassburg.

Carr, D.E. (1966) *Death of the Sweet Waters*. Norton, New York, pp. 37-54.

Claudian, J. (1970) History of the usage of alcohol. In: Tremoiliers, J. (ed.): *International Encyclopedia of Pharmacology and Therapeutics*. Section 20, Vol. 1, Pergamon, Oxford, pp. 3-26.

D'Abernon, L. (1918) *Alcohol: Its Action on the Human Organism*. British Medical Research Council.

Darby, W.J., Ghalioungui, P. and Grivetti, L. (1977) *Food: The Gift of Osiris*. Vols. 1 and 2. Academic Press, New York.

Derry, T.K. and Williams, T.I. (1961) *A Short History of Technology*. Oxford University Press, Oxford.

Forbes, R.J. (1970) *Short History of the Art of Distillation*. E.J. Brill, Leiden.

Hirsch, J. (1953) Historical perspectives on the problem of alcoholism. *Bull. NY Acad. Med.* 29: 961-971.

Jacob, H. (1935) *Coffee*. Translated by E. and C. Paul. Viking, New York.

Keung, W.M. and Vallee, B.L. (1993) *Proc. Natl. Acad. Sci. U.S.A.*, in press.

Keung, W.M. and Vallee, B.L. (1994) this volume.

Lender, M.E. and Martin, J.K. (1987) *Drinking in America*. The Free Press, New York.

Lillywhite, B. (1963) *London Coffee Houses: A Reference Book of Coffee Houses of the Seventeenth, Eighteenth and Nineteenth Centuries*. Allen and Unwin, London.

Lu Yu (~ 800) Ch'a Ching (The Bible of Tea) In: Hu, S.-Y. (ed.): *Gu-Jin-Cha-Shi*. Shanghai Book Store (printed 1985), Shanghai, pp. 1-23.

Marcuse, J. (1899) *Diätetik im Alterthum. Eine Historische Studie*. Verlag von Ferdinand Enke, Stuttgart, pp. 46-49.

O'Brien, J.M. (1980) Alexander and Dionysus: The invisible enemy. *Ann. Scholar.* 1: 31-46.

Wilson, C.A. (1973) *Food and Drink in Britain from the Stone Age to Recent Times*. Anchor Press, London.

Toward a Molecular Basis of Alcohol Use and Abuse
ed. by B. Jansson, H. Jörnvall, U. Rydberg, L. Terenius & B. L. Vallee
© 1994 Birkhäuser Verlag Basel/Switzerland

Reward and its control by dynorphin peptides

Lars Terenius

Department of Clinical Neuroscience, Experimental Alcohol and Drug Addiction Section, Karolinska Institute, S-171 76 Stockholm, Sweden

Summary

Reward is a strong behavioral cue. It is argued that a sense of reward must be strictly controlled in time and magnitude to be functional. Central to the reward system is the mesolimbic dopamine pathway which is titillated by addictive drugs and alcohol. Rat lines or strains that voluntarily drink alcohol have lower dynorphin levels in relevant brain areas than those which avoid alcohol. Reward control may be a protective factor against addiction.

Introduction

Pain and pleasure may be considered emotional antipodes in a superficial sense. Both are extremely important behavioral cues, pain as a sign of tissue injury will protect the individual, pleasure (or reward) acts as a strong motivational impetus. Both are important for learned behaviors and reinforcement of behavior. Pedagogically, alternate use of "the stick and the carrot" is a guiding principle. By quality, pain and pleasure (reward) are, however, different. Pain is accompanied with discomfort and anxiety regardless of intensity. Therefore, pain and pleasure cannot be considered the extremes on a single axis, rather they fall on two different axes.

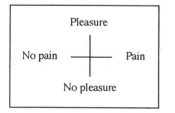

Fig. 1. Pain and pleasure (reward) are separate behavioral cues.

Acute pain is a strong cue for avoidance behavior that protects against further tissue damage. Continued pain can be considered functional only in the sense that it reduces activity and promotes the healing process. However, as pain becomes longlasting, it will lose this function and chronic pain is a devastating condition. Pathologic pain may have its origins in deficient modulation notably by opioid peptides. Such pain may be relieved by artificial stimulation of modulatory systems by acupuncture or transcutaneous electric nerve stimulation (TENS) (Han and Terenius, 1982).

Is there a parallel to excessive reward? Clearly, absence of reward or incapability to experience reward is an unwanted condition. Insensitiveness to pleasure, anhedonia is a symptom or item used to characterize mood disorders. The opposite, excessive feelings of reward, does not have a corresponding place in psychiatric diagnosis. This suggests, that the reward systems in the brain are very strictly controlled. Of course, alcohol and other drugs in recreational use, are artificial stimulants of the reward mechanisms. Physiologically, opposing mechanisms are probably triggered by the reward which explains why reward is only recognized for a short time. A continued feeling of reward would indeed be nonfunctional, as the individual would lose incentive to seek a new reward. Mechanisms that terminate the experience of reward have not been much considered in the past, probably because they have not become equally apparent pathophysiologically. It could be postulated, however, that such mechanisms would be sensitive to change and may be exhausted in drug addiction. This chapter will consider reward modulatory mechanisms, particularly as they may relate to the neuropeptide system, prodynorphin.

Paradoxically, morphine and other opiates are the strongest remedies in severe pain and among the strongest agents producing euphoria and addiction. In contrast to alcohol, the mode of action of opiates is well defined. Areas in the brain where morphine can trigger reward (see Koob et al.; Schwartz et al., this volume) are however, much more restricted than areas where morphine produces analgesia, areas which also include the spinal cord (Basbaum and Fields, 1984). Since morphine is a very frequently used analgesic, interactions between morphine as analgesic and endogenous signal substances have been studied quite extensively. In the brain, several neuropeptides have been described which have antiopioid activity.

Antiopioid peptides

The simplest mechanism for antagonism would be receptor interaction. However, nature rarely operates via receptor antagonists; instead antagonism is induced by counteractivity via separate signal systems and receptors. In the central nervous system this involves several neurons in a network. Antiopioid activity should consequently be possible to observe in a functional system, but not in a receptor preparation. In fact, no one has so far observed a natural opiate receptor antagonist, a naloxone-like compound. Using a functional in vitro system, the electrically stimulated guinea-pig ileum, Wahlström and Terenius (1980) demonstrated the existence of a peptide with functional antiopioid activity. The chemical structure remained unknown. Faris et al. (1983)

showed that the neuropeptide cholecystokinin (CCK) acted as a functional morphine antagonist. Subsequently, Tang *et al.* (1984) and Watkins *et al.* (1984) demonstrated that repeated administration of proglumide, a CCK antagonist, reversed the development of morphine tolerance. This is an important observation, since it suggests that chronic opiate treatment activates endogenous CCK to reduce the opiate effect. Indeed, Pohl *et al.* (1992) showed that CCK mRNA and CCK peptide increased in rat brain and spinal cord after chronic morphine treatment. Also importantly, it could be shown that antimorphine effects could be induced by "safety cues" i.e. when the animal is conditioned to ignore a painful stimulus. Under the tested experimental conditions, CCK was the mediator of the effect (Wiertelak *et al.* 1992).

Another peptide, neuropeptide FF has also been found to modulate the action of morphine (Yang *et al.*, 1985). Recently, it was shown that this peptide acts via mechanisms other than CCK (Magnuson *et al.*, 1990). Its physiological role is unclear. More information is available regarding the dynorphin peptides. These peptides belong to the opioid family and have been extensively studied in our laboratory.

Prodynorphin-derived opioid peptides

Opioid peptides derive from three separate protein precursors. The dynorphin peptides, dynorphin A, dynorphin B and α-neoendorphin are generated from prodynorphin by enzymatic processing. In an analogous manner, β-endorphin derives from proopiomelanocortin and a series of enkephalin peptides from proenkephalin. These peptides are defined as opioid since they act on opioid receptors, which have been defined pharmacologically in three major subgroups, μ which is the classic morphine receptor, κ which was defined by the somewhat different action profile of some ketocyclazocines (Martin *et al.*, 1976) and finally δ defined by its selectivity for enkephalin peptides (Lord *et al.*, 1977). All receptors are characterized by the fact that naloxone acts as an antagonist, although higher doses are required to antagonize actions at the κ- or δ-receptors than at the μ-receptor. With the recent cloning of opioid receptors, it is clear that they are all related structurally and form a separate subfamily with the closest relative in the somatostatin receptor (Evans *et al.*, 1992; Kieffer *et al.*, 1992; Chen *et al.*, 1993; Yasuda *et al.*, 1993). They are all members of the 7-transmembrane family, coupling via G proteins to a second messenger (Table I) and causing membrane hyperpolarization. The receptors are differently distributed with some overlaps of μ- and δ-receptors and a rather different distribution of κ-receptors. This may explain why μ- and δ-agonists are readily self-administered and cause positive reinforcement, whereas κ-agonists are behaviorally neutral or slightly aversive (Mucha and Herz, 1985). The different opioid peptides differ in receptor selectivity (Table I) and the dynorphin peptides are quite selective for κ-receptors. In line with this observation, dynorphin is not self-administered.

Table I. Opioid receptors and their selectivity for opioid peptides. Inhibition of cAMP affects K$^+$-channels.

Receptor	Effector pathway	Naloxone sensitivity	Preferred peptide ligand
μ	cAMP↓	high	(β-endorphin)*
δ	cAMP↓	low	Leu- and Met-enkephalin
κ	Ca^{2+}-channel	intermediate-low	dynorphin A and B

* β-endorphin is unselective and has considerable affinity also for δ- and κ-receptors.

Dynorphin peptides as reward modulators

The pharmacologic actions of dynorphin peptides have been studied in different experimental systems. Although dynorphins act as analgesics, if introduced close to the spinal cord (Herman and Goldstein, 1985), they are not analgesic after injection into the brain. Instead, in the brain, they modulate (reduce) the analgesic actions of systemic morphine (Friedman et al., 1981) and very interestingly potentiate opiate effects in animals made morphine tolerant (Tulunay et al., 1981). In the opiate tolerant animal, dynorphin peptides attenuate withdrawal signs (Green and Lee, 1988; Takemori et al., 1993). The mechanism for the interaction is not known. However, dynorphin is known to interact with both the nigrostriatal dopamine systems and the mesolimbic dopamine system. Since only the latter is considered central in the reward mechanisms (see Koob et al.; Schwartz et al., this volume) it will be discussed here.

The mesolimbic dopamine pathway with cell bodies in the ventral tegmental area (VTA) and terminals in *nucleus accumbens* is known to be a target for addictive drugs (Di Chiara and Imperato,1988). Morphine is a strong reinforcer in the *n. accumbens* (Olds, 1982) and in the VTA (Bozarth and Wise, 1981). If morphine is microinjected into the VTA, dopamine release in the *n. accumbens* is increased. Dynorphin injected into the *n. accumbens* has the opposite effect. This led Spanagel et al. (1992) to propose a model where morphine acts via the μ-receptor on GABA-interneurons in the VTA to stimulate dopamine release and dynorphin acts on κ-receptors located presynaptically on the dopamine nerve terminals in the *n. accumbens* to inhibit the release (Fig. 2). This model generates an hypothesis for the interaction of dynorphins with the reward system.

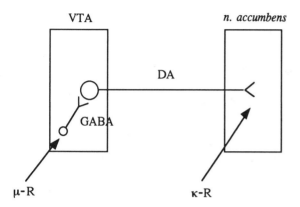

Fig. 2. Opioid influences on the mesolimbic dopamine pathway, believed to be central in the reward system. In the dopamine (DA) cell body area in the ventral tegmental area (VTA) enkephalins interact with μ-receptors on an inhibitory GABA interneuron causing increase of DA activity, whereas in the *n. accumbens*, dynorphin peptides interact with κ-receptors and attenuate DA release. Modified from Spanagel *et al.* (1992).

Testing the hypothesis

To test the hypothesis that dynorphin peptides are reward modulators, several approaches would be possible. We have chosen to compare different rat lines or strains that differ in their propensity to become drug dependent. Two models have been tested, rats that have been selected for voluntary ethanol drinking and genetically inbred rat strains that differ in susceptibility to become opiate dependent.

In experiments to be reviewed here, two rat lines have been compared. These lines were generated by selective outbreeding on the basis of alcohol preference: a line with high alcohol preference (AA) and a line with low preference (ANA). Only rats of the AA strain will drink alcohol spontaneously and voluntarily (Eriksson, 1968). They also differ in other behavioral aspects (Sinclair *et al.*, 1989). Animals of each line (n = 10) had a free choice between tap water and 10% alcohol solution (unsweetened). During the experiment which lasted four weeks, the average alcohol intake in ANA rats was 0.39 ± 0.12 g/kg/day and in AA rats 6.69 ± 0.57 g/kg/day (p < 0.001). After the 4-week period of free choice between alcohol and water, animals were decapitated and analyzed for the content of opioid peptides by radioimmunoassay. Selected data are shown in Table II. In the VTA area enkephalin peptides are rather low. Alcohol-drinking AA rats (but not controls) have a higher level of the peptide marker than ANA rats. In the *n. accumbens* enkephalin peptides are lower in AA control animals than in ANA controls but increase significantly after alcohol intake. Dynorphin peptides in the VTA show no consistent trend. In the *n. accumbens*, however, AA controls have significantly lower levels, particularly of dynorphin B. Alcohol drinking increases levels of dynorphin peptides although dynorphin B is still lower than in ANA rats.

Table II. Effect of voluntary alcohol (alc) drinking in AA rats on opioid peptides in the ventral tegmental area (VTA) and the *n. accumbens* (N. acc.). Non-drinking ANA rats are included for comparison. Met-enkephalin-ArgPhe was measured as an index of proenkephalin, dynorphin A and B as index of prodynorphin activity (Nylander *et al.*, in preparation).

Rat lines	Brain area	Peptide (pmol/g tissue)		
		Met-enk-ArgPhe	Dyn A	Dyn B
ANA	VTA	1.40 ± 0.11	1.6 ± 0.2	1.62 ± 0.37
AA		1.80 ± 0.12	2.2 ± 0.2	1.81 ± 0.22
AAalc		$2.01 \pm 0.19**$	2.6 ± 0.5	1.83 ± 0.30
ANA	N.acc.	35.9 ± 1.8	10.2 ± 0.7	23.2 ± 1.7
AA		$23.7 \pm 2.9*$	$7.4 \pm 1.0*$	$13.6 \pm 1.7***$
AAalc		33.7 ± 2.6	8.4 ± 0.8	$17.9 \pm 1.5*$

* $p < 0.05$; **$p < 0.01$; *** $p < 0.001$ vs ANA. Values for ANA rats offered alcohol are not shown - they were not significantly different from ANA controls.

Alcohol drinking in the AA rats is moderate and the animals never seem intoxicated (Eriksson, 1969). As judged from the change in enkephalin levels, alcohol seems to activate the dopamine pathway via μ-receptors in the VTA, although rather moderately. Differences between the rat lines in relation to dynorphin peptide activity may be larger. The alcohol non-preferring ANA line has clearly higher levels of dynorphin peptides in the *n. accumbens*, suggestive of a stronger modulatory role. It could be speculated that this excess in dynorphin would modulate the consequences of alcohol intake. Experiments where ANA rats were pair-fed alcohol at the rate taken by AA rats voluntarily, would further test the hypothesis.

Inbred strains of rats, which differ in the ease they become morphine tolerant and dependent, also differ with regard to dynorphin peptide levels in mesolimbic areas. A resistent strain has higher levels in the *n. accumbens* (Nylander *et al.*, in preparation). However, this resistence may weaken with chronic treatment. An enzyme fragmenting dynorphin peptides, dynorphin-converting enzyme, is up-regulated in the morphine tolerant rat (Persson *et al.*, 1991).

Outlook

This communication has pointed to the obvious fact that bodily functions always operate inter-actively to create an equilibrium, "le milieu interieur" as described by Claude Bernard. This equilibrium should not be considered static, rather it is dynamic. It has been argued that reward in the physiological sense must be subject to modulation and excesses strictly controlled. Repeated or chronic use of recreational drugs will interfere with these mechanisms. As already indicated they may not be able to respond to sustained activity.

Disposition to become addicted to alcohol or other drugs may involve a large number of factors. In the simplified scheme presented, reward mechanisms or reward modulatory mechanisms have been considered. Promising therapeutic results with the opiate antagonist naltrexone in chronic alcoholism (see O´Brien, this volume) suggest that the reward mechanisms may still be functioning in the addict. Interestingly, patients taking the drug not only have a lower relapse rate, but also if they "slip" into alcohol use, consequences are much less severe. Maybe, and this is of course speculation, these alcoholics have inadequate endogenous modulation of the reward system. If dynorphin peptides are at the center of a modulatory network, inhibitors of dynorphin degradation or synthetic agents acting at the relevant receptors could be therapeutically useful.

Acknowledgements

This work is supported by the Swedish Medical Research Council (3766) and the National Institute on Drug Addiction (NIDA), Washington, D.C.

References

Basbaum, A.I. and Fields, H.L. (1984) Endogenous pain control systems: Brainstem spinal pathways and endorphin circuitry. *Ann. Rev. Neurosci.* 7: 309-338.

Bozarth, M.A. and Wise, R.A. (1981) Intracranial self-administration of morphine into the ventral tegmental area in rats. *Life Sci.* 28: 551-555.

Chen, Y., Mestek, A., Liu, J., Hurley, J.A. and Yu, L. (1993) Molecular cloning and functional expression of a μ-opioid receptor from rat brain. *Mol. Pharmacol.* 44: 8-12.

Di Chiara, G. and Imperato, A. (1988) Drugs abused by humans preferentially increase synaptic dopamine concentrations in the mesolimbic system of freely moving rats. *Proc. Natl. Acad. Sci. USA* 85: 5274-5278.

Eriksson, K. (1968) Genetic selection for voluntary alcohol consumption in the albino rat. *Science* 159: 739-741.

Evans, C.J., Keith Jr., D.E., Morrison, H., Magendzo, K. and Edwards, R.H. (1992) Cloning of a delta opioid receptor by functional expression. *Science* 258: 1952-1955.

Faris, P.L., Komisaruk, B.R., Watkins, L.R. and Mayer, D.J. (1983) Evidence for the neuropeptide cholecystokinin as an antagonist of opiate analgesia. *Science* 219: 310-312.

Friedman, H.J., Jen, M.F., Chang, J.K., Lee, N.M. and Loh, H.H. (1981) Dynorphin: a possible modulatory peptide on morphine or β-endorphin analgesia in the mouse. *Eur. J. Pharmacol.* 69: 351-360.

Green, P.G. and Lee, N.M. (1988) Dynorphin A-(1-13) attenuates withdrawal in morphine-dependent rats: effect of route of administration. *Eur. J. Pharmacol.* 145: 267-272.

Han, J.S. and Terenius, L. (1982) Neurochemical basis of acupuncture analgesia. *Ann. Rev. Pharmacol. Toxicol.* 22: 193-220.

Herman, B.H. and Goldstein, A. (1985) Antinociception and paralysis induced by intrathecal dynorphin-A. *J. Pharmacol. Exp. Ther.* 232: 27-32.

Kieffer, B.L., Befort, K., Gaveriaux-Ruff, C. and Hirth, C.G. (1992) The δ-opioid receptor: Isolation of a cDNA by expression cloning and pharmacological characterization. *Proc. Natl. Acad. Sci. USA* 89: 12048-12052.

Lord, J.A.H., Waterfield, A.A., Hughes, J. and Kosterlitz, H.W. (1977) Endogenous opioid peptides: multiple agonists and receptors. *Nature* 267: 495-499.

Magnuson, D.S., Sullivan, A.F., Simonnet, G., Roques, B.P. and Dickenson, A.H. (1990) Differential interactions of cholecystokinin and FLFQPQRF-NH2 with mu and delta opioid antinociception in the rat spinal cord. *Neuropeptides* 16: 213-218.

Martin, W.R., Eades, C.G., Thompson, J.A., Huppler, R.E. and Gilbert, P.E. (1976) The effects of morphine- and nalorphine-like drugs in the nondependent and morphine-dependent chronic spinal dog. *J. Pharmacol. Exp. Ther.* 197: 517-522.

Mucha, R.F. and Herz, A. (1985) Motivational properties of kappa and mu opioid receptor agonists studied with place and taste preference conditioning. *Psychopharmacology* 86: 274-280.

Nylander, I., Hyytiä, P., Forsander, O. and Terenius, L. (in preparation) Differences between alcohol-preferring (AA) and alcohol-avoiding (ANA) rats in the prodynorphin and pro-enkephalin systems.

Olds, J. (1976) Brain stimulation and the motivation of behavior. *Prog. Brain Res.* 45: 401-426.

Persson, S., Post, C., Alari, L., Nyberg, F. and Terenius, L. (1989) Increased neuropeptide-converting enzyme activities in cerebrospinal fluid of opiate-tolerant rats. *Neurosci. Lett.* 107: 318-322.

Pohl, M., Collin, E., Benoliel, J.J., Bourgoin, S., Cesselin, F. and Hamon, M. (1992) Cholecystokinin (CCK)-like material and CCK mRNA levels in the rat brain and spinal cord after acute or repeated morphine treatment. *Neuropeptides* 21: 193-200.

Sinclair, J.D., Le, A.D. and Kiianmaa, K. (1989) The AA and ANA rat lines, selected for differences in volontary alcohol consumption. *Experientia* 45: 798-805.

Spanagel, R., Herz, A. and Shippenberg, T.S. (1992) Opposing tonically active endogenous opioid systems modulate the mesolimbic dopaminergic pathway. *Proc. Natl. Acad. Sci. USA* 89: 2046-2050.

Takemori, A.E., Loh, H.H. and Lee, N.M. (1993) Suppression by dynorphin A and [des-Tyr¹] dynorphin A peptides of the expression of opiate withdrawal and tolerance in morphine-dependent mice. *J. Pharmacol. Exp. Ther.* 266: 121-124.

Tang, J., Chou, J., Iadarola, M., Yang, H.-Y.T. and Costa, E. (1984) Proglumide prevents and curtails acute tolerance to morphine in rats. *Neuropharmacology* 23: 715-718.

Tulunay, F.C., Jen, M.F., Chang, J.K., Loh, H.H. and Lee, N.M. (1981) Possible regulatory role of dynorphin on morphine- and beta-endorphin-induced analgesia. *J. Pharmacol. Exp. Ther.* 219: 296-298.

Wahlström, A. and Terenius, L. (1980) Factor in human CSF with apparent morphine-antagonistic properties. *Acta Physiol. Scand.* 110: 427-429.

Watkins, L.R., Kinscheck, I.B. and Mayer, D.J. (1984) Potentiation of opiate analgesia and apparent reversal of morphine tolerance by proglumide. *Science* 244: 395-396.

Wiertelak, E.P., Maier, S.F. and Watkins, L.R. (1992) Cholecystokinin antianalgesia: safety cues abolish morphine analgesia. *Science* 256: 830-833.

Yang, H.-Y.T., Fratta, W., Majane, E.A. and Costa, E. (1985) Isolation, sequencing, synthesis, and pharmacological characterization of two brain neuropeptides that modulate the action of morphine. *Proc. Natl. Acad. Sci. USA* 82: 7757-7761.

Yasuda, K., Raynor, K., Kong, H., Breder, C.D., Takeda, J., Reisine, T. and Bell, G.I. (1993) Cloning and functional comparison of κ and δ opioid receptors from mouse brain. *Proc. Natl. Acad. Sci. USA* 90: 6736-6740.

Toward a Molecular Basis of Alcohol Use and Abuse
ed. by B. Jansson, H. Jörnvall, U. Rydberg, L. Terenius & B. L. Vallee

Adaptation of signal transduction in brain

C. Alling, L. Gustavsson, C. Larsson, C. Lundqvist, D. Rodriguez[1] and P. Simonsson

Dept. of Psychiatry and Neurochemistry, Lund University, POB 638, S-220 09 Lund, Sweden, and [1]Departamento de Bioquimica y Biologia Molecular, Facultad de Medicina, Universidad de Salamanca, E-37007 Salamanca, Spain

Summary

Cell culture models were used to study the effects of long-term ethanol exposure on neuronal cells. Effects on phopholipase C and phospholipase D mediated signal transduction were investigated by assaying receptor-binding, G protein function, activities of lipases, formation of second messengers and c-fos mRNA. The signal transduction cascades displayed abnormal activities from 2 to 7 days of exposure which differed from the acute effects. Phosphatidylethanol formed by phospholipase D is an abnormal lipid that may harmfully affect nerve cell function.

Introduction

Ethanol affects most signalling systems in brain (Dietrich *et al.*, 1989). A substantial amount of knowledge has been acquired over the past years concerning the influence of ethanol on release of transmitters (Koob, 1992), but much more is known about the post-synaptic events (Tabakoff and Hoffman, 1993). Ethanol has both hydrophilic and hydrophobic properties and is therefore especially able to interact with both proteins and lipid molecules that constitute the receptor and signal transduction pathways (Hoek and Rubin, 1990; Hoek *et al.*, 1992). There are no specific ethanol receptors. Ethanol acts directly or indirectly on receptors for the signal substances and signal transduction pathways and therefore interferes with normal cell communication. Some receptors, like the NMDA type of glutamate receptors, are sensitive to low concentrations of ethanol and respond quickly to the action of ethanol (Lovinger *et al.*, 1989; Lovinger *et al.*, 1990). Receptors coupled to guanine nucleotide binding proteins (G proteins) are often less sensitive than receptors coupled to ion channels, but ethanol may induce changes in the intracellular response that are more long-lasting via these pathways (Hoffman and Tabakoff, 1990).

Adaptive changes in receptors coupled to G_s-proteins and adenylyl cyclase activity after chronic ethanol exposure have been demonstrated in neuronal cell cultures (Charness *et al.*, 1988; Mochly-Rosen *et al.*, 1988). The results presented here focus on disturbances induced by ethanol in signal transduction pathways mediated by phospholipase C (PLC) and phospholipase D (PLD) with special emphasis on long-term effects, intracellular second messengers and influence on gene expression.

Methods

Culture and stimulation of cells

Neuroblastoma x glioma cells, NG 108-15, (passages 18-32) were cultivated in 30 mm diameter plastic dishes in Dulbecco's modified Eagle's medium with 5% fetal calf serum. The culture dishes were maintained at 37°C in an incubator under a humidified atmosphere containing 10% CO_2/90% air. Medium was changed daily. In experiments studying the effects of long-term exposure of ethanol, varying amounts of ethanol were added to the medium daily. A constant ethanol concentration was obtained by growing the cells in plastic boxes together with an open cell culture dish containing 2% ethanol (Rodriguez *et al.*, 1992b). By equilibrating the atmosphere with ethanol vapour in this way, ethanol concentrations were reduced by less than 5-10% after 24 hours.

Human SH-SY5Y neuroblastoma cells were cultivated in Eagle's minimal essential medium with 10% fetal calf serum in a humidified atmosphere with 5% CO_2/95% air. Medium was changed every 2 days. Ethanol exposure was the same as for NG 108-15 cells.

Assays

Inositol phosphates were separated by anion exchange chromatography (Berridge *et al.*, 1982; Simonsson *et al.*, 1989). Inositol 1,4,5-trisphosphate (IP$_3$) was assayed with a receptor-binding method (Bredt *et al.*, 1989). [^{3}H]-diacylglycerol was isolated from HPTLC plates and radioactivity was analysed in a β-counter. Phosphatidylethanol (PEth) and phosphatidic acid (PA) were assayed either as fatty acid labelled compounds in a scintillation counter or using Coomassie blue staining of HPTLC plates followed by densitometric quantification using an image analysis program (Nakamura and Handa, 1984). Total cellular RNA was isolated (Chomscynski and Sacchi, 1987) and *c-fos* mRNA was assayed by Northern blot analysis using a [^{32}P]-dCTP-labelled cDNA probe (XhoI/NcoI excised fragment of the pc-fos(human)-1 clone -ATCC 41042).

Results and discussion

PLC signal transduction

The effects of ethanol on the PLC transduction system seem to differ from one receptor type to another and from one cell type to another (Hoffman *et al.*, 1986; Gonzales *et al.*, 1986; Smith, 1991; Simonsson *et al.*, 1991; Sun *et al.*, 1993). The present studies give information from defined receptor types in two different clonal cell populations of neuronal origin.

NG 108-15 cells

These cells express bradykinin receptors coupled to phosphatidylinositol 4,5-bisphosphate(PIP_2)-specific PLC. Neither basal nor bradykinin stimulated formation of inositol phosphates were influenced by ethanol in varying concentrations up to 150 mM for time periods up to 15 minutes. However, when cells were grown for 4 days in medium containing 100 mM ethanol, the maximum rate of bradykinin stimulated formation of inositol monophosphate (IP_1), inositol bisphosphate (IP_2) and IP_3 was significantly reduced and the hydrolysis of PIP_2 was inhibited (Simonsson *et al.*, 1989). A time course study indicated that the reduced formation of IP_3 was obtained already after 2 days of ethanol exposure (30% reduction) and was most pronounced after 3 days of exposure (60% reduction).

Theoretically, ethanol could exert its effect directly on any of the three major constituents of the signalling cascade, i.e. receptor protein, G protein or PLC. Experimental methods were therefore set up to approach the question of the molecular mechanisms underlying the inhibition of IP_3 formation. Exposure to 100 mM ethanol for 4 days did not alter the specific binding of [^{3}H]-bradykinin and the PIP_2-specific PLC activity was the same in ethanol exposed cells as in control cells. GTP binding proteins (G proteins) have a pivotal role in cell signalling and have previously been demonstrated to be perturbed by ethanol in adenylyl cyclase-coupled signal transduction (Valverius *et al.*, 1989a, b, c). We tested the G protein function in the NG 108-15 cells by stimulating partly permeabilized cells with the non-hydrolysable GTP analogue GTP[S] and found that cells exposed to ethanol (100 mM) for 4 days showed almost no hydrolysis of PIP_2 and a pronounced reduction (50%) of IP_3 formation (Simonsson *et al.*, 1991). It is therefore tempting to postulate that the G proteins represent the primary targets for long-term ethanol exposure in this signal transduction cascade.

Previous studies on the effect of ethanol on adenylyl cyclase activation have shown that chronic ethanol apparently produces a selective decrease in $G_{s\alpha}$ expression in NG 108-15 cells (Charness *et al.*, 1988; Mochly-Rosen *et al.*, 1988). The G proteins mediating the signal from receptor to PLC are not characterized in detail. Recent studies suggest that a subfamily of G-protein α subunits denoted $G_{q\alpha/11\alpha}$ mediate this effect (Gutowski *et al.*, 1991). Recently,

Williams and Kelly (1993) reported that ethanol (10-200 mM) reduced the concentration of $G_{q\alpha/11\alpha}$ in NG 108-15 membranes for 48 hours as detected by Western blotting. To further investigate the effect of ethanol on the PLC signalling pathway in NG 108-15 cells, ethanol exposure time was extended up to 7 days (100 mM) (Rodriguez *et al.*, 1992a). After this time period the inhibition of agonist-induced activation of PLC observed after 2 and 4 days of ethanol had disappeared. This finding correlated with a change in the bradykinin receptor population after 7 days of treatment towards a larger proportion of high affinity state receptors. The precise molecular mechanism underlying these changes is not fully understood but the alterations in the affinity of the receptor population may be part of a process of adaptation to restore the functional derangement seen after a shorter period of ethanol exposure.

SH-SY5Y cells

The ethanol-induced abnormalities in concentrations of second messengers described above may have serious consequences for the intracellular events further downstream, especially with regard to calcium fluxes, activation of protein kinase C (PKC), phosphorylation of transcription factors and gene expression. It is also known that ethanol exposure of other cell types significantly influences certain PKC isoenzymes (Messing *et al.*, 1991). In our hands NG 108-15 cells were less suitable for experiments that could address these questions. Instead the human neuroblastoma SH-SY5Y cell line was more rewarding. These cells express muscarinic receptors coupled to PIP_2-specific PLC. Calcium and diacylglycerol (DAG) activate protein kinase C in these cells (Larsson *et al.*, 1992) and the PKC induced phosphorylation is counteracted by an okadaic acid sensitive protein phosphatase (Larsson *et al.*, 1993). In SH-SY5Y cells *c-fos* expression can be induced by activation of PKC (Squinto *et al.*, 1989). We have characterized the importance of DAG formation for the muscarinic receptor stimulated and PKC mediated *c-fos* expression in these cells and applied the model to studies on ethanol exposure.

Activation of the M1 receptor with 1 mM carbachol induced a biphasic increase of IP_3 with an initial peak reaching a maximum after 10 seconds and a sustained elevated IP_3 level that was constant between 60-120 seconds. Ethanol (25-100 mM) given acutely reduced the IP_3 formation at 10 seconds by 20% but had no effect on the plateau level. The explanation for the differential effects by ethanol on the two phases of IP_3 formation remains to be elucidated. After 2 days and especially after 4 days of exposure to 100 mM ethanol a significant increase in the formation of

IP$_3$ was observed (140 and 170% increase, respectively). The effect of longterm ethanol exposure on IP$_3$ formation was opposite to the effect of acute ethanol exposure (Table I).

Table I. IP$_3$ levels in SH-SY5Y cells after 4 days of exposure to varying concentrations of ethanol

Ethanol concentration (mM)	IP$_3$ concentration (pmol/mg protein) M $\pm$ s.e.m.
0	109 $\pm$ 4.9
25	155 $\pm$ 10.4
50	212 $\pm$ 19.8
100	270 $\pm$ 21.7

The consequences of a chronic abnormality in this signal transduction pathway on gene expression was tested. It was found that cells exposed to 100 mM ethanol for 4 days displayed a potentiated carbachol stimulated c-fos expression reaching about 40% increase after 30 minutes of carbachol stimulation.

Phospholipase D signal transduction

A number of reports have demonstrated that PLD is a component of signal transduction cascades. Activation of several different receptor complexes, as well as stimulation of PKC, induces increased PLD activity in a variety of cell types (Gustavsson, 1993; Exton 1990; Billah and Anthes 1990; Shukla and Halenda, 1991). Phosphatidic acid (PA) is the normal lipid derived product formed by PLD. However, this enzyme also utilizes short-chain alcohols as substrate leading to the formation of the corresponding phosphatidylalcohol.

In 1984 it was demonstrated that PEth was formed in different organs of ethanol treated rats (Alling *et al.*, 1984) (Fig. 1). The formation of PEth is catalysed by PLD (Gustavsson and Alling, 1987; Kobayashi and Kanfer, 1987). By formation of PEth PLC may contribute to the mechanisms by which ethanol interferes with cell function.

24

PEth formation and reduction of PA levels

Activation of muscarinic acetylcholine receptors, P_2-purinergic receptors, bradykinin receptors and α_1-adrenergic receptors has been proven to activate PLD in the nervous system (Gustavsson *et al.*, 1993). Activation of PKC is another important pathway for the activation of PLD and via this mechanism significant amounts of PEth are formed with ethanol concentrations as low as 25 mM (Gustavsson and Hansson, 1990). Parallel to the formation of PEth,

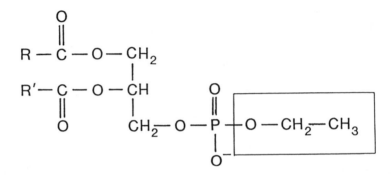

Fig. 1. Structure of PEth.

an inhibition by ethanol of the normal PA formation takes place forming PEth at the expense of PA. Because of the rapid interconversion between PA, formed by PLD, and DAG, formed by PLC, there is a close relationship between these two different signal transduction pathways (Gustavsson *et al.*, 1993). The extent to which PA levels are decreased depends therefore on the activation of other phospholipases which may be induced by the same receptor agonist. Since PA is also formed through phosphorylation of the PLC product, DAG via diacylglycerol kinase, the effects of ethanol are dependent on the relative contribution of PLC and PLD to the total amount of lipid derived second messengers formed (Gustavsson *et al.*, 1993).

Accumulation of PEth

The amount of PEth formed under different conditions and in different tissues varies. Rats injected with ethanol (3 g/kg body weight) formed 200 pmol/mg protein in kidneys (Benthin *et al.*, 1985) whereas NG 108-15 cells exposed for 2 days to ethanol (100 mM) contained 40 pmol/mg protein (Lundqvist *et al.*, 1993). When cells were stimulated with PKC activators, the amount of PEth formed reached 3 nmol/mg protein which corresponds to 3 % of the total phospholipids. In cell types and tissues so far studied, the metabolism of PEth is relatively slow. Degradation of PEth was studied in NG 108-15 cells after incorporation of exogenous PEth to the cell membranes. The disappearance of PEth from NG 108-15 cells that contained 3 % PEth of total lipid phosphoros indicated an average half life of 5 hours and PEth was still detectable 24 hours after removal of the lipid from the medium (Lundqvist *et al.*, 1993).

PEth and cell function

The formation and accumulation of PEth, a phospholipid normally not occurring in nerve cells, may induce changes in membrane function and signal transduction systems. One pathway studied in this context is the formation of IP_3 in NG 108-15 cells (Lundqvist *et al.*, 1993). When these cells were exposed to ethanol or exogenous PEth, bradykinin stimulated formation of IP_3 was not affected but there was a significant increase in the basal level of this second messenger after both treatments. The change in IP_3 was dependent on PEth concentration and also on time exposure (Lundqvist *et al.*, 1993). The effects of exogenously added PEth and ethanol exposure for 2 days on IP_3 levels were not additive. This indicates that the cause of an increased IP_3 concentration may be due to PEth under both conditions. Whether PEth has a stimulatory effect on basal PLC activity or acts through an inhibition of the IP_3 metabolizing

Table II. Effects of PEth

- Changes in membrane fusion properties (Bondeson and Sundler, 1987)
- Activation of protein kinase C (Asaoka *et al.*, 1988)
- Increased membrane fluidity and tolerance to ethanol-induced fluidization
  (Omodeo-Salé *et al.*, 1991)
- Reduction of ethanol-induced increase in Na^+, K^+-ATPase activity
  (Omodeo-Salé *et al.*, 1991)
- Increased resistance to ethanol-induced membrane disordering (Rottenberg
  et al., 1992)

---

enzymes is unknown. A number of other effects of PEth on cell function have previously been reported and are summarized in Table II.

These findings make it highly plausible that PEth plays a role in some of the adaptation phenomenon developing during exposure of ethanol.

Conclusions

Ethanol induces changes in lipase-mediated signal transduction pathways in neuronal cells. The chronic effects on IP_3, DAG and calcium mediated through the PLC system vary among cells and receptor types. Adaptive changes include affinity states of receptor binding, G protein function, IP_3 formation and expression of *c-fos* mRNA. The most remarkable effect of ethanol on these signal transduction pathways is the formation of PEth, an abnormal anionic phospholipid, that can be formed in amounts up to 3% of total phospholipids. PEth, a potentially toxic agent, affects the basal IP_3 level and thereby influences the homeostasis of signal transduction in nerve cells.

Acknowledgement

Financial support has been obtained from the Swedish Medical Research Council (05249 and 10837), Swedish Alcohol Research Fund, Albert Påhlsson Foundation, The Medical Faculty, Lund University, Crafoord Foundation and Magnus Bergvall Foundation.

References

Alling, C., Gustavsson, L., Månsson, J.E., Benthin, G. and Änggård, E. (1984) Phosphatidyl-ethanol formation in rat organs after ethanol treatment. *Biochim. Biophys. Acta* 793: 119-122.

Asaoka, Y., Kikkawa, U., Sekiguchi, K., Shearman, M.S., Kosaka, Y., Nakano, Y., Satoh, T. and Nishizuka, Y. (1988) Activation of a brain-specific protein kinase C subspecies in the presence of phosphatidylethanol. *FEBS Lett.* 231: 221-224.

Benthin, G., Änggård, E., Gustavsson, L. and Alling, C. (1985) Formation of phosphatidyletha-nol in frozen kidneys from ethanol-treated rats. *Biochim. Biophys. Acta* 835: 385-389.

Berridge, M.J., Downes, C.P. and Hanley, M.R. (1982) Lithium amplifies agonist-dependent phosphatidylinositol responses in brain and salivary glands. *Biochem. J.* 206: 587-595.

Billah, M.M. and Anthes, J.C. (1990) The reulgation and cellular functions of phosphatidyl-choline hydrolysis. *Biochem. J.* 269: 281-291.

Bondeson, J. and Sundler, R. (1987) Phosphatidylethanol counteracts calcium-induced membrane fusion but promotes proton-induced fusion. *Biochim. Biophys. Acta* 899: 258-264.

Bredt, D.S., Mourey, R.J. and Snyder, S.H. (1989) A simple sensitive, and specific radio-receptor assay for inositol 1,4,5-trisphosphate in biological tissues. *Biochim. Biophys. Res. Commun.* 159: 976-982.

Charness, M.E., Querimit, L.A. and Henteleff, M. (1988) Ethanol differentially regulates G proteins in neural cultures. *Biochem. Biophys. Res. Commun.* 155: 138-141.

Chomczynski, P. and Sacchi, N. (1987) Single-step method of RNA isolation by acid guanidium thiocyanate-phenol-chloroform extraction. *Anal. Biochem.* 162: 156-159.

Dietrich, R.A., Dunwiddie, T.V., Harris, R.A. and Erwin, V.G. (1989) Mechanism of action of ethanol: Initial central nervous system actions. *Pharmacol. Reviews* 41: 489-537.

Exton, J.H. (1990) Signaling through phosphatidylcholine breakdown. *J. Biol. Chem.* 265: 1-4.

Gonzales, R.A., Theiss, C. and Crews, F.T. (1986) Effects of ethanol on stimulated inositol phospholipid hydrolysis in rat brain. *J. Pharmacol. Exp. Ther.* 237: 92-98.

Gustavsson, L. and Alling, C. (1987) Formation of phosphatidylethanol in rat brain by phospholipase D. *Biochem. Biophys. Res. Commun.* 142: 958-963.

Gustavsson, L. and Hansson, E. (1990) Stimulation of phospholipase D activity by phorbol esters in cultured astrocytes. *J. Neurochem.* 54: 737-742.

Gustavsson, L., Lundqvist, C., Hansson E., Rodriguez, D., Simonsson, P. and Alling, C. (1993) In: Alling, C. and Wood, W.G. (eds.): *Alcohol, Cell Membranes and Signal Transduction in Brain*, Plenum Press, New York, pp. 63-74.

Gutowski, S., Smrcka A., Nowak, L., Wu, D., Simon, M. and Sternweis, P. (1991) Antibodies to the αq subfamily of guanine nucleotide-binding regulatory protein α subunits attenuate activation of phosphatidylinositol 4,5-bisphosphate hydrolysis by hormones. *J. Biol. Chem.* 266: 445-447.

Hoek, J.B. and Rubin, E. (1990) Alcohol and membrane-associated signal transduction. *Alcohol Alcohol* 25: 143-156.

Hoek, J.B., Thomas, A.P., Rooney, T.A., Higashi, K. and Rubin, E. (1992) Ethanol and signal transduction in the liver. *FASEB J.* 6: 2386-2396.

Hoffman, P.L., Moses, F., Luthin, G.R. and Tabakoff, B. (1986) Acute and chronic effects of ethanol on receptor-mediated phosphatidylinositol 4,5-bisphosphate breakdown in mouse brain. *Mol. Pharmacol.* 30: 13-18.

Hoffman, P.L. and Tabakoff, B. (1990) Ethanol and guanine nucleotide binding proteins: a selective interaction. *FASEB J.* 4: 2612-2622.

Kobyashi, M. and Kanfer, J.N. (1987) Phosphatidylethanol formation via transphosphatidylation by rat brain synaptosomal phospholipase D. *J. Neurochem.* 48: 1597-1603.

Koob, G.F. (1992) Drugs of abuse: anatomy, pharmacology and function of reward pathways. *Trends Pharmacol. Sci.* 13: 177-184.

Larsson, C., Saermark, T., Mau, S. and Simonsson, P. (1992) Activation of protein kinase C in permeabilized human neuroblastoma SH-SY5Y cells. *J. Neurochem.* 59: 644-651.

Larsson, C., Alling, C. and Simonsson, P. (1993) An okadaic acid-sensitive protein phosphatase counteracts protein kinase C-induced phosphorylation in SH-SY5Y cells. *Cell Signal* 5: 305-313.

Lovinger, D.M., White, G. and Weight, F.F. (1989) Ethanol inhibits NMDA-activated ion current in hippocampal neurons. *Science* 243: 1721.

Lovinger, D.M., White, G. and Weight, F.F. (1990) Ethanol inhibition of NMDA-activated ion current is not voltage-dependent and ethanol does not interact with other binding sites on the NMDA receptor/ionophore complex. *FASEB J.* 4: A678.

Lundqvist, C., Rodriguez, F.D., Simonsson, P., Alling C. and Gustavsson, L. (1993) Phosphatidylethanol affects inositol 1,4,5-trisphosphate levels in NG 108-15 neuroblastoma xglioma hybrid cells. *J. Neurochem.* 60: 738-744.

28

Messing, R.O., Petersen P.J. and Henrich, C.J. (1991) Chronic ethanol exposure increases levels of protein kinase C δ and ϵ and protein kinase C-mediated phosphorylation in cultured neural cells. *J. Biol. Chem.* 266: 23428-23432.

Mochly-Rosen, D., Chang F-H., Cheever, L., Kim, M., Diamond, I. and Gordon, A.S. (1988) Chronic ethanol causes heterologous desensitization of receptors by reducing α_s messenger RNA. *Nature* 333: 848-850.

Nakamura, K. and Handa, S. (1984) Coomassie brilliant blue staining of lipids on thin layer plates. *Anal. Biochem.* 142: 406-410.

Omodoe-Salé, F., Lindi, C., Palestini, P. and Masserini, M. (1991) Role of phosphatidylethanol in membranes. Effects on membrane fluidity, tolerance to ethanol, and activity of membrane-bound enzymes. *Biochemistry* 30: 2477-2482.

Rodriguez, F.D., Simonsson, P., Gustavsson, L. and Alling, C. (1992a) Mechanisms of adaptation to the effects of ethanol on activation of phospholipase C in NG 108-15 cells. *Neuropharmacol.* 31: 1157-1164.

Rodriguez, F.D., Simonsson, P. and Alling, C. (1992b) A method for maintaining constant ethanol concentrations in cell culture media. *Alcohol Alcohol* 27: 309-313.

Rottenberg, H., Bittman, R. and Hong-Lan, L. (1992) Resistance to ethanol disordering of membranes from ethanol-fed rats is conferred by all phospholipid classes. *Biochim. Biophys. Acta* 1123: 282-290.

Shukla, S.D. and Halenda, S.P. (1991) Phospholipase D in cell signalling and its relationship to phospholipase C. *Life Sci.* 48: 851-866.

Simonsson P., Sun, G.Y., Vécsei, L. and Alling, C. (1989) Ethanol effects on bradykinin-stimulated phosphoinositide hydrolysis in NG 108-15 neuroblastoma-glioma cells. *Alcohol* 6: 475-479.

Simonsson, P., Rodriguez F.D., Loman, N. and Alling, C. (1991) G proteins coupled to phospholipase C: Molecular targets of long-term ethanol exposure. *J. Neurochem.* 56: 2018-2016.

Smith, T.L. (1991) Selective effects of acute and chronic ethanol exposure on neuropeptide and guanine nucleotide stimulated phospholipase C activity in intact N1E-115 neuroblastoma. *J. Pharmacol. Exp. Ther.* 258: 410-415.

Squinto, S.P., Block, A.L., Braquet, P. and Bazan, N.G. (1989) Platelet-activating factor stimulates a fos/jun/AP-1 transcriptional signalling system in human neuroblastoma cells. *J. Neurosci. Res.* 24: 558-566.

Sun, G.Y., Zhang, J. and Lin, T-A. (1993) In: Alling, C. and Wood, W.G. (eds.): *Alcohol, Cell Membranes and Signal Transduction in Brain*, Plenum Press, New York, pp. 205-218.

Tabakoff, B. and Hoffman, P.L. (1993) The neurochemistry of alcohol. *Current Opinion in Psychiatry* 6: 388-394.

Valverius, P., Hoffman, P.L. and Tabakoff, B. (1989a) Hippocampal and cerebellar β-adrenergic receptors and adenylate cyclase are differentially altered by chronic ethanol ingestion. *J. Neurochem.* 52: 492-497.

Valverius, P. Hoffman, P.L. and Tabakoff, B. (1989b) Brain forskolin binding in mice dependent on and tolerant to ethanol. *Brain Res.* 503: 38-43.

Valverius, P., Borg, S., Valverius, M.R., Hoffman, P.L. and Tabakoff, B. (1989c) Beta-adrenergic receptor binding in brains of alcoholics. *Exp. Neurol.* 105: 280-286.

Williams, R.J. and Kelly, E. (1993) Chronic ethanol reduces immunologically detectable $G_{q\alpha/11\alpha}$ in NG108-15 cells. *J. Neurochem.* 61: 1163-1166.

Toward a Molecular Basis of Alcohol Use and Abuse
ed. by B. Jansson, H. Jörnvall, U. Rydberg, L. Terenius & B. L. Vallee
© 1994 Birkhäuser Verlag Basel/Switzerland

Protein kinase C and adaptation to ethanol

R. Roivainen, B. Hundle and R.O. Messing

Ernest Gallo Clinic & Research Center, Department of Neurology, University of California, San Francisco, Building 1, Room 101, 1001 Potrero Avenue, San Francisco, CA 94110, USA

Summary

Adaptation to chronic ethanol exposure results in a decrease in sensitivity to the intoxicating effects of ethanol. Recent evidence indicates that changes in the expression and function of certain proteins involved in signal transduction are important for adaptation to ethanol. Using the neural cell line PC12, we found that chronic exposure to ethanol increases the expression and function of L-type voltage-gated calcium channels and enhances neural differentiation induced by nerve growth factor. Both of these responses to ethanol require protein kinase C (PKC). Chronic ethanol exposure activates PKC-mediated phosphorylation, in part, by increasing the expression of two PKC isozymes, δ and ϵ. The PKC family of enzymes may be important targets for the development of drugs that could modify adaptive and toxic consequences of chronic ethanol exposure.

Introduction

Ethanol-induced changes in voltage-gated Ca^{2+} channels appear to play a role in alcohol dependence and account for several manifestations of alcohol withdrawal. Brief exposure to ethanol inhibits the function of voltage-dependent Ca^{2+} channels (Oakes and Pozos, 1982; Leslie et al., 1983; Messing et al., 1986; Skattebol and Rabin, 1987). In contrast, we (Messing et al., 1986) and others (Skattebol and Rabin, 1987) have found that chronic exposure to 25-200 mM ethanol for 2-6 days increases depolarization-stimulated $^{45}Ca^{2+}$ uptake through dihydropyridine (DHP)-sensitive Ca^{2+} channels in the neural cell line PC12. This increase in uptake remains elevated for several hours following removal of ethanol (Messing et al., 1986) and is associated with an increase in the number of binding sites for DHP Ca^{2+} channel antagonists (Messing et al., 1986; Skattebol and Rabin, 1987). Increases in binding sites for DHPs have also been found in brain membranes from rats made dependent on ethanol (Dolin and Little, 1989), suggesting that chronic ethanol exposure increases the number of DHP-sensitive, voltage-dependent Ca^{2+} channels in brain. The importance of these channels in the pathogenesis of alcohol withdrawal syndromes is supported by evidence that DHPs reduce tremors, seizures, and mortality in alcohol-dependent mice and rats deprived of ethanol (Little et al., 1986; Bone et al.,

1989; Littleton *et al.*, 1990).

Excessive consumption of ethanol can also damage adult and developing nervous systems by interfering with the growth and remodeling of neural processes (neurites). In several brain regions, ethanol inhibits the growth of dendrites (Walker *et al.*, 1981; Hammer, 1986). In other regions, ethanol enhances the development of neural processes (West *et al.*, 1981; Tavares *et al.*, 1986; King *et al.*, 1988; Ferrer *et al.*, 1989; Miller *et al.*, 1990; Pentney and Quackenbush, 1990). Ethanol also enhances neurite outgrowth in cultured cerebellar macroneurons (Zou *et al.*, 1993) and in PC12 cells treated with nerve growth factor (Messing *et al.*, 1991; Wooten and Ewald, 1991). Enhancement of neurite outgrowth could be detrimental in several ways. Increases in dendrite length could place synapses at greater distances from the cell soma, decreasing electrical conduction down dendrites to the soma, as has been proposed in aged neurons (Turner and Deupree, 1991). Enhancement of neurite outgrowth could also interfere with normal elimination of neural processes and rearrangement of synaptic contacts during development and learning (Purves and Lichtman, 1985; Patel and Stewart, 1988). In addition, stimulation of process elongation in some neurons could disrupt the output of neural networks by upsetting the balance of inhibitory and excitatory inputs.

Protein phosphorylation regulates ion channels and increases in calcium channel number have been described following activation of protein kinase C (PKC) in *Aplysia* bag cell neurons (Strong *et al.*, 1987) and cultured chick myocytes (Navarro, 1987). PKC also modulates neural growth and differentiation. Tumor-promoting phorbol esters that activate PKC stimulate the induction of neural cells from neuroectoderm (Otte and Moon, 1992) and induce neurite outgrowth from chick sensory ganglia (Hsu *et al.*, 1984), chick ciliary ganglion neurons (Bixby, 1989), and several neuroblastoma cell lines (Spinelli *et al.*, 1982; Påhlman *et al.*, 1983). PKC is a family of at least ten phospholipid-dependent isozymes which can be subdivided into three structurally-related groups: "conventional" cPKCs (α, β_I, β_{II}, and γ) which are regulated by calcium and diacylglycerols; "novel" nPKCs which are sensitive to diacylglycerols but are calcium-independent (δ, ϵ, η, and Θ); and "atypical" aPKCs (ζ and λ) which are insensitive to calcium and diacylglycerols (Nishizuka, 1992)

We have pursued the following studies to investigate whether PKC isozymes mediate effects of ethanol on calcium channels and neural differentiation. For these studies we used the neural cell line PC12. PC12 cells, derived from a rat pheochromocytoma, express L-type calcium channels and respond to nerve growth factor with dramatic biochemical and morphological differentiation, acquiring a phenotype resembling that of a mature sympathetic neuron (Greene et al., 1987).

Materials and Methods

PC12 cells were cultured as described (Messing et al., 1986). NGF (2.5S) was a gift from William Mobley (UCSF). Phorbol esters were from LC Services (Woburn, MA). [^{3}H](+)PN200-110 was from DuPont NEN. Other radioligands, nitrocellulose and nylon membranes, and the ECL detection kit were from Amersham. Nifedipine, sphingosine, and polyornithine were from Sigma.

^{45}Ca uptake and [^{3}H](+)PN200-110 binding studies were performed as described (Messing et al., 1990). Western analysis was performed with PKC isozyme-selective antibodies from Life Technologies (Gaithersburg, MD), as described (Messing et al., 1991). Neurite lengths were measured from 35 mm photomicrographs (Messing et al., 1991). A neurite was identified as a process greater than one cell body diameter in length and possessing a terminal growth cone.

Levels of mRNA for δ and ϵPKC were measured by slot blot analysis. Poly A$^+$ RNA was isolated using the Micro-FAST TRACK mRNA isolation kit from Invitrogen. Heat-denatured samples of RNA (31-500 ng per slot) were filtered over nylon membranes (Nytran, Schleicher and Schuell) and prehybridized at 42°C for 2-3 hours in 50% formamide, 2 x Denhardt's solution, 2 x SSC, 50 mM MPOS (pH 7.0), 10 mM EDTA, and 1 mg/ml of sheared salmon sperm DNA. Membranes were hybridized overnight in fresh solution containing 100 ng of hexamer-labeled probe (1-5 x 10^8 cpm/μg) (Sambrook et al., 1989). Membranes were then washed twice at 62°C for 20 min in 2 x SSC and 0.1% SDS, and exposed to preflashed Kodak XAR-5 film. Hybridization signals on autoradiograms of blots were quantified by densitometric scanning. Signals were normalized to total poly A$^+$ RNA which was measured from autoradiograms of blots that had been stripped of PKC probe and hybridized with ^{35}S-labeled poly-T (Hollander and Fornace, 1990). PKC probes were made from cDNAs for rat δPKC (obtained from Dr. Peter Parker, Imperial Cancer Research Fund, London, UK), and mouse ϵPKC (obtained from Dr. John Knopf, Genetics Institute, Cambridge, MA). Integrity of the RNA was checked by Northern analysis (not shown) and each probe detected a single band of 3.1 kb (δPKC) or 7.1 kb (ϵPKC), as reported by others (Ono et al., 1988).

Results

To investigate whether PKC participates in up-regulation of L-channels by ethanol, we treated cells with PKC inhibitors. Many of these compounds (e.g. calphostin C, acridine orange, tamoxifen) were toxic to PC12 cells, but sphingosine and polymixin B were not. Both reduced the ability of ethanol to increase K$^+$-stimulated ^{45}Ca^{2+} uptake into PC12 cells (Table I). Sphingosine and polymixin B also inhibit kinases other than PKC. However, we found that the effect of sphingosine on ^{45}Ca^{2+} uptake was reversed by simultaneous exposure to 10 nM phorbol 12,13-dibutyrate (PDBu), which potently activates PKC (Table I). These findings indicate that sphingosine acts in this system by inhibiting PKC. In the absence of ethanol, PDBu did not increase ^{45}Ca^{2+} uptake (not shown), suggesting that activation of PKC alone is insufficient to up-regulate L channels and that additional factors besides PKC are also required for the effect of ethanol.

Table I. ^{45}Ca uptake after treatment with ethanol and PKC modulators for 4 days

Treatment	^{45}Ca uptake, % above control (Mean $\pm$ SE)	N
200 mM ethanol	48 $\pm$ 3	3
200 mM ethanol + 1 μM Sphingosine	10 $\pm$ 3*	3
200 mM ethanol + 1 μM Sphingosine + 10 nM PDBu	55 $\pm$ 3	4
200 mM ethanol + 10 nM PDBu	64 $\pm$ 4*	3

*$p < 0.05$ compared to 200 mM ethanol by ANOVA with Newman Keuls post-hoc test.

We previously found that increased ^{45}Ca^{2+} uptake, following prolonged ethanol exposure, is accompanied by a corresponding increase in the number of binding sites for dihydropyridine (DHP) Ca^{2+} channel antagonists (Messing et al., 1986). We reasoned that if PKC mediates these effects of ethanol, PKC inhibitors should block ethanol-induced increases in DHP binding. Therefore, we cultured cells in the presence or absence of 200 mM ethanol with or without 10 μM sphingosine for 4 days and measured binding of [^{3}H](+)PN200-110 (Messing et al., 1990).

Scatchard analysis of binding demonstrated that the equilibrium dissociation constant was 55 ± 10 pM in control membranes and was not altered by treatment with ethanol (47 ± 6 pM; p = 0.34; N = 4), or ethanol plus sphingosine (40 ± 6 pM; p = 0.47). However, ethanol exposure increased the maximal number of binding sites from 18 ± 2 to 28 ± 1 fmol/mg (p < 0.001; N = 4), whereas the B_{max} in cells treated with ethanol plus sphingosine was not significantly altered (20 ± 3 fmol/mg; p = 0.61; N = 4).

We next asked whether ethanol activates PKC. A fragment of the EGF receptor, KRTLRR, is selectively phosphorylated by PKC *in vitro,* and can be used as a marker for activation of PKC in intact cells, when introduced by detergent permeabilization following activation of PKC (Heasley and Johnson, 1989). Using this method, we found (Messing et al., 1990) that brief (10 min) exposure of PC12 cells to 100 mM ethanol did not alter KRTLRR phosphorylation. However, treatment with 100 mM ethanol for 6 days increased phosphorylation by 32 ± 4% (p < 0.001). Tumor-promoting phorbol esters, unlike ethanol, activate PKC by increasing the affinity of the enzyme for Ca^{2+} and phosphatidylserine (PS) (Nishizuka, 1984). Exposure for 10 min to 1 mM phorbol 12-myristate, 13-acetate (PMA), which maximally activates PKC, increased phosphorylation in cells cultured without (177 ± 22 % above control) and with (205 ± 25 % above control) 100 mM ethanol for 6 days, but the amount of PMA-stimulated phosphorylation was greater in cells pre-treated with ethanol (p < 0.01 compared to cells treated with PMA alone). Therefore, exposure to ethanol for several days increased PKC-mediated phosphorylation. In addition, the effects of ethanol and PMA followed different time courses and were additive, suggesting that these agents stimulate PKC by different mechanisms.

One mechanism by which ethanol could activate PKC is by increasing expression of PKC isozymes. Such a mechanism has been proposed to explain activation of PKC by 1,25-dihydroxyvitamin D_3 in HL-60 cells (Obeid et al., 1990). To investigate this possibility, we measured kinase activity in whole cell extracts under conditions that maximally activate PKC (Messing et al., 1991). Treatment of cells with 25-200 mM ethanol for 6 days caused a concentration-dependent increase in PKC activity, which was significant (21 ± 2%) after exposure to 50 mM ethanol (p < 0.004; N = 4) and maximal (32 ± 2%) after exposure to 100 mM ethanol (Messing et al., 1991). This effect was time-dependent and was significant after exposure to 200 mM ethanol for 2 days (18 ± 5 % increase; p < 0.012; N = 6). The time course for the increase in PKC activity resembled the time course for up-regulation of L-

34

channels by ethanol (Messing *et al.*, 1986). The concentration-dependence of both were also quite similar, although a maximal increase in PKC activity was achieved with 100 mM ethanol, whereas L-channel up-regulation was increased further by 200 mM ethanol (Messing *et al.*, 1986). This may suggest that additional, PKC-independent mechanisms, stimulated by concentrations of ethanol > 100 mM, also contribute to up-regulation of L-channels.

Tumor-promoting phorbol esters bind with high affinity to PKC isozymes. Therefore, to investigate whether ethanol increased total PKC activity by increasing levels of PKC, we measured saturation binding of [³H]PDBu to homogenates of PC12 cells (Messing *et al.*, 1991). Scatchard analysis showed a single population of binding sites with no significant difference in the K_D in control (11 $\pm$ 2 nM) and ethanol-treated cells (14 $\pm$ 7 nM; p = 0.64; N = 3). However, ethanol treatment (200 mM for 6 days) increased the B_{max} from 163 $\pm$ 30 to 297 $\pm$ 32 pmol/mg (p = 0.037; N = 3). These findings suggest that chronic exposure to ethanol increases PKC activity and phorbol ester binding by increasing levels of PKC.

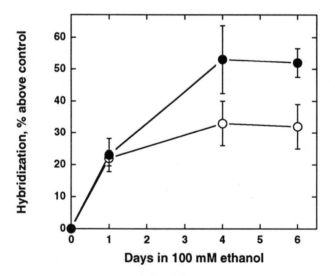

Fig. 1. Ethanol increases mRNA levels for δPKC and εPKC in PC12 cells. Cells were treated with 100 mM ethanol for 0-4 days and 31-500 μg samples of poly A+ RNA were analyzed on slot blots with ³²P-labeled cDNA probes to δPKC (●) or εPKC (○). Data are mean $\pm$ SEM values from three experiments.

We next used PKC isozyme-selective antibodies to investigate which isozymes are regulated by ethanol. Treatment with 100 mM ethanol for 6 days increased δPKC and εPKC immunoreactivity by 46-48%, without significantly altering α, β, or ζPKC immunoreactivity

(Messing *et al.*, 1991). As shown in Fig. 1, treatment with 100 mM ethanol for 1-4 days also increased the abundance of mRNA for δ and εPKC. Thus chronic ethanol exposure appears to increase levels of δ and εPKC by enhancing transcription or decreasing degradation of their mRNAs. These results suggest that chronic ethanol exposure increases PKC activity by increasing expression of δ and εPKC.

Since PKC is important for neural differentiation and ethanol regulates PKC, we investigated the role of PKC in enhancement of neurite outgrowth by ethanol in PC12 cells treated with NGF (Roivainen *et al.*, 1993). Like ethanol, 3-10 nM PMA greatly enhanced NGF-induced neurite outgrowth (Fig. 2). In contrast, concentrations of PMA $\geq$ 100 nM, which depleted cells of β, δ, and εPKC, but not α or ζPKC (Roivainen *et al.*, 1993), completely prevented enhancement of neurite outgrowth by ethanol (Fig. 2). These findings suggest that β, δ, or εPKC are required for enhancement of neurite outgrowth by ethanol.

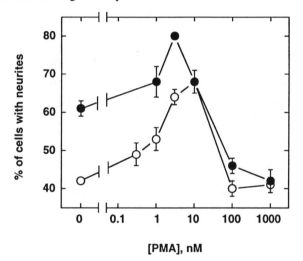

Fig. 2. Neurite outgrowth in PC12 cells treated with PMA. Cells were cultured with (●) or without (○) 100 mM ethanol for 4 days and then PMA was added to the cultures. After 2 days, 50 ng/ml of NGF was added and neurite outgrowth was scored 4 days later. Data shown are mean ± SEM values from 4-12 experiments. (Adapted from Roivainen *et al.*, 1993)

Discussion

The results of our studies indicate that the PKC family of enzymes plays a role in two adaptive responses to chronic ethanol exposure, up-regulation of L-type calcium channels, and enhancement of NGF-induced neurite outgrowth in PC12 cells. Since chronic ethanol exposure

36

increases the expression of δ and εPKC, it is likely that these isozymes mediate the effects of ethanol on L channels and neurites. However, activation of PKC is unlikely to fully account for up-regulation of L channels by ethanol since activation of PKC by phorbol esters does not increase L channel density. In addition, it is possible that enhancement of neurite outgrowth by ethanol involves a process that requires βPKC, rather than δ or εPKC. At present there are no compounds available to selectively activate or inhibit individual PKC isozymes (Roivainen and Messing, 1993). It is our hope that ongoing studies with antisense DNA and RNA will clearly identify which PKC isozymes mediate these adaptive responses to ethanol.

Conclusion

Chronic exposure to ethanol up-regulates the number of neuronal L-type calcium channels and enhances neurite outgrowth in some regions of the brain. Our studies in PC12 cells demonstrate that the effects of ethanol on L channel density and neurite outgrowth are PKC-dependent. Two isozymes of PKC, δPKC and εPKC are up-regulated by ethanol and are therefore likely mediators of the effects of ethanol, but proof of this awaits further study. Identification of specific PKC isozymes that mediate responses to chronic ethanol exposure could lead to the development of selective PKC antagonists that might be useful in the treatment of alcohol-related neurologic disorders.

Acknowledgements

This work was supported by grants from the National Institute on Alcohol Abuse and Alcoholism and from the Alcoholic Beverage Medical Research Foundation to R.O.M. The authors acknowledge permission from Elsevier Science Publishers BV for Fig. 2.

References

Bixby, J.L. (1989) Protein kinase C is involved in laminin stimulation of neurite outgrowth. *Neuron* 3: 287-297.

Bone, G.H., Majchrowicz, E., Martin, P.R., Linnoila, M. and Nutt, D.J. (1989) A comparison of calcium antagonists and diazepam in reducing ethanol withdrawal tremors. *Psychopharmacology* 99: 386-388.

Dolin, S.J. and Little, H.J. (1989) Are changes in neuronal calcium channels involved in ethanol tolerance? *J. Pharmacol. Exp. Ther.* 250: 985-91.

Ferrer, I., Galofre, E., Fabregues, I. and Lopez-Tejero, D. (1989) Effects of chronic ethanol consumption beginning at adolescence: increased numbers of dendritic spines on cortical pyramidal cells in the adulthood. *Acta Neuropathol.* 78: 528-532.

Greene, L.A., Aletta, J.M., Rukenstein, A. and Green, S.H. (1987) PC12 pheochromocytoma cells: culture, nerve growth factor treatment, and experimental exploitation. *Methods Enzymol.* 147: 207-216.

Hammer, R.P., Jr. (1986) Alcohol effects on developing neuronal structure. In: West, J.R. (ed.): *Alcohol and Brain Development,* Oxford University Press, New York, pp. 184-203.

Heasley, L.E. and Johnson, G.L. (1989) Regulation of protein kinase C by nerve growth factor, epidermal growth factor, and phorbol esters in PC12 pheochromocytoma cells. *J. Biol. Chem.* 264: 8646-8652.

Hollander, M.C. and Fornace, A.J. (1990) Estimation of relative mRNA content by filter hybridization to a polythymidylate probe. *BioTechniques* 9: 174-179.

Hsu, L., Natyzak, D. and Laskin, J.D. (1984) Effects of tumor promoter 12-o-tetradecanoyl phorbol-13-acetate on neurite outgrowth from chick embryo sensory ganglia. *Cancer Res.* 44: 4607-4614.

King, M.A., Hunter, B.E. and Walker, D.W. (1988) Alterations and recovery of dendritic spine density in rat hippocampus following long-term ethanol ingestion. *Brain Res.* 459: 381-385.

Leslie, S.W., Barr, E., Chandler, J. and Farrar, R.P. (1983) Inhibition of fast- and slow-phase depolarization-dependent synaptosomal calcium uptake by ethanol. *J. Pharmacol. Exp. Ther.* 225: 571-575.

Little, H.J., Dolin, S.J. and Halsey, M.J. (1986) Calcium channel antagonists decrease the ethanol withdrawal syndrome. *Life Sci.* 39: 2059-2065.

Littleton, J.M., Little, H.J. and Whittington, M.A. (1990) Effects of dihydropyridine calcium channel antagonists in ethanol withdrawal; doses required, stereospecificity and actions of Bay K 8644. *Psychopharmacology* 100: 387-92.

Messing, R.O., Carpenter, C.L. and Greenberg, D.A. (1986) Ethanol regulates calcium channels in clonal neural cells. *Proc. Natl. Acad. Sci. USA* 83: 6213-6215.

Messing, R.O., Henteleff, M. and Park, J.J. (1991) Ethanol enhances growth factor-induced neurite formation in PC12 cells. *Brain Res.* 565: 301-311.

Messing, R.O., Petersen, P.J. and Henrich, C.J. (1991) Chronic ethanol exposure increases levels of protein kinase C δ and ε and protein kinase C-mediated phosphorylation in cultured neural cells. *J. Biol. Chem.* 266: 23428-23432.

Messing, R.O., Sneade, A.B. and Savidge, B. (1990) Protein kinase C participates in up-regulation of dihydropyridine-sensitive calcium channels by ethanol. *J. Neurochem.* 55: 1383-1389.

Miller, M.W., Nicholas, N.L. and Rhoades, R.W. (1990) Intracellular recording and injection study of corticospinal neurons in the rat somatosensory cortex: effect of prenatal exposure to ethanol. *J. Comp. Neurol.* 297: 91-105.

Navarro, J. (1987) Modulation of [3H]dihydropyridine receptors by activation of protein kinase C in chick muscle cells. *J. Biol. Chem.* 262: 4649-4652.

Nishizuka, Y. (1984) The role of protein kinase C in cell surface signal transduction and tumor promotion. *Nature* 308: 693-698.

Nishizuka, Y. (1992) Intracellular signaling by hydrolysis of phospholipids and activation of protein kinase C. *Science* 258: 607-614.

Oakes, S.G. and Pozos, R.S. (1982) Electrophysiologic effects of acute ethanol exposure. II. Alterations in the calcium component of action potentials from sensory neurons in dissociated culture. *Dev. Brain Res.* 5: 251-255.

Obeid, L.M., Okazaki, T., Karolak, L.A. and Hannun, Y.A. (1990) Transcriptional regulation

38

of protein kinase C by 1,25-dihydroxyvitamin D3 in HL-60 cells. *J. Biol. Chem.* 265: 2370-2374.

Ono, Y., Fujii, T., Ogita, K., Kikkawa, U., Igarashi, K. and Nishizuka, Y. (1988) The structure, expression, and properties of additional members of the protein kinase C family. *J. Biol. Chem.* 263: 6927-6932.

Otte, A.P. and Moon, R.T. (1992) Protein kinase C isozymes have distinct roles in neural induction and competence in Xenopus. *Cell* 68: 1021-1029.

Påhlman, S., Ruusala, A.-I., Abrahamsson, L., Odelstad, L. and Nilsson, K. (1983) Kinetics and concentration effects of TPA-induced differentiation of cultured neuroblastoma cells. *Cell Differentiation* 12: 165-170.

Patel, S.N. and Stewart, M.G. (1988) Changes in the number and structure of dendritic spines 25 hours after passive avoidance training in the domestic chick, *Gallus domesticus. Brain Res.* 449: 34-36.

Pentney, R.J. and Quackenbush (1990) Dendritic hypertrophy in Purkinje neurons of old Fischer 344 rats after long-term ethanol treatment. *Alcohol. Clin. Exp. Res.* 14: 878-886.

Purves, D. and Lichtman, J.W. (1985) *Principles of Neural Development.* Sinauer, Sunderland, MA.

Roivainen, R., MacMahon, T. and Messing, R.O. (1993) Protein kinase C isozymes that mediate enhancement of neurite outgrowth by ethanol and phorbol esters in PC12 cells. *Brain Res.* 624: 85-93.

Roivainen, R. and Messing, R.O. (1993) The phorbol derivatives thymeleatoxin and 12-deoxyphorbol-13-*O*-phenylacetate-10-acetate cause translocation and down-regulation of multiple protein kinase C isozymes. *FEBS Lett.* 319: 31-34.

Sambrook, J., Fritsch, E.F. and Maniatis, T. (1989) *Molecular Cloning.* Cold Spring Harbor Press, Cold Spring Harbor.

Skattebol, A. and Rabin, R. (1987) Effects of ethanol on $^{45}Ca^{2+}$ uptake in synaptosomes and in PC12 cells. *Biochem. Pharmacol.* 36: 2227-2229.

Spinelli, W., Sonnenfeld, K.H. and Ishii, D.N. (1982) Effects of phorbol ester tumor promoters and nerve growth factor on neurite outgrowth in cultured human neuroblastoma cells. *Cancer Res.* 42: 5067-5073.

Strong, J.A., Fox, A.P., Tsien, R.W. and Kaczmarek, L.K. (1987) Stimulation of protein kinase C recruits covert calcium channels in Aplysia bag cell neurons. *Nature* 325: 714-717.

Tavares, M.A., Paula-Barbosa, M.M. and Volk, B. (1986) Chronic alcohol consumption induces plastic changes in granule cell synaptic boutons of the rat cerebellar cortex. *J. Submicrosc. Cytol.* 18: 725-730.

Turner, D.A. and Deupree, D.L. (1991) Functional elongation of CA1 hippocampal neurons with aging in Fischer 344 rats. *Neurobiol. Aging* 12: 201-210.

Walker, D.W., Hunter, B.E. and Abraham, W.C. (1981) Neuroanatomical and functional deficits subsequent to chronic ethanol administration in animals. *Alcohol. Clin. Exp. Res.* 5: 267-282.

West, J.R., Hodges, C.A. and Black, A.C.J. (1981) Prenatal exposure to ethanol alters the organization of hippocampal mossy fibers in rats. *Science* 211: 957-959.

Wooten, M.W. and Ewald, S.J. (1991) Alcohols synergize with NGF to induce early differentiation of PC12 cells. *Brain Res.* 550: 333-339.

Zou, J.-Y., Rabin, R.A. and Pentney, R.J. (1993) Ethanol enhances neurite outgrowth in primary cultures of rat cerebellar macroneurons. *Dev. Brain Res.* 72: 75-84.

Toward a Molecular Basis of Alcohol Use and Abuse
ed. by B. Jansson, H. Jörnvall, U. Rydberg, L. Terenius & B. L. Vallee

Molecular control of neuronal survival in the chick embryo

Stefano Biffo, Georg Dechant, Hitoshi Okazawa, and Yves-Alain Barde

Department of Neurobiochemistry, Max-Planck Institute for Psychiatry, 82152 Martinsried, Germany.

Summary

Neurotrophins are structurally related proteins which promote the survival and differentiation of specific neuronal populations during the development of vertebrate embryos. Like many growth factors, the neurotrophins mediate their actions by binding to membrane proteins that have a ligand-activated tyrosine kinase activity. The interactions of the neurotrophins with their neuronal receptors have been mostly studied using chick embryonic neurons. These neurons are also extensively used to characterise biological responses to neurotrophins in physiologically relevant systems. We have recently cloned and expressed the chick homologue of *trk*B (c*trk*B), thought to be a receptor for BDNF, and examined by *in situ* hybridisation the pattern of expression of the c*trk*B gene during development of the chick embryo. We found that whereas the sequence of c*trk*B shows a high degree of conservation with the mammalian homologues in the intracellular tyrosine kinase domain, the extracellular binding domain is less well conserved. As in mammals, c*trk*B mRNAs appear to exist in differentially spliced forms that result in a full length and a truncated receptor lacking the tyrosine kinase domain. These two forms are differentially expressed in neurons and non-neuronal cells respectively. The binding characteristics of c*trk*B expressed in a transfected cell line are similar, but not identical to those of the BDNF binding sites on primary chick neurons, specially with regard to the affinity of BDNF.

Introduction

The process of programmed cell death, often referred to as *apoptosis* since the original suggestion by Kerr, Wylie and Currie (1972), plays a fundamental role in regulating cell numbers and shaping the organisation of the vertebrate body. In the nervous system, the large scale elimination of neurons by apoptosis generally closely follows the establishment of the first contact between the post mitotic neurons and their target cells. There is strong evidence that neuronal targets play a pivotal role in controlling neuronal survival during this crucial developmental period of a neuron, and it has been proposed that the target cells secrete limiting amounts of trophic molecules that are necessary to support the survival of embryonic neurons. According to this model, these secretory molecules bind to specific neuronal receptors, and are somehow able to stop programmed cell death in those neurons that have been able to successfully compete for them. It is likely that this target-dependent-selection serves some important function as neurons are exquisitely dependent on target-derived molecules at the time when a permanent relationship is established, but not before and much less (if at all) afterwards. This tightly developmentally regulated process might serve the purpose of quantitative matching between two populations that have been generated independently of each other, and it remains to be seen if it is of general importance in eliminating neurons projecting to inappropriate targets. At the molecular level, various proteins have been shown, mainly *in vitro*, to be able to block the process of apoptosis in embryonic neurons. Some of these proteins belong to a family of structurally related factors, the neurotrophins, which are small, basic and secretory proteins (Barde, 1990). At least for some neurotrophins, there is evidence that they are produced by target cells before and during the process of neuronal death, and the finding that some of them can also rescue damaged neurons, either during development after axotomy or even in the adult brain has spurred interest both in the basic mechanisms underlying their action and potential clinical applications (Thoenen *et al.,* 1987). In addition to their role in regulating neuronal survival, neurotrophins are very likely to play additional, important functions in the developing and adult nervous system. These include mitotic effects on neuronal precursors (Cattaneo and McKay, 1990; Kalcheim *et al.,* 1992), instructive differentiation effects on neural crest cells (Sieber-Blum, 1991), stimulation of neurite outgrowth (Lindsay, 1988) and regulation of neurotransmitter release (Lohof *et al.,* 1993). One of the greatest hopes is that the neurotrophins might help preventing the degeneration of adult neurons in neurodegenerative diseases, for example of motoneurons in amyotrophic lateral sclerosis or help in the re-growth of axons in the CNS after lesion once the inhibitory actions of oligodendrocyte-bound membrane components are neutralised (for review, see Schwab *et al.,* 1993).

So far, 4 members of the neurotrophin family have been discovered: nerve growth factor (NGF), brain-derived neurotrophic factor (BDNF), neurotrophin-3 (NT-3) and neurotrophin-4/5 (NT-4/5). Except for the last of these, the neurotrophins are highly conserved throughout vertebrate evolution. Within the neurotrophin family, each member is characterised by its unique pattern of expression and its action on specific sub-classes of neurons (Barde, 1990).

The first step in the action of neurotrophins is their recognition and binding by specific cell surface receptors. Binding studies with NGF, BDNF, and NT-3 using embryonic chick neurons indicate the existence of low affinity (K_d 10^{-9}M) as well as of high affinity (K_d 10^{-11}M) binding sites. In addition to their affinity, the high affinity binding sites exhibit a much higher selectivity for each neurotrophin compared with the low affinity sites. Also, they are thought to represent the signal transducing elements (Rodríguez-Tébar *et al.*, 1991). At present, three distinct genes coding for functional neurotrophin receptors have been characterised in mammals: *trk*, *trk*B and *trk*C. They belong to a sub-family of the receptor tyrosine kinases and are composed of an extracellular binding region, a single transmembrane domain and a tyrosine kinase domain followed by a very short C-terminal stretch of aminoacids (Barbacid *et al.*, 1991). Studies with recombinant rodent receptors on fibroblasts cell lines have shown that the interactions between the *trk*s and neurotrophins are complex (Fig. 1A). *Trk* has been shown to bind preferentially NGF, but also NT-3 and NT-4/5, *trk*B binds BDNF and NT-4/5 equally well and to a lesser extent NT-3, *trk*C appears to be a specific receptor for NT-3 (for review, see Meakin and Shooter, 1992).

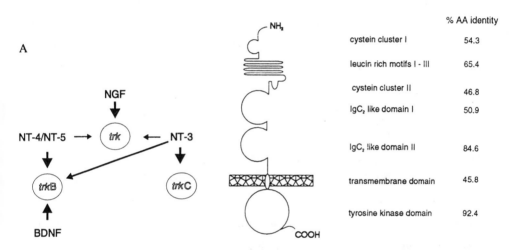

Fig. 1. A. Scheme of the interactions between neurotrophins and their *trk* receptors.
    B. Sequence homolgies in the extracellular domain of chick *trk*B.

While two different binding sites with low and high affinities for the same ligand can be discriminated on cell lines expressing recombinant *trk* receptors, in all cases reported so far the low affinity binding sites are by far the predominant ones. The proportion of receptors with high affinity on the transfected cell lines ranges from not detectable to about 10 percent of the total receptor population. Whether these high affinity receptors are structurally related or even identical to the high affinity neurotrophin binding sites on responsive primary neurons remains to be established. The molecular complexity of interactions between neurotrophins and *trk* receptors is further complicated by the existence of alternatively spliced forms of all *trk* receptors. For *trk,* two isoforms are predicted from cDNA clones, which differ by the absence or presence of a short stretch of six amino acids in the extracellular domain close to the transmembrane region (Barker *et al.,* 1993). For *trk*B, three forms have been described in the rodent brain: a full length form with a tyrosine kinase domain, as well as two forms which are truncated intracellularly and which are characterised by short, specific amino acid sequences at their carboxyterminal ends (Klein *et al.,* 1990; Middlemas *et al.,* 1991). The most complex *trk* locus appears to be *trk*C for which one truncated form, as well as probably more than two forms with amino acid insertions in the tyrosine kinase domain have been described recently (Valenzuela *et al.,* 1993; Tsoulfas *et al.,* 1993; Lamballe *et al.,* 1993). The binding of neurotrophins to *trk* tyrosine kinase receptors is followed by a rapid phosphorylation of tyrosine residues, presumably as a result of ligand induced dimerisation and receptor cross-phosphorylation. Depending on the cellular context, activation of a cascade of intracellular signals may result in the stimulation of mitosis with fibroblasts or in cellular differentiation with PC12 cells (Ip *et al.,* 1993).

So far, all data characterising the interactions between the neurotrophins and the *trk*s have been obtained with mammalian *trk*s. However, most studies characterising the action of neurotrophins on embryonic neurons and their interactions with neuronal receptors have made use of neurons isolated form the chick (*Gallus gallus*) embryo. The analysis of how the structure, localisation and interaction with the neurotrophins of the *trk*-family of receptors have been conserved in non-mammalian models may provide insight into the way neurotrophins can regulate neuronal properties including the control of neuronal survival. Furthermore, while the recent results summarised above all indicate that the *trk* genes code for functional neurotrophin receptors on neurons, some questions remained unresolved. These include the precise molecular composition of the neuronal high affinity receptors for neurotrophins on responsive neurons, as well as the significance for the *in vivo* situation of the low affinity *trk* binding sites found on transfected fibroblasts. As some of the discrepancies between *trk receptors* on fibroblasts and neurotrophin binding sites on responsive primary neurons might simply result from species differences, we decided to approach this question by using chick *trk*B (c*trk*B) as a model (Dechant *et al.,* 1993).

Sequence comparison between chick *trk*B and mammalian *trk* receptors

The open reading frame of the c*trk*B encoding cDNA predicts a protein of 818 amino acids. The comparison of this amino acid sequence with the known mammalian *trk* sequences reveals that the greatest similarity is found with rat *trk*B (76.8% identity) followed by rat *trk*C (57.1% identity) and human *trk* (51%). The homology between mammalian *trk*B and chick *trk*B (c*trk*B) is unevenly distributed along the molecule: while the tyrosine kinase domain is highly conserved (92.4%), the extracellular and the transmembrane domains have undergone more extensive variation during vertebrate evolution. The high conservation of the tyrosine kinase domains is very likely due to the high evolutionary pressure to maintain functional cytoplasmic domains. In this part of the molecule, c*trk*B contains sites of phosphorylation related to its enzymatic activity, as well as those implicated in the interactions with intracellular signal transducing molecules (Schlessinger and Ullrich, 1992). In contrast, the transmembrane and the extracellular binding domain of ctrkB are, respectively, only 45.8 and 61.2% identical with the corresponding rat *trk*B sequences. This is somewhat unexpected in view of the high degree of conservation of BDNF (identical in all mammals and 92.4% between chick and mammals). The extracellular sequence can be further subdivided into several sub domains, which are characterised by their relatedness to extracellular domains of other proteins. Chicken *trk*B contains two immuno-globulin-like domains, three leucine-rich motifs (LRM), and two cysteine clusters (see Fig. 1B). Two structural elements were found to be highly conserved: the three LRM elements, as well as the second immunoglobulin-like domain. The function of the LRM elements is not clearly understood, but such motifs have been described in several cell adhesion and receptor molecules, and they are thought to participate in homophilic or heterophilic protein-protein interactions (Schneider and Schweiger, 1991). In *trk*B (as well as in the other *trk*s) they might favour dimer formation. The highest degree of conservation in the extracellular domain of *trk*B (85.4% identity) is found in the second immunoglobulin-like domain. Since a similar domain has been reported to be involved in specific ligand binding of the FGF receptor-4 tyrosine kinase (Vainikka *et al.*, 1992), it is conceivable that this part of the molecule might be important in the binding of BDNF.

Chick c*trk*B is a receptor for BDNF, NT-4 and NT-3 but not for NGF

In binding studies with radioiodinated BDNF as a ligand, we compared the binding properties of c*trk*B expressed on a cell line (A293-c*trk*B) with the BDNF binding sites expressed on DRG neurons freshly isolated from the chick embryo (Dechant *et al.*, 1993). These experiments confirmed that c*trk*B is indeed a BDNF receptor, and that the BDNF binding sites on A293-c*trk*B cells and chick sensory neurons do show some common features with regard to ligand specificity. Thus, for example, the binding of BDNF to both A293-c*trk*B cells and sensory neurons receptors can be efficiently prevented by NT-4, less so by NT-3, and not at all for c*trk*B by NGF. Despite these general similarities, differences were also observed. In particular, in steady state binding experiments, the only binding site for BDNF on A293-ctrkB exhibited a low affinity of essentially 10^{-9}M, as opposed to the neuronal receptors displaying an approximately 50-fold higher affinity. A nanomolar affinity for BDNF, like on A293-c*trk*B cells, has also been reported for cell lines transfected with rodent *trk*B (Soppet *et al.*, 1991), indicating that the molecular differences between rodent and c*trk*B are not sufficient to explain the differences in the affinity between recombinant receptors and the neuronal binding sites. Also, whereas NT-4 completely and efficiently blocked the binding of BDNF to A293-c*trk*B cells, a sub population of neuronal BDNF binding sites was insensitive to even high concentrations of NT-4. The latter finding indicates heterogeneity amongst the neuronal BDNF high affinity binding sites. Taken together, these results indicate that while c*trk*B is highly likely to be part of the BDNF high affinity binding sites observed on peripheral chick sensory neurons, it probably exists in a modified form in these neurons.

Expression of full length c*trk*B in the peripheral nervous system
Northern Blot analysis showed that in the developing chick embryo, several transcripts coding for c*trk*B can be detected and that one class of transcripts lacks the tyrosine kinase domain. In general, a good correlation was observed between the neuronal structures that expressed full length c*trk*B mRNA *in vivo*, and those known to contain BDNF responsive neurons. Thus, c*trk*B mRNA was found in BDNF-responsive ganglia of cranial nerves such as the trigeminal, nodose or stato-acoustic ganglia. In the trigeminal and dorsal root ganglia, morphologically discernible subsets of neurons expressed c*trk*B mRNA, namely the large neurons located in the ventral portion of the ganglion. In other BDNF responsive ganglia, as the ganglion nodosum, c*trk*B expressing neurons were found amongst unlabelled neurons (Dechant *et al.*, 1993). Of interest is the observation that in all structures examined and known to respond to BDNF *in vivo* or *in*

vitro (with the exception of the retinal ganglion cells), c*trk*B mRNA expression largely preceded the onset of developmentally controlled neuronal death, and was maintained after this period.

c*trk*B is abundantly expressed in the developing brain

Like the brain of all vertebrates, the chick encephalon develops first as a series of thin vesicles surrounded by a neuroepithelium. In the developmental period between E3 and E14 cell division, cell death and cell migration shape the morphological appearance of the adult brain. At E5, the most prominent structure of the chick brain is the optic tectum. Subsequently, the optic tectum can be schematically divided into a superficial region that receives primary afferent fibres from the retina (stratum opticum), a nuclear and fiber layer containing various classes of interneurons and dendrites from the underlying layer (stratum griseum and fibrosum superficiale), as well as a nuclear layer containing the cell bodies of the ganglion cells (stratum griseum centrale). These ganglion cells represent the main output of the optic tectum and they convey the visual information to other structures mainly through the tecto-bulbar and tecto-thalamic tracts. In the developing optic tectum, the localisation of full length c*trk*B mRNA was almost exclusively in the ganglion cells. These large neurons were found to express c*trk*B message at E5, concomitant with their migration on glial processes (Fig. 2). It is interesting to observe that in the mammalian brain, the visual cortex, a brain structure of different embryological origin, but with a similar function to the avian optic tectum, also contains high levels of *trk*B receptor. As in the peripheral nervous system, the expression of c*trk*B mRNA remained strong throughout the developmental period studied. In addition, chick *trkB* mRNA was found in a variety of nuclei in the developing brain. In the chick optic tectum itself, mRNA coding for the truncated form of chick c*trk*B was also observed: mRNA coding for truncated receptors was detected in the ependymal layer lining the ventricles (Fig. 2), and in lepto-meningeal cells surrounding the optic tectum. The functional significance of these *trk*B variants, also described in mammals with similar patterns of expression is currently unclear. It appears likely that in lepto-meningeal cells, the *trk*B variants lacking the tyrosine kinase domain are expressed (at least in some cells) without the simultaneous expression of the transcripts coding for the tyrosine kinase domain. Curiously, in view of the widespread and often high levels expression of c*trk*B mRNA in the CNS, the retinal ganglion cells did not show detectable levels of c*trk*B mRNA until E14. This is also surprising in view of the known BDNF responsiveness of these cells *in vitro*.

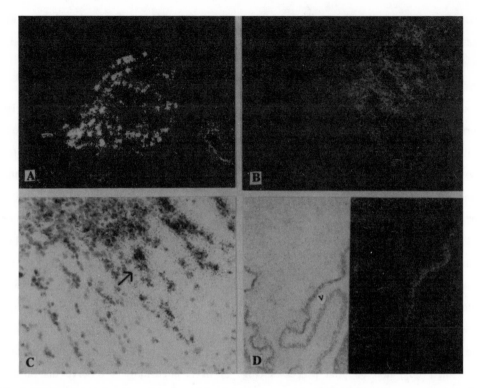

Fig. 2. Full length *trk*B is expressed in the peripheral nervous system (top left, E14 dorsal root ganglion), in the developing tectum (B,C embryonic day 5: dark and bright field) and later in ganglion cells (B).The truncated receptor is expressed in non-neuronal ependymal cells lining the ventricles (D).

Conclusion

In developing embryos, it has been shown that the administration of BDNF can prevent neuronal death in peripheral ganglia, both in the neural-crest derived DRG, as well as in the (NGF-insensitive) placode-derived neurons of the nodose ganglion (Hofer and Barde, 1988). The lack of antibodies specifically blocking the biological activity of BDNF has prevented the converse (and crucial) experiment to be done. However, the observation that c*trk*B both binds BDNF, as first shown in rodent systems, and is expressed on the neurons known to respond to BDNF lend support to the suggestion that BDNF is a molecule that is physiologically relevant and prevents the elimination of certain neuronal populations during normal development. The early expression of c*trk*B suggests other roles for BDNF during development beyond the block of *apoptosis*, and the expression pattern in the CNS, as previously observed with rodents, also suggest that in fact numerous CNS neurons might respond to BDNF. The abundant and early expression of truncated forms of c*trk*B, specially in the lepto-meninges is very intriguing. In this context, it is perhaps relevant to note that c*trk*B is a receptor that binds with substantial affinity to at least three different ligands, BDNF, NT-4/5 and NT-3, but not NGF, the expression of which is far more restricted than that of the other neurotrophins. Possibly, the truncated form of this receptor serves the function of preventing diffusion of several neurotrophins and helps to create boundaries.

References

Barbacid, M., Lamballe, F., Pullido, D. and Klein, R. (1991) The trk family of tyrosine kinase receptors. *BBA Rev. Cancer* 1072: 115-127.

Barde, Y.-A. (1990) The Nerve Growth Factor Family. *Prog. Growth Factor Res.* 2: 237-248.

Barker, P.A., Lomen-Hoerth, C., Gensch, E.M., Meakin, S.O., Glass, D.J. and Shooter, E.M. (1993) Tissue-specific alternative splicing generates two isoforms of the *trkA* receptor. *J. Biol. Chem.* 268: 15150-15157.

Cattaneo, E. and McKay, R. (1990) Proliferation and differentiation of neuronal stem cells regulated by nerve growth factor. *Nature* 347: 762-765.

Dechant, G., Biffo, S., Okazawa, H., Kolbeck, R., Pottgiesser, J. and Barde, Y.-A. (1993) Expression and binding characteristics of the BDNF receptor chick trkB. Development 119: 545-558.

Hofer, M.M. and Barde, Y.-A. (1988) Brain-derived neurotrophic factor prevents neuronal death *in vivo*. *Nature* 331: 261-262.

Ip, N.Y., Stitt, T.N., Tapley, P., Klein, R., Glass, D.J., Fandl, J., Greene, L.A., Barbacid. M. and Yancopoulos, G.D. (1993) Similarities and differences in the way neurotrophins interact with the Trk receptors in neuronal and nonneuronal cells. *Neuron* 10: 137-149.

Kalcheim, C., Carmeli, C. and Rosenthal, A. (1992) Neurotrophin 3 is a mitogen for cultured neural crest cells. *Proc. Natl. Acad. Sci. USA* 89: 1661-1665.

Kerr, J.F.R., Wyllie, A.H. and Currie, A.R. (1972) Apoptosis: a basic biological phenomenon with wide-ranging implications in tissue kinetics. *Br. J. Cancer* 26: 239-244.

Klein, R., Conway, D., Parada, L.F. and Barbacid, M. (1990) The trkB tyrosine protein kinase gene codes for a second neurogenic receptor that lacks the catalytic kinase domain. *Cell* 61: 647-656.

Lamballe, F., Tapley, P. and Barbacid, M. (1993) *trkC* encodes multiple neurotrophin-3 receptors with distinct biological properties and substrate specificities. *EMBO J.* 12: 3083-3094.

Lindsay, R.M. (1988) Nerve growth factors (NGF, BDNF) enhance axonal regeneration but are not required for survival of adult sensory neurons. *J. Neurosci.* 8: 2394-2405.

Lohof, A.M., Ip, N.Y. and Poo, M. (1993) Potentiation of developing neuromuscular synapses by the neurotrophins NT-3 and BDNF. *Nature* 363: 350-353.

Meakin, S.O. and Shooter, E.M. (1992) The nerve growth factor family of receptors. *Trends Neurosci.* 15: 323-331.

Middlemas, D.S., Lindberg, R.A. and Hunter, T. (1991) trk B, a neural receptor protein-tyrosine kinase: evidence for a full-length and two truncated receptors. *Mol. Cell. Biol.* 11: 143-153.

Rodriguez-Tébar, A., Dechant, G. and Barde, Y.-A. (1991) Neurotrophins: structural related ness and receptor interactions. *Phil. Trans. R. Soc. Lond.* 331: 255-258.

Schlessinger. J. and Ullrich, A. (1992) Growth factor signaling by receptor tyrosine kinases. *Neuron* 9: 383-391.

Schneider, R. and Schweiger, M. (1991) A novel modular mosaic of cell adhesion motifs in the extracellular domains of the neurogenic *trk* and *trkB* tyrosine kinase receptors. *Oncogene* 6: 1807-1811.

Schwab, M.E., Kapfhammer, J.P. and Bandtlow, C.E. (1993) Inhibitors of neurite growth. *Annu. Rev. Neurosci.* 16: 565-595.

Sieber-Blum, M. (1991) Role of the neurotrophic factors BDNF and NGF in the commitment of pluripotent neural crest cells. *Neuron* 6: 1-20.

Soppet, D., Escandon, E., Maragos, J., Middlemas, D.S., Reid, S.W., Blair, J., Burton, L.E., Stanton, B.R., Kaplan, D.R., Hunter, T., Nicolics, K. and Parada, L.F. (1991) The neurotrophic factors Brain-Derived Neurotrophic Factor and Neurotrophin-3 are ligands for the trkB tyrosine kinase receptor. *Cell* 65: 21.

Thoenen, H., Bandtlow, C. and Heumann, R. (1987) The physiological function of nerve growth factor in the central nervous system: comparison with the periphery. *Rev. Physiol. Biochem. Pharmacol.* 109: 145-178.

Tsoulfas, P., Soppet, D., Escandon, E., Tessarollo, L., Mendoza-Ramirez, J.-L., Rosenthal, A., Nikolics, K. and Parada, L.F. (1993) The rat *trkC* locus encodes multiple neurogenic receptors that exhibit differential response to neurotrophin-3 in PC12 cells. *Neuron* 10: 975-990.

Vainikka, S., Partanen, J., Bellosta, P., Coulier, F., Basilico, C., Jaye, M. and Alitalo, K. (1992) Fibroblast growth factor receptor-4 shows novel features in genomic structure, ligand binding and signal transduction. *EMBO J.* 11: 4273-4280.

Valenzuela, D,M., Maisonpierre, P.C., Glass, D.J., Rojas, E., Nuñez, L., Kong, Y., Gies, D.R., Stitt, T.N., Ip, N.Y. and Yancopoulos, G.D. (1993) Alternative forms of rat TrkC with different functional capabilities. *Neuron* 10: 963-974.

Toward a Molecular Basis of Alcohol Use and Abuse
ed. by B. Jansson, H. Jörnvall, U. Rydberg, L. Terenius & B. L. Vallee
© 1994 Birkhäuser Verlag Basel/Switzerland

Effects of alcohol on gene expression in neural cells

Norbert Wilke, Michael Sganga, Steven Barhite and Michael F. Miles

Ernest Gallo Clinic and Research Center and the Department of Neurology, University of California at San Francisco, Building 1, Room 101, San Francisco General Hospital, San Francisco, CA 94110, USA

Summary

Our studies in the NG108-15 neuroblastoma x glioma cell line previously showed that the molecular chaperonin, Hsc70, is an ethanol-responsive gene (EtRG) regulated at the level of transcription by ethanol. We recently identified two related molecular chaperonins, GRP94 and GRP78, as EtRGs with GRP94 mRNA abundance being induced by ethanol more than three-fold vs. control. Stable transfection studies show that GRP78 transcription is also regulated by ethanol and that ethanol also potentiates GRP78 induction by classical inducing agents such as tunicamycin. Recently, we have found that ethanol induction of Hsc70 may require cis-acting promoter sequences recognized by the DNA-binding protein Sp1. Chronic ethanol exposure does not alter Sp1 DNA-binding activity, thus suggesting a possible ethanol-induced post-translational modification that activates Sp1 function. We predict that the molecular mechanisms underlying ethanol regulation of Hsc70, GRP94 and GRP78 may be similar since they have related functions. GRP94 and GRP78 (GRP94/78) are known to be induced by agents which inhibit glycoprotein processing or deplete endoplasmic reticulum stores of calcium. In turn, induction of GRP78 expression is known to selectively alter the transport of glycoproteins and produce "tolerance" to depletion of sequestered intracellular calcium. The regulation of these genes by ethanol could thus relate to the known effects of ethanol on calcium homeostasis and protein trafficking. The actions of ethanol on chaperonin gene expression may have important mechanistic implications for CNS adaptation to ethanol, particularly if other EtRGs share the same regulatory mechanisms.

Introduction

Alcoholics have been reported to survive or even remain sober with blood ethanol levels above 200 mM (Lindblad and Olsson, 1976). In contrast, blood ethanol levels above 65 mM are usually associated with severe central nervous system (CNS) depression or death in naive individuals. This suggests that alterations in brain function provide a large role in the adaptation to chronic ethanol use. It is our working hypothesis that ethanol-induced alterations in CNS gene expression could underlie the development of tolerance and dependence.

We and several other laboratories have recently documented that ethanol can cause increases (Parent et al., 1987; Charness et al., 1988; Kolber et al., 1988; Gayer et al., 1991; Miles et al., 1991) or decreases (Dave et al., 1986; Mochly-Rosen et al., 1988; Gulya et al., 1991) in the abundance of specific mRNAs or proteins in cultured neural cells or in the CNS of intact animals. Ethanol-responsive genes (EtRGs) that have been identified to date suggest that such ethanol-induced changes could have profound functional consequences. Tyrosine hydroxylase (Gayer *et al.*, 1991), the rate-limiting enzyme in catecholamine biosynthesis, and the GTP-binding protein subunits, $G_{s\alpha}$ (Mochly-Rosen *et al.*, 1988) and $G_{i\alpha}$ (Charness *et al.*, 1988) are such examples of EtRGs which suggest that ethanol-induced changes in gene expression could lead to important adaptive changes in CNS function. This chapter will summarize our recent work on the identification, regulation and function of EtRGs.

Materials and Methods

Drosophila embryonic Schneider line 2 (SL2) cells were cultured in Schneider's *Drosophila* Medium (Gibco) supplemented with 10% heat-inactivated fetal calf serum at 25°C without CO_2 in 25 cm^2 T-flasks (Corning) and transfected as described by Courey *et al.* (1989). All cell culturing, transfection and molecular biology procedures were done as described previously or according to standard techniques (Gayer *et al.*, 1991; Miles *et al.*, 1991; 1993).

Results

We have recently used a modified subtractive hybridization procedure to isolate EtRGs (Miles, Wilke and Elliot, manuscript in revision). Seven EtRGs have been identified to date, as verified by Northern blot analyses. Gene inductions were 1.6-3.0 fold following 24 hour treatment with 100 mM ethanol. DNA sequence analysis identified 4 EtRGs as an intracisternal A-type particle gene, an insulin-responsive growth associated protein, malic enzyme and GRP94, a member of the "glucose-responsive" subgroup of stress proteins (see below). The DNA sequence of another EtRG showed an open reading frame with a 65% homology to phosducin, a retinal phospho-protein that is known to regulate G protein function (Bauer *et al.*, 1992). The remaining genes had no significant homology in DNA sequence data bases. Ongoing studies are aimed at further identifying the structure and function of the novel EtRGs and the physiological role of these gene inductions in CNS adaptation to ethanol.

Identification of a novel subset of stress proteins as EtRGs
The identity of different genes that have been shown to respond to ethanol provides little clue

as to underlying mechanisms of regulation or even the overall impact on cellular phenotype. However, as noted above, subtractive hybridization cloning identified the molecular chaperonin, GRP94, as an EtRG. Recently, we found that the highly related gene, GRP78, is also regulated by ethanol (Sganga *et al.*, 1992). Thus, these two molecular chaperonins, together with the chaperonin Hsc70, which we have previously identified as an EtRG, are coordinately regulated by ethanol. Hsc70, GRP78 and GRP94 are constitutively expressed members of the molecular chaperonin gene family and are thought to provide crucial functions in protein trafficking.

Hsc70 is a molecular chaperonin which binds to nascent polypeptide chains on polyribosomes and is required for their transport across the endoplasmic reticulum or mito-chondrial membrane (Deshaies *et al.*, 1988; Beckman *et al.*, 1990). Hsc70 has also been shown to act as an uncoating enzyme for the removal of clathrin triskelia from clathrin-coated vesicles (Chappell *et al.*, 1986). Thus, this protein plays a crucial role in protein and vesicular trafficking. We previously showed that Hsc70 is induced by chronic ethanol in NG108-15 neuro-blastoma x glioma cells through an increase in Hsc70 gene transcription (Miles *et al.*, 1991).

GRP94 and GRP78 respond to ethanol even more prominently than Hsc70 (Fig. 1). GRP78 and GRP94 are targeted to the endoplasmic reticulum (ER) by presence of a "KDEL" sequence (Munro and Pelham, 1986) and are thought to function in the trafficking and processing of glycoproteins. GRP78 is an intraluminal protein while GRP94 contains a single membrane-spanning domain and is thought to exist as a transmembrane ER protein.

Unlike the classic heat shock gene Hsp70, Hsc70, GRP78 and GRP94 show little or no response to heat shock. Hsc70 is known to be induced by agents which activate proliferation in lymphocytes (Hansen *et al.*, 1991). GRP78 and GRP94 are usually coordinately regulated with GRP78 being the most responsive to any stimuli studied to date. The promoter regions of GRP78 and GRP94 show striking similarities (Chang *et al.*, 1989). These genes are induced by glucose deprivation, calcium ionophore (e.g. A23187) and agents, such as tunicamycin, AlF_4- and brefeldin A, which inhibit normal processing or trafficking of glycoproteins (Chang *et al.*, 1989). The underlying molecular event regulating GRP induction by all of these factors is unknown but it has been suggested that either depletion of intracellular stores of calcium (Drummond *et al.*, 1987) and/or accumulation of malfolded proteins in the ER (Dorner *et al.*, 1989) are the instigating factors. The former mechanism is particularly emphasized by the

52

finding that a selective modulator of intracisternal calcium, thapsigargin, is sufficient to induce GRP78/94. Thapsigargin produces depletion of ER calcium stores by inhibiting a specific calcium-ATPase present in the ER membrane (Thastrup *et al.*, 1990).

We found that ethanol produces a unique induction pattern of GRP94 > GRP78 = Hsc70 (Fig. 1). Other classical GRP inducers, such as tunicamycin, an inhibitor of core glycosylation, induce GRP78 most prominently (Fig. 1). Ethanol treatment is the first instance where GRP equals or even exceeds the induction level seen with GRP78. Interestingly, addition of ethanol along with tunicamycin consistently produced more than additive responses with saturating levels of tunicamycin (Fig. 1). These results suggest that ethanol may induce GRP expression through a mechanism(s) distinct from that of classical inducers.

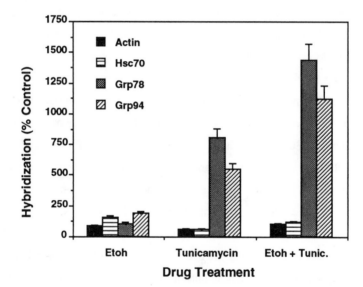

Fig. 1. Northern blot analysis of ethanol effects on GRP94, GRP78 and Hsc70 mRNA abundance.
NG108-15 cells were exposed to ethanol (100 mM), tunicamycin (0.5 mg/ml) or a combination of the two drugs for 24 hours. RNA levels for the indicated genes were then determined by Northern blot hybridization. Results are expressed as percent of control (untreated) and represent the mean ± S.D. for quadruplicate determinations. Similar results were obtained in experiments repeated twice. Results show that the increase in GRP78 and GRP94 mRNAs by tunicamycin + ethanol is much greater than additive results of the two drugs in isolation. Ethanol similarly potentiates the brefeldin A or A23187 induction of GRP78 and GRP94 (not shown).

Identification of ethanol-responsive cis-acting sequences in the Hsc70 promoter

We have recently studied how ethanol induces increased mRNA abundance for the Hsc70 and

GRP78 genes. Nuclear runoff analysis with the Hsc70 gene showed that ethanol increased the transcription rate of this gene while showing no effect on the rate of β-actin or lactate dehydrogenase (LDH) transcription (Miles *et al.*, 1991). This was the first direct evidence that showed that ethanol could regulate gene transcription.

Stable transfection analyses with the Hsc70 gene showed that the proximal 2500 base pairs of the promoter contain ethanol-responsive cis-acting elements (Miles *et al.*, 1991). Transient transfection studies have been used to finely map the ethanol-responsive regions of the Hsc70 promoter. Ethanol concentrations of 100-200 mM were generally used in deletion experiments to achieve more reproducible responses. Fig. 2 shows the results of deletion and point mutation analysis with the Hsc70 promoter.

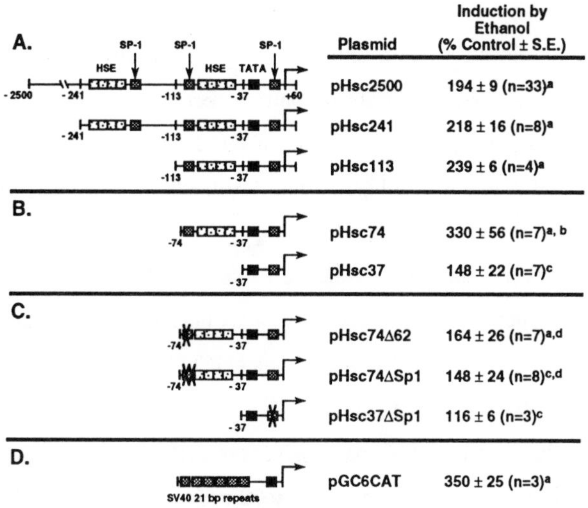

Fig. 2. Identification of ethanol-responsive cis-acting sequences. Deletions or point mutations in the Hsc70 promoter were made by progressive exonuclease digestion (panel A) or using PCR (panels B & C). Plasmid pGC6CAT transfected into NG108-15 cells by electroporation, treated with 200 mM ethanol for 24 h and analyzed for CAT activity. Results are expressed as percent of CAT activity in control (mock ethanol treatment) cells transfected in the same cuvette as ethanol-treated cells. Assays were done on triplicate wells o cells and results are presented as mean ± S.E. from experiments repeated as indicated. Statistical significance was determined by single group t-test analysis (for differences from control cells) or ANOVA with Scheffe post-hoc analysis to determine significant differences between various constructs: [a]p<0.05 vs control (t-test), [b]p<0.05 vs. pHsc2500 (ANOVA, Scheffe post-hoc), [c]p<0.05 vs. pHsc74 (ANOVA, Scheffe post-hoc) and [d]p<0.05 vs. control (t-test).

The positions of various known cis-acting elements are identified. Fig. 2 (panel A) shows that deletion constructs extending from -2500 to -113 of the Hsc70 promoter all showed significant inductions by ethanol (200 mM) that were of very similar magnitudes.

Further deletions showed that pHsc74 (endpoint at -74) was actually more ethanol responsive than the full length pHsc2500. Point mutations in the Sp1 site at -67 to -61 (pHsc74ΔSp1, pHsc74Δ62) caused a large decrease (p < 0.05, ANOVA) in ethanol-responsiveness compared to the parent construct pHsc74 (Fig. 2, panel C). Construct pHsc74Δ62 was identical to pHsc74 except for a single base mutation which changed the Sp1 consensus site (-67 to -61) from GGGGCGG to GGGGCTG. Plasmid pHsc74ΔSp1 contained two point mutations in this Sp1 site (GGGGCGG to GGGGCTT). This more extensive mutation reduced the ethanol-response to below statistical significance (148±24; p=0.81, one-sample t-test, n=7). Deletion of the entire distal Sp1 site and HSE (pHsc37) caused no further decrease in ethanol-responsiveness (Fig. 2, panel B). Mutation of a potential Sp1 site at -9 (pHsc37ΔSp1) produced an even lower mean response to ethanol but did not reach a statistically significant difference from pHsc37 (116±6 vs. 148+22, respectively). When corrected for transfection efficiency, by use of an internal standard plasmid expressing β-galactosidase, all of the deletion and point mutations had levels of basal activity at least 3-fold above background levels obtained with a "promoter-less" CAT coding region (data not shown).

The ability of an Sp1 consensus sequence to confer ethanol-responsiveness was further confirmed using a construct (pGC6CAT) containing six tandem Sp1 binding sites from the SV40 21 bp repeat region. Fig. 2 (Panel D) shows that pGC6CAT respnded to ethanol with CAT activity increasing to 350% of control values.

To more directly correlate Sp1 responsiveness with ethanol induction of the Hsc70 promoter, we assayed the response of various Hsc70 promoter constructs to Sp1 in *Drosophila* SL2 cells, which lack endogenous Sp1 (Courey and Tjian, 1988). Hsc70 promoter CAT constructs were cotransfected into SL2 cells together with an Sp1 expression plasmid (pPacSp1) or an empty expression vector (pPac0). The highly ethanol-responsve plasmids, pHsc2500 and pHsc74, have striking responses to Sp1 in SL2 cells (Sganga and Miles, manuscript in preparation). Similar results were observed when responses to Sp1 were measured by *in vitro* transcription in *Drosophila* embryo lysates that were obtained from a commercial vendor (Promega) (M. Miles, manuscript in preparation).

The SL2 cell data, together with the results from Fig. 2, suggests that the distal SP1 consensus binding site at -67 to -61 is a strong Sp1-dependent transcriptional activator and that this activity is required for ethanol-responsiveness of Hsc70. The more proximal Sp1 site at -9 produced minimal ethanol responsiveness. This suggests that either context of the Sp1 site is also important in conferring ethanol-responsiveness or, that synergism between Sp1 sites is needed for ethanol induction of transcription. All of the Hsc70 promoter constructs had basal CAT expression levels of at least 3 to 4-fold above background levels.

To determine how ethanol might modulate Sp1-dependent transcription, gel mobility shift assays (GMSA) were performed to determine whether ethanol alters Sp1 binding activity. Using the -40/-74 region of the Hsc70 promoter as probe, a single major band and several minor more rapidly migrating bands were seen on GMSA. Appropriate controls with cold competitor oligo-nucleotides and commercially available Sp1 (Promega) identified the major band as an Sp1-like protein. Control and ethanol-treated (200 mM, 24 hours) protein lysates from three separate experiments had no significant difference in Sp1-binding activity on GMSA using non-saturating amounts of protein lysate. These results suggest that chronic ethanol treatment modulates Hsc70 expression without altering Sp1 binding activity.

Ethanol induction of GRP78 transcription

We have recently initiated studies on the GRP78 and GRP94 promoter regions to determine if these genes are also transcriptionally regulated by ethanol using the same cis-acting elements as seen with Hsc70. Stable tranfection experiments using the rat GRP78 promoter coupled to a CAT reporter gene (generous gift of Dr. Amy Lee, University of Southern California) suggest that ethanol regulates GRP78 transcription, as seen with Hsc70. Cells transfected with a construct containing 1.25 kb of the rat GRP78 promoter (Resendez *et al.*, 1985) showed CAT activity increase to $153 \pm 12\%$ of control ($p < 0.001$, t-test, n=13) following 24 hours treatment with 100 mM ethanol. Moreover, when ethanol induction was compared to thapsigargin, which induces GRP expression by causing a selective depletion of endoplasmic reticulum calcium (Thastrup *et al.*, 1990), there was a potentiation of the thapsigargin induction of GRP78 by ethanol (Fig. 3). Thus, across the entire thapsigargin dose response range, cells treated with ethanol (100 mM) + thapsigargin show a 2-fold increase in CAT activity compared to thapsigargin alone.

56

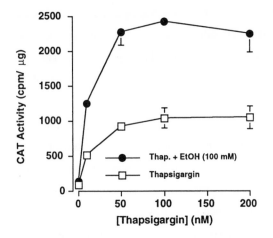

Fig. 3. Ethanol potentiates induction of GRP78 promoter activity by thapsigargin. NG108-15 cells transfected with GRP78-CAT (NG78CAT.25 cells) were treated for 24 hours with the indicated concentrations of thapsigargin in the presence or absence of 100 mM ethanol. CAT activity is expressed as the mean $\pm$ S.D. of results from triplicate wells of cells. Points withou error bars have S.D. smaller than the symbol size. Ethanol caused a 2.0-2.5-fold increase in the thapsigargin response at all concentrations of thapsigargin. Induction by ethanol alone was 50% above control levels (129 ± 11 cpm/μg vs. 83 ± 6 cpm/μg, respectively).

When the ethanol concentration-response curve was studied in the presence or absence of 100 nM thapsigargin, ethanol concentrations as low as 25 mM produced a near doubling of the thapsigargin induction (Miles, Wilke and Harris, manuscript in preparation). This potentiation of GRP induction by ethanol was also seen with calcium ionophore (A23187) treatment (not shown). Thus, although ethanol alone causes an induction of GRP78 promoter activity, an even greater magnitude of ethanol response is seen with the potentiation of thepsigargin or ionophore-induced GRP78 expression. This data is consistent with the Northern blot studies on GRP78 and GRP94 mRNA which showed ethanol potentiation of tunicamycin inductions (Fig. 1). This potentiation by ethanol occurs at saturating levels of tapsigargin, tunicamycin or A23187, suggesting that ethanol induction of GRP promoter activity is separate from the pathway involved in classical inducer action.

Discussion

Our own recent studies and the work of other investigators have clearly documented specific changes in neuronal gene expression with chronic exposure to ethanol. In particular, we have identified ethanol regulation of three related genes all involved in protein trafficking. GRP94/78, in addition to "sensing" protein trafficking, are also closely regulated by intracellular stores of calcium. Thus, ethanol induction of these genes possibly either could reflect or cause alterations in cellular protein trafficking and calcium homeostasis. These EtRGs thus may be indicative of a major aspect of cellular adaptation to ethanol. Definition of the mechanisms underlying these ethanol-induced changes in gene expression is a major focus of our labratory.

Our preliminary studies have suggested a requirement for the trans-acting factor Sp1 in ethanol induction of Hsc70 transcription. Although clearly more work is required to firmly establish the role of Sp1 in ethanol regulation of Hsc70, involvement of Sp1 in gene regulation by ethanol is an attractive hypothesis since this would allow the coordinate regulation of a number of genes by ethanol, thus producing the pleiotropic changes which are the hallmark of this drug. Sp1 could, nonetheless, mediate highly selective changes in gene expression since not all genes have GC box motifs. Furthermore, only a subset of promoters with consensus Sp1 sites would be expected to bind Sp1 or use bound Sp1 as a rate-limiting factor. For example, the proper spacing of Sp1 sites with each other or the TATA box have been shown to be crucial for Sp1 induction of transcription (Segal and Berk, 1991). Promoter context, i.e. the other DNA-binding proteins in the direct vicinity of a potential Sp1 binding site together with factors which might interact positively or negatively with Sp1, could also influence the ethanol-responsiveness of a Sp1-containing promoter.

How ethanol could regulate SP1 function remains to be determined. However, Sp1 has been shown to be both O-glycosylated and phosphorylated. Jackson and Tjian demonstrated that Sp1 bears O-linked N-acetylglucosamine monosaccarides and that this modification significantly increased Sp1 functional activity without altering DNA binding (Jackson and Tjian, 1988).

Given our findings to date, we now need to determine whether ethanol regulation of GRP94 and GRP78 proceed through the same cis-acting element(s) as Hsc70. It is tempting to speculate that Sp1 sites in the GRP promoter regions may mediate ethanol-induced increases in basal GRP94/78 transcription and that Sp1 is capable of interacting with other transcription factors which are responsible for "classical" GRP-inducing agent action. Such interactions are well known for SP1 (Pascal and Tjian, 1991) and could explain the potentiation responses we see with GRP induction by ethanol + thapsigargin vs. thapsigargin or ethanol alone (Fig. 3). Thus, ethanol could potentially modulate a variety of transcriptional regulatory elements through

58

such indirect actions. This would greatly increase the complexity of ethanol action on CNS function. The identification of such central processes in CNS adaptation to ethanol may provide new therapeutic insights for alcoholism or ethanol-related CNS injury.

Acknowledgements

This work was supported by grants from the NIAAA, the Alcoholic Beverage Medical Research Foundation and by intramural funding from the Ernest Gallo Clinic and Research Center.

References

Bauer, P.H., Muller, S., Puzicha, M., Pippig, S., Obermaier, B., Helmrich, E.J.M. and Lohse, M.J. (1992) Phosducin is a protein kinase A-regulated G-protein regulator. *Nature* 358: 73-76.

Beckmann, R.P., Mizen, L.A. and Welch, W.J. (1990) Interaction of Hsp70 with newly syn thesized proteins: Implications for protein folding and assembly. *Science* 248: 850-854.

Chang, S.C., Erwin, A.E. and Lee, A.S. (1989) Glucose-regulated protein (GRP94 and GRP78) genes share common regulatory domains and are coordinately regulated by common *trans*-acting factors. *Mol. Cell. Biol.* 9: 2153-2162.

Chappell, T.G., Welch, W.J., Schlossman, D.M., Palter, K.B., Schlesinger, M.J. and Rothman, J.E. (1986) Uncoating ATPase is a member of the 70 kilodalton family of stress proteins. *Cell* 45: 3-13.

Courey, A.J., Holtzman, D.A., Jackson, S.P. and Tjian, R. (1989) Synergistic activation by the glutamine-rich domains of human transcription factor Sp1. *Cell* 59: 827-836.

Courey, A.J. and Tjian, R. (1988) Analysis of Sp1 in vivo reveals multiple transcriptional domains, a novel glutamine-rich activation motif. *Cell* 55: 887-896.

Dave, J.R., Eiden, L.E., Karaman, J.W. and Eskay, R.L. (1986) Ethanol exposure decreases pituitary corticotropin-releasing factor binding, adenylate cyclase activity, proopio-melanocortin biosynthesis and plasma β-endorphin levels in the rat. Endocrinoogy 118: 280-286.

Deshaies, R.J., Koch, B.D., Werner-Washburne, M., Craig, E.A. and Schekman, R. (1988) A subfamily of stress proteins facilitates translocation of secretory and mitochondrial precursor polypeptides. *Nature* 332: 800-805.

Dorner, A.J., Wasley, L.C. and Kaufman, R.J. (1989) Increased synthesis of secreted proteins induces expression of glucose-regulated proteins in butyrate-treated Chinese hamster ovary cells. *J. Biol. Chem.* 264: 20602-20607.

Drummond, I.A.S., Lee, A.S., Resendez, E. and Steinhardt, R.A. (1987) Depletion of intra cellular calcium stores by calcium ionophore A23187 induces the genes for glucose-regulated proteins in hamster fibroblasts. *J. Biol. Chem.* 262: 12801-12805.

Gayer, G.G., Gordon, A. and Miles, M.F. (1991) Ethanol increases tyrosine hydroxylase gene expression in N1E-115 neuroblastoma cells. *J. Biol. Chem.* 266: 22279-22284.

Gulya, K., Dave, J.R. and Hoffman, P.L. (1991) Chronic ethanol ingestion decreases vasopressin mRNA in hypothalamic and extrahypothalamic nuclei of mouse brain. *Brain Res.* 557: 129-135.

Hansen, L.K., Houchins, J.P. and O'Leary, J.J. (1991) Differential regulation of HSC70, HSP70, HSP90α, and HSP90β mRNA expression by mitogen activation and heat shock in human lymphocytes. *Exp. Cell Res.* 192: 587-596.

Jackson, S.P. and Tjian, R. (1988) O-glycosylation of eukaryotic transcription factors: impli cations for mechanisms of transcriptional regulation. *Cell* 55: 125-133.

Kolber, M.A., Walls, R.M., Hinner, M.L. and Singer, D.S. (1988) Evidence of increased class I MHC expression on human peripheral blood lymphocytes during acute ethanol intoxication. *Alcoholism: Clin. Exp. Res.* 12: 820-823.

Lindblad, B. and Olsson, R. (1976) Unusually high levels of blood alcohol? *JAMA* 236: 1600-1602.

Miles, M.F., Barhite, S., Sganga, M. and Elliott, M. (1993) Phosducin-like protein (PhLP): an ethanol-responsive potential modulator of G-protein function. *Proc. Natl. Acad. Sci. USA*, in press.

Miles, M.F., Diaz, J.E. and DeGuzman, V.S. (1991) Mechanisms of neuronal adaptation to ethanol: Ethanol induces Hsc70 gene transcription in NG108-15 neuroblastom x glioma cells. *J. Biol. Chem.* 266: 2409-2414.

Mochly-Rosen, D., Chang, F.-H., Cheever, L., Kim, M., Diamond, I. and Gordon, A.S. (1988) Chronic ethanol causes heterologous desensitization by reducing α_s mRNA. *Nature* 333: 848-850.

Munro, S. and Pelham, H.R.B. (1986) An Hsp70-like protein in the ER: Identity with the 78 kd glucose-regulated protein and immunoglobulin heavy chain binding protein. *Cell* 46: 291-300.

Parent, L.J., Ehrlich, R., Matis, L. and Singer, D.S. (1987) Ethanol: an enhancer of major histocompatibility complex antigen expression. *FASEB J.* 1: 469-473.

Pascal, E. and Tjian, R. (1991) Different activation domains of Sp1 govern formation of multi-mers and mediate transcriptional synergism. *Genes Dev.* 5: 1646-1656.

Resendez, E., Attenello, J.W., Grafsky, A., Chang, C.S. and Lee, A.S. (1985) Calcium iono-phore A23187 induces expression of glucose-regulated genes and their heterologous fusion genes. *Mol. Cell. Biol.* 5: 1212-1219.

Segal, R. and Berk, A.J. (1991) Promoter activity and distance constraints of one *versus* two Sp1 binding sites. *J. Biol. Chem.* 266: 20406-20411.

Sganga, M.W., Wilke, N., Gayer, G., Chin, W., Barhite, S. and Miles, M.F. (1992) Regula-tion of ethanol-responsive genes in neural cells. *Soc. Neurosci. Abstr.* 18: 1335.

Thastrup, O., Cullen, P.J., Drobak, B.K., Hanley, M.R. and Dawson, A.P. (1990) Thapsigar-gin, a tumour promoter discharges intracellular Ca^{2+} stores by specific inhibition of the endo-plasmic reticulum Ca^{2+}-ATPase. *Proc. Natl. Acad. Sci. USA* 87: 2466-2470.

Toward a Molecular Basis of Alcohol Use and Abuse
ed. by B. Jansson, H. Jörnvall, U. Rydberg, L. Terenius & B. L. Vallee
© 1994 Birkhäuser Verlag Basel/Switzerland

The role of the NMDA receptor in ethanol withdrawal

Paula L. Hoffman and Boris Tabakoff

University of Colorado Health Sciences Center, Department of Pharmacology, C236, Denver, Colorado 80262, USA

Summary

The function of the N-methyl-D-aspartate (NMDA) subtype of glutamate receptor is very sensitive to acute inhibition by ethanol. Because of the role of this receptor in processes such as synaptic plasticity and neuronal development, it may contribute to the acute cognitive deficits caused by ethanol, or to the deleterious effects of ethanol during gestation. Excessive stimulation of the NMDA receptor is believed to be involved in the generation of epileptiform seizure activity as well as in excitotoxic cell death. Our studies have demonstrated that there is an adaptive "up-regulation" of NMDA receptor function in brains of chronically ethanol-treated animals and in cultured cells that have been exposed chronically to ethanol. This up-regulation appears to contribute to ethanol withdrawal seizure activity, since withdrawal seizures can be attenuated by specific NMDA receptor antagonists, and the time course of the change in receptor number parallels the time course of withdrawal seizures. In addition, cells exposed chronically to ethanol are significantly more susceptible to glutamate-induced cell death, which is mediated by the NMDA receptor, indicating a key role of the NMDA receptor in the well-characterized neuronal damage that is observed after chronic ethanol exposure and withdrawal in animals and humans. Understanding the basis for withdrawal seizures and withdrawal-induced neurotoxicity provides for the development of specific and selective therapeutic agents to ameliorate these consequences of chronic ethanol exposure and withdrawal.

Introduction

Physical dependence on ethanol (alcohol) is defined by the occurrence of a characteristic withdrawal syndrome following the cessation of chronic ethanol intake and the elimination of ethanol from an individual. The symptoms of ethanol withdrawal are, for the most part, opposite to the signs of acute intoxication (Tabakoff and Rothstein, 1983), and include seizure activity. Ethanol withdrawal seizures have some characteristics of grand mal seizures, but there is no seizure focus, and the EEG during sober periods is normal, in contrast to the abnormal EEG found in epileptics (Tabakoff and Rothstein, 1983). In addition, approximately half of patients with alcohol withdrawal seizures demonstrate myoclonus or convulsive seizures following photic stimulation, a phenomenon that does not occur in epileptic patients (Tabakoff and Rothstein, 1983). Although ethanol withdrawal seizures can therefore be distinguished from epileptic seizures, withdrawal seizures are more likely to occur in epileptics, since ethanol withdrawal lowers the seizure threshold (Tabakoff and Rothstein, 1983). Another predictor of the occurrence of ethanol withdrawal seizures is a history of ethanol withdrawal. It has been postulated that the generation of ethanol withdrawal seizures may be likened to a kindling process, since repeated episodes of ethanol withdrawal result in more rapid development of physical dependence and a more severe ethanol withdrawal syndrome (Ballenger and Post, 1978; Becker and Hale, 1993). This hypothesis suggests that the neurochemical mechanisms that underlie ethanol withdrawal seizures may be similar to those that are involved in kindling.

The search for mechanisms of ethanol withdrawal seizures and other withdrawal symptoms has been influenced by the fact that the signs of withdrawal are opposite to those of acute intoxication. It has been suggested that an adaptation of the central nervous system (CNS) to the initial effects of ethanol or other drugs would lead to withdrawal signs when the drug was removed (e.g., Goldstein and Goldstein, 1968). Although for many years the acute effects of ethanol were attributed to its ability to enter into and "fluidize" neuronal cell membranes (Hunt, 1985), more recently it has been recognized that certain membrane-bound proteins, including receptor-coupled ion channels and signal transduction systems, are very sensitive to perturbation by ethanol. These systems have been called "receptive elements" for ethanol (Tabakoff and Hoffman, 1987), and one such receptive system is the N-methyl-D-aspartate (NMDA) subtype of glutamate receptor.

Glutamate is believed to be the major excitatory neurotransmitter in brain, and it can interact with several different ionotropic and metabotropic receptor subtypes that have been characterized on the basis of their sensitivity to specific agonists and, more recently, by their molecular properties (Collingridge and Lester, 1989; Nakanishi, 1992). The NMDA receptor has several unusual features that have led to its extensive study (Collingridge and Lester, 1989). The activation of this receptor-coupled channel is voltage-dependent, meaning that there is greater activation as the cell becomes depolarized. The voltage-dependence is a consequence of the fact that Mg^{2+} binds to a site within the NMDA receptor-coupled channel and blocks it; the Mg^{2+} is released when the cell is depolarized. The importance of this phenomenon is that the response produced by the neurotransmitter, glutamate, at the NMDA receptor is dependent on the level of postsynaptic depolarization produced in a cell by glutamate (acting at other receptors) and/or other neurotransmitters. These conditions are similar to those that have been postulated to be necessary for the synaptic strengthening (e.g., long-term potentiation) that is believed to be important for learning and memory (see Cotman et al., 1989). In fact, there is substantial evidence implicating the NMDA receptor in the induction of long-term potentiation and in certain aspects of learning (Collingridge and Lester, 1989). Another unusual aspect of NMDA receptor function is an absolute requirement for glycine, which acts at a strychnine-insensitive site on the receptor, as a co-agonist. In the absence of glycine, glutamate cannot activate the NMDA receptor. Within the NMDA receptor-coupled ion channel there is a binding site for the dissociative anesthetics, ketamine and phencyclidine, and for dizocilpine, a dibenzylcyclo-heptenimine. All of these compounds act as uncompetitive antagonists at the NMDA receptor, and, when radioactively labeled, are often used to assess the number of NMDA receptors in tissue. One of the most important properties of the NMDA receptor is that, when the receptor is activated by agonist (e.g., glutamate or NMDA), the associated ion channel becomes permeable not only to monovalent cations, but also to Ca^{2+} (Collingridge and Lester, 1989). The response to NMDA is relatively slow (compared, for example, to the response to glutamate at other receptor subtypes), and as a result a large amount of Ca^{2+} can enter the cell through the activated NMDA receptor. It is this influx of Ca^{2+} which is believed to contribute to the involvement of the NMDA receptor in synaptic plasticity and development. However, when the receptor is excessively stimulated, the influx of Ca^{2+} can lead to epileptiform seizure activity and cell death (Cotman *et al.*, 1989; Collingridge and Lester, 1989).

Over the past few years, it has been repeatedly demonstrated that NMDA receptor function is very sensitive to acute inhibition by ethanol (see Hoffman, 1994). Initial studies showed that, in cerebellar granule cells in primary culture, low concentrations of ethanol, which are attainable *in vivo* (10 - 25 mM), inhibited NMDA-stimulated Ca^{2+} uptake, and ethanol was much less potent as an inhibitor of the response to agonists acting at other glutamate receptors (e.g., kainate). This selectivity of ethanol-induced inhibition of glutamate receptor responses was also observed in electrophysiological studies of cultured hippocampal neurons. In addition, ethanol was reported to inhibit NMDA-stimulated neurotransmitter release from brain slices. Following these studies, ethanol was shown to inhibit NMDA receptor function in *in vitro* biochemical and electrophysiological studies using cultured embryonic neurons, neurons and brain slices from adult animals and *Xenopus* oocytes expressing brain mRNA (see Hoffman, 1994; Dildy-Mayfield and Harris, 1992). Ethanol, administered systemically, has also been shown to inhibit the response of certain medial septal neurons, hippocampal neurons and locus coeruleus neurons to iontophoretically applied NMDA (see Hoffman *et al.*, 1994; Simson *et al.*, 1993).

The mechanism of ethanol-induced inhibition of NMDA receptor function is still under investigation. Ethanol does not appear to interfere with the action of phencyclidine or Mg^{2+}, or to directly interact with the agonist binding site on the receptor. In cerebellar granule cells, however, the inhibitory effect of ethanol could be overcome by high concentrations of glycine, and glycine also reversed the effect of ethanol on NMDA-stimulated dopamine release in striatal slices. Recently, we have demonstrated that an effect of ethanol in cerebellar granule cells is to reduce the potency (increase the EC_{50}) of glycine to act as a co-agonist at the NMDA receptor (Snell et al., 1994a). Thus, at low glycine concentrations, ethanol inhibition may be a result of a decreased response to glycine; however, at higher glycine concentrations, ethanol inhibition is no longer apparent (Snell et al., 1994a). Ethanol does not directly affect glycine binding to the NMDA receptor (Snell et al., 1993). However, we also found that, in cerebellar granule cells, activation of protein kinase C by phorbol 12-myristate, 13-acetate (PMA) had an effect similar to ethanol on the response to glycine. That is, protein kinase C activation inhibited NMDA responses in the presence of low glycine concentrations and increased the EC_{50} for glycine (Snell et al., 1994b). Furthermore, both the effects of the phorbol ester and of ethanol could be blocked by inhibitors of protein kinase C activity (Snell et al., 1994a,b). These results suggested that, in cerebellar granule cells, protein kinase C is involved in the inhibitory effect of ethanol on NMDA receptor function. Whether this mechanism will apply to other cell types, which may express different subunits of the NMDA receptor (see Nakanishi, 1992), is under investigation.

Whether or not the molecular mechanism of ethanol's effects is identical in all cells, it is clear that ethanol acutely inhibits NMDA receptor function in most systems tested. Based on the known roles of the NMDA receptor, these findings suggest that the receptor may be involved in producing the cognitive deficits observed after ethanol ingestion (Lister et al., 1987), and may play a role in the abnormal CNS development associated with the Fetal Alcohol Syndrome (Clarren and Smith, 1978). In addition, chronic ethanol exposure might result in an adaptation in NMDA receptor function as a response to the acute inhibitory effect of ethanol. Since the NMDA receptor plays a key role in the generation of epileptiform seizures and kindling, an adaptation involving an increase in receptor function might be expected to contribute to ethanol withdrawal seizures and other pathological effects of ethanol. This adaptive process has been studied both in animals and in cultured cells.

Materials and Methods

To assess the effect of chronic ethanol exposure on the characteristics of NMDA receptors in brain, male C57BL/6 mice were fed ethanol in a liquid diet for seven days, while control mice were pair-fed a liquid diet in which sucrose equicalorically replaced the ethanol (Ritzmann and Tabakoff, 1976). This regimen produces physical dependence on ethanol in the ethanol-fed mice, which can be assessed by monitoring withdrawal signs including handling-induced seizures. The animals were used to determine the effects of NMDA receptor antagonists on ethanol withdrawal symptoms, and to quantitate NMDA receptors in brain by ligand binding analyses (see Hoffman, 1994). To assess NMDA receptor *function* after chronic ethanol exposure and withdrawal, primary cultures of cerebellar granule cells were exposed to ethanol for two-four days in culture (Iorio *et al.*, 1992). NMDA-induced increases in intracellular Ca^{2+} were measured using the fluorescent dye fura-2. Glutamate-induced cytotoxicity in control and ethanol-exposed cerebellar granule cells was determined 24 hours after exposure of the cells to glutamate, with the use of the fluorescent dye fluorescein diacetate (Iorio *et al.*, 1993).

Results and Discussion

Chronic ethanol exposure results in an increase in NMDA receptor number in brain. In early studies, Michaelis *et al.* (1978) reported increased glutamate binding in brains of rats after chronic ethanol treatment. More recently, our results showed that, in mice, there was an increase in brain dizocilpine binding after chronic ethanol ingestion, as measured both in membrane binding studies and by quantitative autoradiography (see Hoffman, 1994). In the hippocampus of these animals, there was also an increase in NMDA-sensitive glutamate binding, but no change in glycine binding or in the binding of the NMDA receptor antagonist, CGS-19755. This pattern of changes suggests the possibility that chronic ethanol treatment results in altered characteristics of the glutamate receptor. Recently, a number of NMDA receptor subunits have been cloned. These include the NR1 and ζ subunits (from rat and mouse, respectively), and a family of NR2 (rat) and ε (mouse) subunits (see Nakanishi, 1992). The NR1 subunit is widely distributed in brain, while the NR2 subunits have more discrete localization. Expression of NR1 alone in *Xenopus* oocytes produces a glutamate receptor with many of the characteristics of the NMDA receptor, but which demonstrates a low response to NMDA. However, expression of NR1 with the various NR2 subunits (which alone do not generate an electrophysiological response to agonist) results in more robust responses. Each of the NR2 subunits confers somewhat different characteristics on the NMDA receptor response, in terms of agonist vs. antagonist affinities, effects of Mg^{2+} and response to glycine. In addition to these subunits, another glutamate binding protein has been cloned that was suggested to represent one subunit of a multisubunit NMDA receptor (Kumar *et al.*, 1991). In this case, each of the subunits was reported to bind a different ligand that acts at the NMDA receptor (e.g., glycine, antagonist). These models of NMDA receptor structure postulate heteromeric receptor-channel complexes. Thus, it is possible to speculate that chronic ethanol treatment may alter the subunit composition of the NMDA receptor, resulting in changes in binding of some ligands and not others.

Changes in ligand binding, however, do not necessarily predict a change in the functional capacity of the receptor, and to evaluate receptor function we determined the effect of NMDA on intracellular Ca^{2+} levels in cerebellar granule cells that had been exposed chronically to ethanol (100 mM for 2-4 days or 20 mM for three days or longer). There was an increased maximal response to NMDA (in the presence of glycine) in the ethanol-exposed cells, compatible with an increased number of receptors. There appeared to be no difference in sensitivity to various antagonists of the NMDA receptor in the ethanol-treated cells, compared to control cells, also consistent with the interpretation of an increased number, rather than a change in properties of the receptor. The mechanism of this NMDA receptor "up-regulation" remains to be determined. In preliminary studies, however, we have found a significant increase in the mRNA coding for the glutamate binding protein that was suggested to be a subunit of an NMDA receptor, as well as in glutamate binding protein levels *per se*, in cerebellar granule cells that were treated chronically with ethanol. A smaller increase in the level of mRNA coding for NR1 was also found.

The increased number and/or function of NMDA receptors in brains of chronically ethanol-treated animals appears to play a role in ethanol withdrawal seizures. Our analysis of the time course of the change in hippocampal dizocilpine binding in ethanol-fed mice revealed that the number of receptors was increased at the time of ethanol withdrawal, remained increased at eight hours after withdrawal, the time of peak withdrawal seizures, and had returned to control levels by 24 hours after withdrawal, when seizures had dissipated. This time course of changes in the quantity of NMDA receptors (dizocilpine binding sites) is consistent with a contribution of excessive stimulation of NMDA receptors to ethanol withdrawal seizures. The fact that seizures do not occur at the time the animals are withdrawn from ethanol, although receptor levels are elevated, can be attributed to the high brain levels of ethanol that are found in the animals at this time (Ritzmann and Tabakoff, 1976). Since ethanol acutely inhibits NMDA receptor function, the animals are protected from seizures until the ethanol has been eliminated. Further evidence for a role of NMDA receptors in ethanol withdrawal seizures came from studies showing that seizures were attenuated by the administration of competitive and non-competitive NMDA receptor antagonists, while administration of NMDA, at a dose that had no effect in control animals, exacerbated the withdrawal seizures (see Hoffman, 1994). In addition, we found a higher number of dizocilpine binding sites in the hippocampus of replicate lines of mice that were selectively bred to be susceptible to ethanol withdrawal seizures (WSP mice) than in brains of mice that were bred to be resistant to ethanol withdrawal seizures (WSR mice). After these lines of mice were fed ethanol chronically, dizocilpine binding was increased in hippocampus from mice of both lines, such that the relative difference was maintained (i.e., WSP mice still had a higher number of hippocampal NMDA receptors than WSR mice) (Valverius et al., 1990). The finding of a biochemical difference between lines of mice that have been selectively bred for a behavioral response strongly implicates that biochemical characteristic in mediating the selected response (Phillips et al., 1989).

The demonstration that the NMDA receptor plays a key role in ethanol withdrawal seizures provides the opportunity to develop therapeutic strategies to treat these seizures. For example, the drug ADCI ((±)-5-aminocarbonyl-10,11-dihydro-5H-dibenzo[a,d]cyclohepten-5,10-imine), which is a low-affinity blocker of the NMDA receptor-coupled channel, suppressed ethanol withdrawal seizures and was less toxic than dizocilpine when tested for its incoordinating effects (Grant et al., 1992). This drug, and other related compounds, could prove to be useful in treating ethanol withdrawal seizures and possibly other aspects of the ethanol withdrawal syndrome (e.g., tremors) (Grant et al., 1992).

It should be pointed out, however, that although the data strongly implicate the NMDA receptor in ethanol withdrawal seizures, other neurochemical systems are also likely to contribute to these and other withdrawal symptoms. For example, there is evidence for "down-regulation" of $GABA_A$ receptor function after chronic ethanol treatment (e.g., Allan and Harris, 1987), as well as an increase in voltage-sensitive Ca^{2+} channels (VSCC) (Dolin et al., 1987). In fact, the administration of VSCC antagonists has been reported to reduce the severity of ethanol withdrawal seizures (Little et al., 1988), and benzodiazepines (albeit at high doses) are the most commonly used drug for treatment of ethanol withdrawal (Litten and Allen, 1991).

In addition to playing a role in ethanol withdrawal seizures, the increase in NMDA receptor function observed after chronic ethanol treatment appears to contribute to neuronal damage, consistent with the key role of the NMDA receptor in glutamate-induced excitotoxic cell death (Choi, 1992). Primary cultures of cerebellar granule cells (Iorio et al., 1993) or cerebral cortical cells (Chandler *et al.*, 1993) were exposed to 100 mM ethanol for several days, and then ethanol was removed and the cells were exposed to glutamate or NMDA, respectively. In both cell types, delayed neurotoxicity was significantly increased in the ethanol-exposed cells, as compared to the controls. In our studies of cerebellar granule cells, selective receptor antagonists were used to show that the glutamate-induced toxicity was mediated by the NMDA receptor. Because ethanol, acutely, has been shown to protect cells from glutamate-induced cell death (Takadera *et al.*, 1990), the enhanced neurotoxicity observed in these studies is apparently a result of ethanol withdrawal. Thus, as for withdrawal seizures, the presence of ethanol in the cell culture medium (or brain) would be expected to protect the cells from the consequences of NMDA receptor up-regulation, which would appear after ethanol was eliminated or removed. The role of NMDA receptors in withdrawal-induced neuronal damage is also supported by a study in which rats that had been chronically treated with ethanol were given intrahippocampal injections of NMDA. The ethanol-treated rats were more sensitive to the effects of NMDA on enzyme activities and to NMDA-induced mortality than the control rats, and these differences were suggested to reflect enhanced NMDA receptor function in the ethanol-withdrawn animals (Davidson *et al.*, 1993).

Conclusion

Overall, the results indicate a significant role of increased NMDA receptor function in both ethanol withdrawal seizures and in the neuronal damage that is associated with chronic ethanol exposure and withdrawal. The increase in NMDA receptor function observed in ethanol-treated animals and cells may represent an adaptive response to the initial inhibition of NMDA receptor function by ethanol. Regardless of the mechanism by which receptor function is increased (e.g., changes in receptor subunit composition), the data support the contention that ethanol withdrawal hyperexcitability may be treated with drugs that are antagonists at the NMDA receptor. This treatment is indicated because of the evidence for a worsening of ethanol withdrawal seizures following multiple episodes of withdrawal (e.g., Becker and Hale, 1993), and the fact that most alcoholics undergo withdrawal numerous times during their lives. In addition, it is likely that the repeated "up-regulation" of NMDA receptors is a key factor in the well-characterized neuronal damage that is observed in alcoholics (Charness, 1993). Treatment with NMDA receptor antagonists could provide an efficacious means to simultaneously attenuate withdrawal seizures and alleviate or prevent withdrawal-induced neuronal damage.

69

Acknowledgements

This work was supported in part by the National Institute on Alcohol Abuse and Alcoholism, USPHS (AA 9005; AA 3527) and by the Banbury Foundation.

References

Allan, A.M. and Harris, R.A. (1987) Acute and chronic ethanol treatments alter GABA receptor-operated chloride channels. *Pharmacol. Biochem. Behav.* 27: 665-670.

Ballenger, J.C. and Post, R.M. (1978) Kindling as a model for alcohol withdrawal syndromes. *Br. J. Psych.* 133: 1-14.

Becker, H.C. and Hale, R.L. (1993) Repeated episodes of ethanol withdrawal potentiate the severity of subsequent withdrawal seizures: an animal model of alcohol withdrawal "kindling". *Alcoholism: Clin. Exp. Res.* 17: 94-98.

Chandler, L.J., Newsom, H., Sumners, C. and Crews, F.T. (1993) Chronic ethanol exposure potentiates NMDA excitotoxicity in cerebral cortical neurons. *J. Neurochem.* 60: 1578-1581.

Charness, M.E. (1993) Brain lesions in alcoholics. *Alcoholism: Clin. Exp. Res.* 17: 2-11.

Choi, D.W. (1992) Excitotoxic cell death. *J. Neurobiol.* 23: 1261-1276.

Clarren, S.K. and Smith, D.W. (1978) The fetal alcohol syndrome. *N. Engl. J. Med.* 298: 1063-1067.

Collingridge, G.L. and Lester, R.A.J. (1989) Excitatory amino acid receptors in the vertebrate central nervous system. *Pharmacol. Rev.* 40: 143-210.

Cotman, C.W., Bridges, R.J., Taube, J.S., Clark, A.S., Geddes, J.W. and Monaghan, D.T. (1989) The role of the NMDA receptor in central nervous system plasticity and pathology. *J. NIH Res.* 1: 65-74.

Davidson, M.D., Wilce, P. and Shanley, B.C. (1993) Increased sensitivity of the hippocampus in ethanol-dependent rats to toxic effect of N-methyl-D-aspartic acid *in vivo*. *Brain Res.* 606: 5-9.

Dildy-Mayfield, J.E. and Harris, R.A. (1992a) Comparison of ethanol sensitivity of rat brain kainate, DL-α-amino-3-hydroxy-5-methyl-4-isoxaloneproprionic acid and N-methyl-D-aspartate receptors expressed in *Xenopus* oocytes. *J. Pharmacol. Exp. Ther.* 262: 487-494.

Dolin, S., Hudspith, M., Pagonis, C., Little, H. and Littleton, J. (1987) Increased dihydropyridine-sensitive Ca^{2+} channels in rat brain may underlie ethanol physical dependence. *Neuropharmacol.* 26: 275-279.

Goldstein, A. and Goldstein, D.B. (1968) Enzyme expansion theory of drug tolerance and physical dependence. *Res. Pub.- Assoc. Res. Nerv. Mental Disease* 46: 265-267.

Grant, K.A., Snell, L.D., Rogawski, M.I., Thurkauf, A. and Tabakoff, B. (1992) Comparison of the effects of the uncompetitive N-methyl-D-aspartate antagonist ($\pm$)-5-aminocarbonyl-10,11-dihydro-5H-dibenzo[a,d]cyclohepten-5,10-imine (ADCI) with its structural analogs dizocilpine (MK-801) and carbamazepine on ethanol withdrawal seizures. *J. Pharmacol. Exp. Ther.* 260: 1017-1022.

Hoffman, P.L. (1994) The effects of alcohol on excitatory amino acid receptor function. In: Kranzler, H. (ed): *Handbook of Experimental Pharmacology: The Pharmacology of Alcohol Abuse*. Springer-Verlag, Berlin, in press.

Hunt, W.A. (1985) *Alcohol and biological membranes*. The Guilford Press, New York.

Iorio, K.R., Reinlib, L., Tabakoff, B. and Hoffman, P.L. (1992) Chronic exposure of cerebellar granule cells to ethanol results in increased NMDA receptor function. *Mol. Pharmacol.* 41:

1142-1148.

Iorio, K.R., Tabakoff, B. and Hoffman, P.L. (1993) Glutamate-induced neurotoxicity is increased in cerebellar granule cells exposed chronically to ethanol. *Eur. J. Pharmacol.* 248: 209-212.

Kumar, K.N., Tilakaratne, N., Johnson, P.S., Allen, A.E. and Michaelis, E.K. (1991) Cloning of cDNA for the glutamate-binding subunit of an NMDA receptor complex. *Nature* 354: 70-73.

Lister, R.G., Eckardt, M.J. and Weingartner, H. (1987) Ethanol intoxication and memory. Recent developments and new directions. In: Galanter, M. (ed): *Recent Developments in Alcoholism, Vol. 5.* Plenum Press, New York, pp. 111-126.

Litten, R.Z. and Allen, J.P. (1991) Pharmacotherapies for alcoholism: promising agents and clinical issues. *Alcoholism: Clin. Exp. Res.* 15: 620-633.

Little, H.J., Dolin, S.J. and Halsey, M.J. (1988) Calcium channel antagonists decrease the ethanol withdrawal syndrome. *Life Sci.* 39: 2059-2065.

Michaelis, E.K., Michaelis, M.L. and Freed, W.K. (1978) Effects of acute and chronic ethanol intake on synaptosomal glutamate binding activity. *Biochem. Pharmacol.* 27: 1685-1691.

Nakanishi, S. (1992) Molecular diversity of glutamate receptors and implications for brain function. *Science* 258: 597-603.

Phillips, T.J., Feller, D.J. and Crabbe, J.C. (1989) Selected mouse lines, alcohol and behavior. *Experientia* 45: 805-827.

Ritzmann, R.F. and Tabakoff, B. (1976) Body temperature in mice: A quantitative measure of alcohol tolerance and dependence. *J. Pharmacol. Exp. Ther.* 199: 158-163.

Simson, P.E., Criswell, H.E. and Breese, G.R. (1993) Inhibition of NMDA-evoked electro physiological activity by ethanol in selected brain regions: evidence for ethanol-sensitive and ethanol-insensitive NMDA-evoked responses. *Brain Res.* 607: 9-16.

Snell, L.D., Tabakoff, B. and Hoffman, P.L. (1993) Radioligand binding to the N-methyl-D-aspartate receptor/ionophore complex: alterations by ethanol *in vitro* and by chronic *in vivo* ethanol ingestion. *Brain Res.* 602: 91-98.

Snell, L.D., Tabakoff, B. and Hoffman, P.L. (1994a) Involvement of protein kinase C in ethanol-induced inhibition of NMDA receptor function in cerebellar granule cells. *Alcoholism: Clin. Exp. Res.,* in press.

Snell, L.D., Iorio, K.R., Tabakoff, B. and Hoffman, P.L. (1994b) Protein kinase C activation attenuates N-methyl-D-aspartate induced increases in intracellular calcium in cerebellar granule cells. *J. Neurochem.,* in press.

Tabakoff, B. and Hoffman, P.L. (1987) Biochemical pharmacology of alcohol. In: Meltzer, H.Y. (ed): *Psychopharmacology - The Third Generation of Progress.* Raven Press, New York, pp. 1521-1526.

Tabakoff, B. and Rothstein, J.D. (1983) Biology of tolerance and dependence. In: Tabakoff, B., Sutker, P.B. and Randall, C.L. (eds): *Medical and Social Aspects of Alcohol Abuse.* Plenum Press, New York, pp. 187-220.

Takadera, T., Suzuki, R. and Mohri, T. (1990) Protection by ethanol of cortical neurons from N-methyl-D-aspartate-induced neurotoxicity is associated with blocking calcium influx. *Brain Res.* 537: 109-115.

Valverius, P., Crabbe, J.C., Hoffman, P.L. and Tabakoff, B. (1990) NMDA receptors in mice bred to be prone or resistant to ethanol withdrawal seizures. *Eur. J. Pharmacol.* 184: 185-189.

Toward a Molecular Basis of Alcohol Use and Abuse
ed. by B. Jansson, H. Jörnvall, U. Rydberg, L. Terenius & B. L. Vallee
© 1994 Birkhäuser Verlag Basel/Switzerland

Molecular diversity of glutamate receptors and their physiological functions

S. Nakanishi, M. Masu, Y. Bessho, Y. Nakajima, Y. Hayashi and R. Shigemoto[1]

Institute for Immunology, and [1]Department of Morphological Brain Science, Kyoto University Faculty of Medicine, Yoshida, Sakyo-ku, Kyoto 606, Japan

Summary

Glutamate receptors play an important role in many integrative brain functions and in neuronal development. We report the molecular diversity of NMDA receptors and metabotropic glutamate receptors on the basis of our studies of molecular cloning and characterization of the diverse members of these receptors. The NMDA receptors consist of two distinct types of subunits. NMDAR1 possesses all properties characteristic of the NMDA receptor-channel complex, whereas the four NMDAR2 subunits, termed NMDAR2A-2D, show no channel activity but potentiate the NMDAR1 activity and confer functional variability by different heteromeric formations. The NMDA receptor subunits are considerably divergent from the other ligand-gated ion channels, and the structural architecture of these subunits remains elusive. The mGluRs form a family of at least seven different subtypes termed mGluR1-mGluR7. These receptor subtypes have seven transmembrane segments and possess a large extracellular domain at their N-terminal regions. The seven mGluR subtypes are classified into three subgroups according to their sequence similarities, signal transduction mechanisms and agonist selectivities: mGluR1/mGluR5, mGluR2/mGluR3 and mGluR4/mGluR6/mGluR7.

On the basis of our knowledge of the molecular diversity of the NMDA receptors and mGluRs, we have studied the physiological roles of individual receptor subunits or subtypes. We have shown that K^+-induced depolarization or NMDA treatment in primary cultures of neonatal cerebellar granule cells induces the functional NMDA receptor and specifically up-regulates NMDAR2A mRNA among the multiple NMDA receptor subunits through the increase in resting intracellular Ca^{2+} concentrations. Our study demonstrates that the regulation of the specific NMDA receptor subunit mRNA governs the NMDA receptor induction that is thought to play an important role in granule cell survival and death. Analysis of an agonist selectivity and an expression pattern of mGluR6 has indicated that mGluR6 is responsible for synaptic neurotransmission from photoreceptor cells to ON-bipolar cells in the visual system. We have also investigated the function of mGluR2 in granule cells of the accessory olfactory bulb by combining immunoelectron-microscopic analysis with slice-patch recordings on the basis of the identification of a new agonist selective for this receptor subtype. Our results demonstrate that mGluR2 is present at the presynaptic site of granule cells and modulates inhibitory GABA transmission from granule cells to mitral cells. This finding indicates that the mGluR2 activation relieves excited mitral cells from GABA inhibition but maintains the lateral inhibition of unexcited mitral cells, thus resulting in enhancement of the signal-to-noise ratio between the excited mitral cells and their neighboring unexcited mitral cells.

Introduction

Glutamate receptors mediate excitatory neurotransmission and play an important role in neuronal plasticity and neurotoxicity in the central nervous system (Nakanishi, 1992; Nakanishi and Masu, 1994). These receptors are essential for inducing long-lasting changes in neuronal responsiveness which are thought to underlie learning, memory, and neuronal development. They also play a critical role in pathophysiological processes such as epilepsy and ischemic neuronal cell death. The diverse functions of glutamate neurotransmission are mediated by a variety of glutamate receptors that are classified into two major groups termed ionotropic and metabotropic receptors (Nakanishi, 1992; Nakanishi and Masu, 1994). The former receptors can be subdivided into N-methyl-D-aspartate (NMDA) receptors and α-amino-3-hydroxy-5-methyl-4-isoxazolepropionate (AMPA)/kainate receptors, both of which contain glutamate-gated, cation-specific ion channels. The latter receptors (mGluRs) are coupled to intracellular signal transduction through G proteins. For the past few years, we have studied the properties, functions and regulation of the NMDA receptors and mGluRs, which were molecularly isolated by using our novel receptor cloning strategy that combined electrophysiology and a *Xenopus* oocyte expression system (Masu *et al.*, 1987; Masu *et al.*, 1991; Moriyoshi *et al.*, 1991). This article deals with the molecular nature and functions of the diverse members of the NMDA receptors and mGluRs and discusses some of the physiological roles of different glutamate receptors.

Results and Discussion

Molecular diversity of NMDA receptors and metabotropic glutamate receptors

Our molecular cloning studies demonstrated that the NMDA receptors consist of two distinct types of subunits, one termed NMDAR1 and the other four termed NMDAR2A-2D (Moriyoshi *et al.*, 1991; Sugihara *et al.*, 1992; Ishii *et al.*, 1993). NMDAR1 possesses all properties characteristic of the NMDA receptor-channel complex, including agonist and antagonist selectivity, glycine modulation, voltage-dependent Mg^{2+} blockade, Ca^{2+} permeability and Zn^{2+} inhibition (Moriyoshi *et al.*, 1991; Sugihara *et al.*, 1992; Karp *et al.*, 1993). NMDAR2A-2D show no channel activity in their homomeric structures but potentiate the NMDA receptor activity in combined expression with NMDAR1 (Ishii *et al.*, 1993). These subunits also confer functional variability by different heteromeric subunit configurations (Ishii *et al.*, 1993). The NMDAR1 mRNA is ubiquitously expressed throughout the brain regions, whereas individual NMDAR2 mRNAs are distinctly distributed in different brain regions (Moriyoshi *et al.*, 1991; Ishii *et al.*, 1993). The functional heterogeneity of the NMDA receptors in different neuronal cells is thus produced by the functional and anatomical differences of the NMDAR2 subunits. All of the NMDA receptor subunits, like the other ligand-gated ion channels, were initially thought to comprise four membrane-spanning domains preceded and followed by extracellular domains at both the N-terminal and C-terminal sides. However, the NMDA receptors are by far divergent from the other ligand-gated ion channels in their amino acid sequences, and several lines of recent evidence suggested that the C-terminal tail is located on the intracellular side rather than on the extracellular side. A transmembrane model of the NMDA receptors is thus illustrated in Fig. 1 under the assumption that the N-terminal and C-terminal portions are located

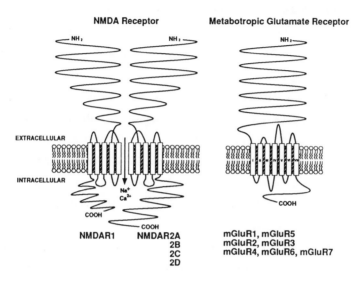

Fig. 1 Transmembrane models of the NMDA receptors and metabotropic receptors. It has generally been accepted that the mGluR subtypes comprise seven membrane-spanning domains with a large extracellular N-terminus, whereas the number of transmembrane segments of the NMDA receptors and their topology relative to the membrane remain unsettled.

on the extracellular and intracellular sides, respectively. However, the number of transmembrane segments and the transmembrane topology of the NMDA receptors still remain elusive. The mutational analysis of NMDAR1 indicated that asparagine at the second transmembrane segment is within a channel pore and is responsible for governing a high Ca^{2+} permeability and the blockades of Mg^{2+} and other channel blockers (Sakurada et al., 1993).

The mGluRs form a family of at least seven different subtypes termed mGluR1-mGluR7 (Masu et al., 1991; Tanabe et al., 1992, 1993; Abe et al., 1992; Nakajima et al., 1993; Okamoto et al., 1994). These receptor subtypes have seven transmembrane segments common to the other members of G protein-coupled receptors but possess a large extracellular domain at their N-terminal regions and thus represent a novel family of G protein-coupled receptors (Fig. 1). The seven mGluR subtypes differ in their agonist selectivities and signal transduction mechanisms as analyzed in CHO cells by DNA transfection (Masu et al., 1991; Tanabe et al.,

1992, 1993; Aramori *et al.*, 1992; Abe *et al.*, 1992; Nakajima *et al.*, 1993; Okamoto *et al.*, 1994). mGluR1 and mGluR5 are coupled to the IP_3/Ca^{2+} signal transduction and efficiently respond to quisqualate. The other five are all linked to the inhibitory cyclic AMP cascade, but their agonist selectivities are totally different between the mGluR2/mGluR3 subgroup and the mGluR4/mGluR6/mGluR7 subgroup. The former effectively interacts with trans-1-aminocyclo-pentane-1, 3-dicarboxylate (trans-ACPD), whereas the latter potently reacts with L-2-amino-4-phosphonobutyrate (L-AP4). The seven subtypes of the mGluR family are thus classified into three subgroups according to their sequence similarities, signal transduction mechanisms and agonist selectivities: mGluR1/mGluR5, mGluR2/mGluR3 and mGluR4/mGluR6/mGluR7 (Nakanishi, 1992; Nakanishi and Masu, 1994). All but mGluR6 mRNA are widely but distinctly distributed in various brain regions (Tanabe *et al.*, 1992, 1993; Abe *et al.*, 1992; Shigemoto *et al.*, 1992; Ohishi *et al.*, 1993a, 1993b; Okamoto *et al.*, 1994). These findings strongly indicate that the individual mGluR subtypes have their own functions by specializing the signal trans-duction and expression patterns in different nerve cells. The chimeric experiments between mGluR1 and mGluR2 demonstrated that the amino-terminal extracellular domain serves as a glutamate binding site of mGluR (Takahashi *et al.*, 1993). Thus, the mode of agonist binding of mGluR is different from that of the other G protein-coupled receptors for small molecule transmitters.

Physiological functions of different glutamate receptors

In order to understand the physiological role of glutamate transmission in integrative brain functions and development, it is essential to characterize the functions and regulation of individual glutamate receptors on the basis of our knowledge of the diverse members of these receptors. The NMDA receptors are known to cause neuronal degeneration and neuronal cell death but are also required for survival of certain neuronal cells. The involvement of the NMDA receptors in neuronal cell death and survival has been well characterized in primary cultures of neonatal cerebellar granule cells (Balázs *et al.*, 1992). K^+ depolarization or NMDA treatment promotes the survival of cultured granule cells, and these cells become sensitive to NMDA toxicity after prolonged K^+ depolarization (Cox *et al.*, 1990; Balázs *et al.*, 1992). The above treatments are thought to produce an influence that mimics the physiological stimulation of immature granule cells during cerebellar development. Primary cultures of cerebellar granule cells thus provide a useful system to study the role of the glutamate receptors in neuronal functions. We investigated the function and regulation of the glutamate receptors in cultured cerebellar granule cells (Bessho *et al.*, 1993; Bessho *et al.*, 1994).

Primary cultures of cerebellar granule cells were prepared from 6-day-old rats, and these cells were treated with high KCl or NMDA. Both treatments induced functional NMDA receptors as assessed by fura-2 fluorescence analysis of NMDA receptor-mediated intracellular Ca^{2+} increase. When the effects of K^+ depolarization and NMDA treatment on mRNA levels of individual NMDA receptor subunits were examined by Northern blot analysis, both treatments were found to specifically up-regulate the NMDAR2A mRNA among the multiple NMDA receptor subunits. Furthermore, antisense oligonucleotides specific for NMDAR2A mRNA prevented the induction of functional NMDA receptors, thus confirming the contribution of the specific up-regulation of NMDAR2A mRNA in the induction of the NMDA receptor. When Ca^{2+} influx was blocked during K^+ depolarization by the Ca^{2+} channel blocker nifedipine or during NMDA treatment by the NMDA receptor antagonist D-2-amino-5-phosphonovalerate, this block abolished both the NMDA receptor induction and the up-regulation of NMDAR2A mRNA. Thus, our investigation demonstrates that the increase in intracellular Ca^{2+} governs the up-regulation of the specific NMDA receptor subunit mRNA that is responsible for the NMDA receptor induction. It has been supposed that modestly elevated intracellular Ca^{2+} promotes neuronal cell survival, whereas substantially elevated intracellular Ca^{2+} leads to neuronal cell death (Franklin and Johnson, 1992). Thus, it is very likely that the induction of the NMDA receptor is involved in the survival of granule cells by maintaining moderate levels of intracellular Ca^{2+}, but excess stimulation of the elevated NMDA receptor results in substantial increase in intracellular Ca^{2+} and causes granule cell death. However, we also found that NMDA is very toxic after prolonged (5 days) K^+ depolarization, but not so after short-term (1 day) depolarization, although the NMDA receptor markedly increases under both conditions. It is thus plausible that the NMDAR2A mRNA up-regulation is necessary, but may not be sufficient, for the NMDA toxicity in granule cells after K^+ depolarization.

Because studies of mGluRs have been initiated only recently, the physiological roles of this receptor family largely remain unknown. One interesting system of mGluRs is glutamate-mediated synaptic transmission within the visual system. Recent electrophysiological studies indicated that the L-AP4-sensitive mGluR is responsible for synaptic transmission between photoreceptor cells and ON-bipolar cells through the coupling to the cyclic GMP cascade (Nawy and Jahr, 1990; Shiells and Falk, 1990). To investigate the mGluR in the visual system, we screened a retinal cDNA library. We identified mGluR6 from the retinal library and characterized its properties and expression in detail (Nakajima *et al.*, 1993). When the agonist selectivity of mGluR6 was examined in CHO cells after DNA transfection, the selective and potent response to L-AP4 and L-serine-O-phosphate was found, thus consistent with the property of the mGluR reported in ON-bipolar cells. The expression of mGluR6 mRNA is restricted in the retina. Furthermore, *in situ* hybridization signals of mGluR6 mRNA were exclusively observed in the inner nuclear layer where ON-bipolar cells are known to be distributed in the retina. When the cellular localization of mGluR6 was examined by immunocytochemical analysis using a polyclonal antibody raised against the mGluR6 protein, the mGluR6 immunoreactivity was confined to the boundary between the photoreceptor cell layer and the bipolar cell layer, indicating that mGluR6 is located at the postsynaptic site. It is, however, to be noted that mGluR6 expressed in CHO cells is coupled to the inhibitory cyclic AMP cascade and is thus different in signal transduction from the cyclic GMP-coupled mGluR in ON-bipolar cells. However, it is also known that when a preferred G protein or subsequent effectors are absent in transfected cells, the receptor expressed in a heterologous system is capable of modulating different second messengers. It is thus plausible that mGluR6 corresponds to the L-AP4 recepor in ON-bipolar cells.

On the basis of the above findings of mGluR6, together with the electrophysiological characterization of mGluR-mediated signal transduction in ON-bipolar cells (Nawy and Jahr, 1990; Shiells and Falk, 1990), the following model can be discussed for synaptic transmission between photoreceptor cells and ON-bipolar cells. When light activates the photoreceptor, intracellular concentrations of cyclic GMP decrease as a result of the stimulation of phosphodiesterase through the activation of transducin. The decrease in cyclic GMP concentrations leads to the closure of the cyclic GMP-gated ion channel and hyperpolarizes the photoreceptor cells. This hyperpolarization reduces glutamate release. Under this situation, the mGluR6-G protein-phosphodiesterase system in ON-bipolar cells becomes inactive, and high concentrations of cyclic GMP are maintained in ON-bipolar cells and stimulate the cyclic GMP-gated ion channel, thus resulting in depolarization of ON-bipolar cells. Glutamate release from ON-bipolar cells is augmented and in turn excites ganglion cells, and this excitation is transmitted to the brain. Our investigation thus raises an interesting possibility that a specific mGluR subtype plays a critical role in synaptic transmission in the visual system.

We also defined the physiological role of mGluR2 in sensory transmission of the accessory olfactory bulb (AOB) by combining immunoelectron microscopy and slice-patch recording on the basis of the identification of a new selective agonist for this receptor subtype (Hayashi *et al.*, 1993). Our previous study indicated that two extended forms of 2-(carboxycyclopropyl)glycines (CCGs) are potent and selective agonists for the mGluR family (Hayashi *et al.*, 1992). Recently, Ohfune and co-workers developed a new CCG derivative, (2S,1'R,2'R,3'R)-2-(2,3-dicarboxy-cyclopropyl)glycine, abbreviated as DCG-IV (Ishida *et al.*, 1993). We tested the agonist potency and selectivity of this compound for different mGluR subtypes expressed in CHO cells. DCG-IV was found to be a very potent and specific agonist for mGluR2/mGluR3 among the mGluR family members. DCG-IV is also capable of binding to the NMDA receptor, but this potency is much less than that for the activation of mGluR2/mGluR3. DCG-IV has no effects on the AMPA/kinate receptors. DCG-IV is thus a very useful agonist for mGluR2.

Because the prominent expression of mGluR2 mRNA in granule cells of the AOB was observed by *in situ* hybridization analysis, the subcellular localization of mGluR2 was examined immunoelectron-microscopically with the aid of a polyclonal antibody raised against the mGluR2 protein expressed in *E. coli*. This analysis indicated that an immunoreactivity of mGluR2 is confined to dendrites of granule cells of the AOB. In the AOB, mitral cells receive afferent inputs from the vomeronasal nerve and transmit excitatory outputs to various brain regions. Granule cells are inhibitory interneurons that form typical dendrodendritic synapses with mitral

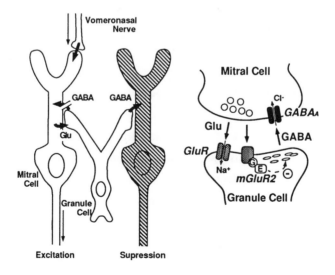

Fig. 2. A model of a regulatory role of mGluR2 in the olfactory transmission of the accessory olfactory bulb. GABA$_A$, GABA$_A$ receptor; GluR, ionotropic glutamate receptor. For detailed explanation, see the text.

cells. These synapses undergo reciprocal regulation, in which the granule cell is excited by glutamate from the mitral cell and exerts an inhibition onto the mitral cell by GABA (Fig. 2). We investigated the role of mGluR2 in synaptic transmission between the mitral cell and granule cell by examining the effect of a newly identified agonist DCG-IV on the GABA transmission from the granule cell to the mitral cell (Hayashi *et al.*, 1993). Whole cell recording of a mitral cell was performed by the slice-patch method, and extracellular stimuli were applied to a granule cell. As expected, when a granule cell was electrically stimulated, GABA-mediated inhibitory postsynaptic currents (IPSCs) were evoked in a mitral cell. When mGluR2 was activated by the addition of DCG-IV in this system, IPSCs were markedly reduced in a reversible manner.

These observations as well as others lead to the conclusion that glutamate released from the mitral cell activates mGluR2 at the presynaptic site of the granule cell and relieves a GABA-mediated inhibition onto the mitral cell (Fig. 2). The granule cell lacks an axon and forms divergent synaptic contacts with not only the original mitral cell but also a large number of neighboring mitral cells. The excitation of mitral cells thus evokes lateral inhibition in the neighboring mitral cells through the divergent synaptic formations with trans-synaptically excited granule cells. Under the mechanism revealed in our study, however, it can be postulated that the GABA inhibition is relieved in excited mitral cells by the activation of mGluR2. Furthermore, this activation is thought to be confined to the synapses of the excited mitral cells and would thus maintain the lateral inhibition of unexcited neighboring mitral cells. This mechanism would evidently enhance the signal-to-noise ratio between the excited mitral cells and their neighboring mitral cells (Fig. 2). It is thus tempting to speculate that the mGluR2-mediated modulation in the microcircuitry between the mitral cell and granule cell plays an important role in discrimination and resolution of the olfactory sensory transmission in the AOB.

It has now been revealed that both NMDA receptors and mGluRs are more diversified than previously envisioned. Our investigations discussed here demonstrate that different glutamate receptors have their own functions and play important roles in the regulation of glutamate or other transmissions involved in integrative brain functions and neuronal development. Further study of not only the physiological roles of multiple glutamate receptors but also the cooperative functions of different glutamate receptors will undoubtedly be interesting and important for understanding glutamate-mediated brain functions.

Acknowledgements

We thank our colleagues at the Institute for Immunology, Department of Physiology, and Department of Morphological Brain Sciences, Kyoto University Faculty of Medicine for productive collaborations. This work was supported by research grants from the Ministry of Education, Science and Culture of Japan and the Ministry of Health and Welfare.

References

Abe, T., Sugihara, H., Nawa, H., Shigemoto, R., Mizuno, N. and Nakanishi, S. (1992) Molecular characterization of a novel metabotropic glutamate receptor mGluR5 coupled to inositol phosphate/Ca^{2+} signal transduction. *J. Biol. Chem.* 267: 13361-13368.

Aramori, I., and Nakanishi, S. (1992) Signal transduction and pharmacological characteristics of a metabotropic glutamate receptor, mGluR1, in transfected CHO cells. *Neuron* 8: 757-765.

Balázs, R., Hack, N. and Jørgensen, O.S. (1992) Cerebellar granule cells and the neurobiology of excitatory amino acids. In: Llinás, R. and Sotelo, C. (eds.): *The Cerebellum Revisited*, Springer-Verlag, New York, pp. 56-71.

Bessho, Y., Nawa, H. and Nakanishi, S. (1993) Glutamate and quisqualate regulate expression of metabotropic glutamate receptor mRNA in cultured cerebellar granule cells. *J. Neurochem.* 60: 253-259.

Bessho, Y., Nawa, H. and Nakanishi, S. (1994) Selective up-regulation of an NMDA receptor subunit mRNA in cultured cerebellar granule cells by K^+-induced depolarization and NMDA treatment. *Neuron*, in press.

Cox, J.A., Felder, C.C. and Henneberry, R.C. (1990) Differential expression of excitatory amino acid receptor subtypes in cultured cerebellar neurons. *Neuron* 4: 941-947.

Franklin, J.L. and Johnson Jr., E.M. (1992) Suppression of programmed neuronal death by sustained elevation of cytoplasmic calcium. *Trends Neurosci.* 15: 501-508.

Hayashi, Y., Tanabe, Y., Aramori, I., Masu, M., Shimamoto, K., Ohfune, Y. and Nakanishi, S. (1992) Agonist analysis of 2-(carboxycyclopropyl)glycine isomers for cloned metabotropic glutamate receptor subtypes expressed in Chinese hamster ovary cells. *Br. J. Pharmacol.* 107: 539-543.

Hayashi, Y., Momiyama, A., Takahashi, T., Ohishi, H., Ogawa-Meguro, R., Shigemoto, R., Mizuno, N. and Nakanishi, S. (1993) Role of a metabotropic glutamate receptor in synaptic modulation in the accessory olfactory bulb. *Nature* 366: 687-690.

Ishida, M., Saitoh, T., Shimamoto, K., Ohfune, Y. and Shinozaki, H. (1993) A novel metabotropic glutamate receptor agonist - Marked depression of monosynaptic excitation in the newborn rat isolated spinal cord. *Br. J. Pharmacol.* 109: 1169-1177.

Ishii, T., Moriyoshi, K., Sugihara, H., Sakurada, K., Kadotani, H., Yokoi, M., Akazawa, C., Shigemoto, R., Mizuno, N., Masu, M. and Nakanishi, S. (1993) Molecular characterization of the family of the N-methyl-D-aspartate receptor subunits. *J. Biol. Chem.* 268: 2836-2843.

Karp, S., Masu, M., Eki, T., Ozawa, K. and Nakanishi, S. (1993) Molecular cloning and chromosomal localization of the key subunit of the human N-methyl-D-aspartate receptor. *J. Biol. Chem.* 268: 3728-3733.

Masu, M., Tanabe, Y., Tsuchida, K., Shigemoto, R. and Nakanishi, S. (1991) Sequence and expression of a metabotropic glutamate receptor. *Nature* 349: 760-765.

Masu, Y., Nakayama, K., Tamaki, H., Harada, Y., Kuno, M. and Nakanishi, S. (1987) cDNA cloning of bovine substance-K receptor through oocyte expression system. *Nature* 329: 836-838.

Moriyoshi, K., Masu, M., Ishii, T., Shigemoto, R., Mizuno, N. and Nakanishi, S. (1991) Molecular cloning and characterization of the rat NMDA receptor. *Nature* 354: 31-37.

Nakajima, Y., Iwakabe, H., Akazawa, C., Nawa, H., Shigemoto, R., Mizuno, N. and Nakanishi, S. (1993) Molecular characterization of a novel retinal metabotropic glutamate receptor mGluR6 with a high agonist selectivity for L-2-amino-4-phosphonobutyrate. *J. Biol. Chem.* 268: 11868-11873.

Nakanishi, S. (1992) Molecular diversity of glutamate receptors and implications for brain function. *Science* 258: 597-603.

Nakanishi, S. and Masu, M. (1994) Molecular diversity and functions of glutamate receptors. *Annu. Rev. Biophys. Biomol. Struct.*, in press.

Nawy, S. and Jahr, C.E. (1990) Suppression by glutamate of cGMP-activated conductance in retinal bipolar cells. *Nature* 346: 269-271.

Ohishi, H., Shigemoto, R., Nakanishi, S. and Mizuno, N. (1993a) Distribution of the messenger RNA for a metabotropic glutamate receptor, mGluR2, in the central nervous system of the rat. *Neuroscience* 53: 1009-1018.

Ohishi, H., Shigemoto, R., Nakanishi, S. and Mizuno, N. (1993b) Distribution of the mRNA for a metabotropic glutamate receptor (mGluR3) in the rat brain: An in situ hybridization study. *J. Comp. Neurol.* 335: 252-266.

Okamoto, N., Hori, S., Akazawa, C., Hayashi, Y., Shigemoto, R., Mizuno, N. and Nakanishi, S. (1994) Molecular characterization of a new metabotropic glutamate receptor mGluR7 coupled to inhibitory cyclic AMP signal transduction. *J. Biol. Chem.*, in press.

Sakurada, K., Masu, M. and Nakanishi, S. (1993) Alteration of Ca^{2+} permeability and sensitivity to Mg^{2+} and channel blockers by a single amino acid substitution in the N-methyl-D-aspartate receptor. *J. Biol. Chem.* 268: 410-415.

Shiells, R.A. and Falk, G. (1990) Glutamate receptors of rod bipolar cells are linked to a cyclic GMP cascade via a G-protein. *Proc. R. Soc. Lond.* B 242: 91-94.

Shigemoto, R., Nakanishi, S. and Mizuno, N. (1992) Distribution of the mRNA for a metabotropic glutamate receptor (mGluR1) in the central nervous system: an in situ hybridization study in adult and developing rat. *J. Comp. Neurol.* 322: 121-135.

Sugihara, H., Moriyoshi, K., Ishii, T., Masu, M. and Nakanishi, S. (1992) Structures and properties of seven isoforms of the NMDA receptor generated by alternative splicing. *Biochem. Biophys. Res. Commun.* 185: 826-832.

Takahashi, K., Tsuchida, K., Tanabe, Y., Masu, M. and Nakanishi, S. (1993) Role of the large extracellular domain of metabotropic glutamate receptors in agonist selectivity determination. *J. Biol. Chem.* 268: 19341-19345.

Tanabe, Y., Masu, M., Ishii, T., Shigemoto, R. and Nakanishi, S. (1992) A family of metabotropic glutamate receptors. *Neuron* 8: 169-179.

Tanabe, Y., Nomura, A., Masu, M., Shigemoto, R., Mizuno, N. and Nakanishi, S. (1993) Signal transduction, pharmacological properties, and expression patterns of two metabotropic glutamate receptors, mGluR3 and mGluR4. *J. Neurosci.* 13: 1372-1378.

Toward a Molecular Basis of Alcohol Use and Abuse
ed. by B. Jansson, H. Jörnvall, U. Rydberg, L. Terenius & B. L. Vallee
© 1994 Birkhäuser Verlag Basel/Switzerland

Multiple dopamine receptors: The D_3 receptor and actions of substances of abuse

J.-C. Schwartz, J. Diaz[1], N. Griffon, D. Levesque, M.-P. Martres and P. Sokoloff

Unité de Neurobiologie et Pharmacologie (U.109) de l'INSERM, Centre Paul Broca, 2ter rue d'Alésia, 75014 Paris, France and [1]Laboratoire de Physiologie, Faculté de Pharmacie, Université René Descartes, Paris - France

Summary

Our knowledge of dopamine receptor diversity has markedly increased during the past few years as a result of discovery of five distinct genes, splice variants and polymorphic receptors. The genes can be classified in two subfamilies: the intronless genes that encode the D_1 and D_5 receptors positively linked to adenylyl cyclase and genes with introns that encode the two isoforms of the D_2 receptor and the D_3 and D_4 receptors. The various dopamine receptor subtypes can be distinguished by their sequence, intracellular signalling systems, pharmacology and localisation. The localisation of the D_3 receptor in the shell of *nucleus accumbens* suggests its participation in brain reward circuits and actions of substances of abuse.

Introduction

The participation of dopaminergic systems, particularly those projecting to the ventral striatal complex as a common final pathway in the mediation of a variety of brain reward processes, alcohol and drug abuse seems well established now. This conviction arises mainly from observations showing that these dopaminergic systems are activated during such processes and can be manipulated pharmacologically through interferences with the metabolism of or responses to dopamine. For a long time responses to dopamine were thought to be mediated by two receptor subtypes, the D_1 and D_2 receptors (Spano *et al.*, 1978; Kebabian *et al.*, 1979). With the application of molecular biology in the field, the picture has been modified considerably, opening new possibilities mainly in the fields of therapy and genetics (reviewed in Schwartz *et al.*, 1992).

Molecular biology reveals multiple dopamine receptors

The first cloning of a dopamine receptor gene, that of the D_2 receptor, was largely achieved by serendipity: it was the result of a search for genes displaying sequences similar to those of the ß-adrenoreceptors. Thus the $ß_2$-adrenoreceptor coding sequence was initially used as a hybridization probe to screen a rat genomic library under low stringency conditions, leading finally to the isolation of a cDNA which was shown to encode a protein exhibiting characteristic D_2 receptor-binding activity (Bunzow et al., 1988). The human homologue of the rat D_2 receptor was subsequently cloned and shown to exhibit 96% amino-acid sequence identity (Grandy et al., 1989; Dal Toso et al., 1989).

The determination of the D_2 receptor sequence paved the way for the cloning of a series of other dopamine receptor genes, based on the significant sequence homology between these receptors and the D_2 receptor. As anticipated, the D_1 receptor, which is nearly as abundant as the D_2 receptor in brain, was the first to follow (Zhou et al., 1990; Dearry et al., 1990; Sunahara et al., 1990). After that, the genes of a series of less abundant forms were established, constituting less expected receptors that markedly extend the dopamine receptor family: the rat and human D_3 (Sokoloff et al., 1990; Giros et al., 1990), the human D_4 (Van Tol et al., 1991) and human D_5 receptors (Sunahara et al., 1991), as well as the rat counterpart of the D_5, termed D_{1B} (Tiberi et al., 1991). The amino-acid sequences of all these receptors, as deduced from their established nucleotide sequences, reveal that they belong to a larger superfamily, that of receptors with seven transmembrane domains and coupled to their intracellular transduction system by a GTP-binding (G) protein.

All these receptors comprise a pattern of seven stretches of 20 to 25 hydrophobic amino-acids postulated to form transmembrane α-helices, connected by alternating extracellular and cytoplasmic loops composed of hydrophilic residues. The transmembrane helices form the ligand-binding domain, particularly three amino-acid residues thought to interact with catecholamines: an aspartate in the third transmembrane domain that forms an ion pair with the protonated amine group of dopamine and two serines in the fifth transmembrane domain that presumably form a hydrogen bonding interaction with the two phenol groups of dopamine. This latter interaction, specific for dopamine and its agonists, could cause a conformational change in the helix that would be transmitted to the third cytoplasmic loop.

The cytoplasmic domains, particularly the third, exhibit the largest sequence dissimilarity among the various dopamine receptor subtypes, which may reflect selective interaction of each subtype with one member of the large family of G proteins, leading to distinct intracellular signals.

Dopamine receptors can be subdivided by the length of the third cytoplasmic loop (Fig.1).

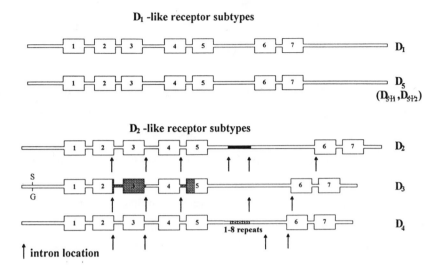

D₁ -like receptor subtypes

D₂ -like receptor subtypes

↑ intron location

Fig. 1. The family of dopamine receptor genes and their various transcripts. The numbered rectangles correspond to the seven putative transmembrane domains. Arrows indicate the position of introns and shaded areas to alternative exons or polymorphic variants in humans.

Thus, the D_1 and D_5 receptors are characterized by a short third cytoplasmic loop and a long C-terminal tail and are coupled to Gs proteins. On the other hand D_2, D_3 and D_4 receptors, which display a long third cytoplasmic loop and a short C-terminal tail, might be coupled to a Gi protein (or another G protein termed Go). This subdivision extends to the absence of introns in the genes for D_1 and D_5 and their presence in the genes for D_2, D_3 and D_4 (see Fig. 1).

One additional source of diversity is in the process of alternative splicing which can take place in those genes (D_2, D_3, D_4) in which the coding sequence is contained in discontinuous DNA segments (exons) interspersed among introns.

For the D_2 receptor, alternative splicing has been shown to occur in several animal species, potentially producing two receptors that differ by a stretch of 29 amino acids in the third cytoplasmic loop, these variants, or isoforms, are called $D_{2(444)}$ (or D_{2L} for D_2 long, or D_{2A}) and $D_{2(415)}$ (or D_{2S}, for D_2 short, or D_{2B}). They display identical pharmacology but have different patterns of expression among cerebral areas, although $D_{2(444)}$ predominates (Dal Toso *et al.*, 1989; Giros *et al.*, 1989; Monsma *et al.*, 1989). In addition, treatment with haloperidol, a D_2 antagonist, predominantly enhances the abundance of the smaller isoform in brain (Martres *et al.*, 1992) and pituitary (Arnauld *et al.*, 1991).

Taken together these observations suggest that the two isoforms are variously regulated and differ functionally in some fashion, perhaps at the second messenger level, through coupling to distinct G proteins, because such coupling probably involves the third cytoplasmic loop. The two isoforms, however, do not differ significantly in either pharmacology or signalling properties. Various truncated forms of the D_3 receptor mRNA, generated by alternative splicing, have also been detected in rat and human brain which do not correspond to a functional receptor (Giros *et al.*, 1991). Hence the physiological significance of the process of alternative splicing remains elusive, even more for the D_3 than for the D_2 receptor.

Another source of diversity is the existence of polymorphisms in the coding sequence of some dopamine receptor genes. These occur at the level of one amino acid of the N-terminal tail of the D_3 receptor (Lannfelt *et al.*, 1992) and of the third cytoplasmic loop of the D_4 receptor, where a variable number of repeats are present in the human gene (Van Tol *et al.*, 1992).

A final source of diversity is the occurrence of pseudogenes encoding for truncated, presumably non-functional D_5 receptors (Grandy *et al.*, 1991).

Dopamine receptor subtypes differ by their pharmacology and signalling systems

The expression of each of the various dopamine receptor subtypes in separate cells has allowed the establishment of their pharmacological profiles which confirm the existence of two subfamilies, based upon this criterion (Tables I and II). It is also clear that, whereas available compounds distinguish the two subfamilies, they are generally unable to distinguish members of each subfamily, an observation which constitutes a considerable challenge to medicinal chemists (Table III).

Before the advent of molecular biology approaches, the D_1 and D_2 receptors were defined by their positive and negative coupling to adenylate cyclase, respectively. With the heterologous expression of receptor subtypes in recipient cells this distinction was extended to the two D_1 and D_2 subfamilies, with one exception, however, that of the D_3 receptor. The latter does not couple to any identified signaling system in several transfected cell types but, recently, was found to trigger c-fos expression and mitogenesis in transfected NG 108-15 cells (Sokoloff *et al.*, 1993). The observation that D_3 receptor stimulation triggers c-fos expression contrasts with opposite effects found in the case of the D_2 receptor in brain and suggests that D_2 and D_3 receptors might

Table I. Synopsis of dopamine receptor subtypes: the D_1-like subfamily

	D_1	$D_5(D_{1B})$
Coding sequence	446 a.a.	477 a.a.
Chromosome	5 q31-q34	4 p 16.3
Highest brain densities	Neostriatum	Hypothalamus, Hippocampus
Pituitary	No	No
DA neurons (A9, A10)	No	?
Affinity for DA	Micromolar	Submicromolar
Characteristic agonist	SKF-38393	SKF-38393
Characteristic antagonist	SCH-23390	SCH-23390
Adenylyl cyclase	Activates	Activates

a.a., amino acids

Table II. Synopsis of dopamine receptor subtypes: the D_2-like subfamily

	D_2	D_3	D_4
Coding sequence	D_{2A} = 443 a.a. D_{2B} = 414 a.a.	400 a.a.	387 a.a.
Chromosome	11 q22-23	3 q13.3	11 p
Highest brain densities	Neostriatum	Paleostriatum Archicerebellum	Medulla Frontal cortex
Pituitary	Yes	No	Yes
DA neurons (A9, A10)	Yes	Yes	No
Affinity for DA	Micromolar	Nanomolar	Submicromolar
Characteristic agonist	Bromocriptine	7-OH-DPAT	?
Characteristic antagonist	Haloperidol	UH 232	Clozapine
Adenylyl cyclase	Inhibits	No effect	Inhibits
Phospholipase A_2	Activates	No effect	?
Mitogenesis	Activates	Activates	?

a.a., amino acids; 7-OH-DPAT, 7-hydroxy dipropylaminotetralin

mediate opposite behavioral responses.

In addition, novel signaling pathways were detected for the D_2 receptor, i.e. activation of phospholipases A_2 and C (Vallar et al., 1990; Piomelli et al., 1991) (Table II).

Two major conclusions arise from these studies: i) dopamine may affect target cells through three major signalling pathways i.e. adenylyl cyclase, phospholipase A_2 and phospholipase C products, ii) the nature of responses to dopamine depends not only on the receptor subtype involved but also on the cell type in which it is co-expressed with a given set of G proteins and effector systems.

Dopamine receptor subtypes are differently expressed among dopaminoceptive areas: selective localisation of the D_3 receptor in ventral striatum

Northern blot and *in situ* hybridization studies have established that the genes for the D_1 and D_2 receptors are the most abundantly expressed in brain, the corresponding mRNAs being detected easily in all main dopamine responsive areas. These localizations are essentially consistent with the results of membrane binding studies and autoradiography.

Both D_1 and D_2 receptors are the most abundant in the caudate-putamen, *nucleus accumbens* and olfactory tubercle. In contrast, D_3, D_4 and D_5 receptors are much less abundant and are selectively expressed in a few, restricted, dopaminoceptive areas. For instance the D_5 receptor mRNA is restricted mainly to lateral mammillary nuclei, anterior pretectal nuclei and some hippocampal layers (Tiberi *et al.*, 1991).

The expression of the D_3 receptor gene is mainly restricted to some areas receiving dopamine inputs from the A_{10} cell group, e.g. the *nucleus accumbens*, islands of Calleja, bed nucleus of the stria terminalis and other limbic areas such as septal or mammillary nuclei; the D_3 receptor gene is also highly expressed in cerebellar lobules 9 and 10 (Bouthenet *et al.*, 1991; Levesque *et al.*, 1992; Schwartz *et al.*, 1993). In most of these areas, the distributions of D_2 and D_3 receptor mRNAs do not seem to overlap, indicating that distinct cell populations are involved.

The distribution of the D_3 receptor and its mRNA were recently studied in some detail in the *nucleus accumbens*, in view of the well established role of this brain structure in processes of reward and drug abuse (Fig.2).

The D_3 receptor was mainly expressed in the rostral and shell divisions of *nucleus accumbens*. The former division, which also abundantly expresses D_2 receptors has mixed connections, whereas the shell division, which displays little D_2 receptor expression, selectively receives inputs from widespread subcortical neurons in the bed nucleus of stria terminalis, medial amygdala, lateral hypothalamus, brainstem reticular formation and projects to the hypothalamus and the ventral prefrontal and entorhinal cortex via the ventromedial pallidum and mediodorsal

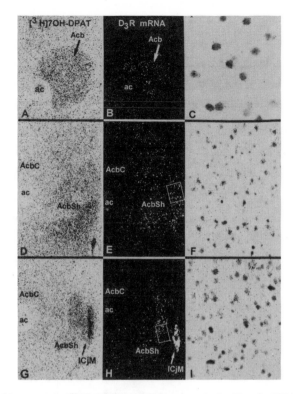

Fig. 2. Distribution of D₃ receptors ([³H]7-OH-DPAT binding sites, A, D, G) and mRNAs (B, E, H) at various rostrocaudal anterior areas of the rat *nucleus accumbens*. High magnification bright-field microphotographs (C, F, I) of positive cells in the shell part of *nucleus accumbens* are also shown.
Abbreviations: ac, anterior commissura; Acb, *nucleus accumbens*; AcbC, core part of *nucleus accumbens*; AcbSh, shell part of nucleus accumbens; ICjM, island of Calleja major.

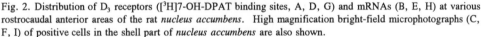

thalamus. In contrast, the core division of *nucleus accumbens*, which expresses D₂ but not D₃ receptors receives afferents mainly from the dorsal prefrontal cortex, subiculum and basal amygdala and projects to premotor and primary motor cortex via the ventrolateral pallidum, subthalamic nucleus, ventromedial and ventrolateral thalamus (Zahm and Brog, 1992).

Hence the connectivity of the core is consistent with this area being involved in motor controls of premotor and supplementary motor areas of the cortex via pathways similar to those of the dorsal striatum. In contrast, the connectivity of the shell shows its association with the extended amygdala, lateral hypothalamus and prefrontal cortical areas lacking direct access to the primary motor cortex. The two subdivisions of the *nucleus accumbens* are further differentiated by their dopaminergic innervation: i) the feedback loop controlling dopaminergic neuron activity involves *substantia nigra* and the ventral tegmental area for the core and shell respectively, ii) dopaminergic axons are more resistant to the neurotoxin 6-hydroxydopamine in the shell than in the core, perhaps because they are less enriched in dopamine transporters, iii) at the ultrastructural level, these axons less frequently contact small spiny neurons (at the level of dendritic spines) in the shell than their counterparts in the core.

From all these points of view, the rostral pole of the *nucleus accumbens* appears as a composite area in which typical neuroanatomical aspects of the shell and the core are combined. Hence it appears that D_2 and D_3 receptor distributions define sub-areas in the *nucleus accumbens* which were already distinguished by a variety of other neuroanatomical approaches. Even more importantly, these observations suggest that it will be possible to manipulate selectively a single dopaminergic subsystem when adequate pharmacological tools become available. One preliminary example is that of studies showing that (relatively) "D_3 selective agonists" (as defined from binding data) may decrease self-administration of cocaïne in rats (Caine and Koob, 1993).

Dopamine autoreceptors belong to diverse subtypes

Dopamine autoreceptors are a functional class of receptors found in dendrites, cell bodies or axons of dopamine neurons and mediating inhibition of the activity of these neurons or of dopamine synthesis and release. These autoreceptors had been identified pharmacologically as D_2 receptors. Accordingly, dopamine neurons in the *substantia nigra pars compacta* and ventral tegmental area express D_2 mRNA (Le Moine *et al.*, 1990; Mansour *et al.*, 1990; Bouthenet *et al.*, 1991), as confirmed by 6-hydroxydopamine lesions (Sokoloff *et al.*, 1990), a localization which is likely to reflect autoreceptor biosynthesis. The diversity of dopamine autoreceptors was first shown by the demonstration that both $D_{2(444)}$ and $D_{2(415)}$ isoforms are expressed in rat *substantia nigra* (Giros *et al.*, 1989; Martres *et al.*, 1992).

In the *substantia nigra pars compacta*, particularly the lateral part, and the ventral tegmental area, dopamine neurons were shown to express D_3 receptor mRNA, indicating that the D_3 receptor is also an autoreceptor (Sokoloff *et al.*, 1990). Other dopaminergic neurons (A_{13}, A_{14}, A_{16}) may also be endowed with D_3 autoreceptors (Bouthenet *et al.*, 1991).

In front of this diversity it remains to establish what is, among the various roles attributed to autoreceptors in the control of dopamine neuron activity, the role of each receptor subtype (or isoform). It would also be important to establish whether a single dopamine neuron expresses several receptor subtypes or isoforms.

Conclusion

The recent discovery of so many different dopamine receptor subtypes suggests that heterogeneity provides a means of increasing the complexity of messages transmitted by a single messenger molecule, dopamine.

First, the various subtypes that differ in their affinity for dopamine, e.g. D_2 and D_3 and, to a lesser extent, D_1 and D_5 receptors, may also differ in the time scales of their activation cycles or in the distance at which they may be activated by released dopamine. Second, the diverse amino acid sequences of the intracellular loops, mainly the third, may ensure differential interactions of the various dopamine receptor subtypes with members of the family of G proteins, new members of which are still being discovered. As G proteins couple in diverse ways to various intracellular signalling pathways this, in turn, is likely to correspond to the generation of a range of intracellular signals. Obviously many more biochemical and electrophysiological studies are required to analyse all possible signalling pathways activated by the various dopamine receptor subtypes.

The discovery of such a range of molecular targets for dopamine agonists and antagonists also opens unique opportunities for developing new classes of drugs that more clearly define the role of the dopamine receptor subtypes and provide more specific and effective treatments for several major neurological and psychiatric diseases. This is particularly clear in the case of the D_3 receptor: with selective ligands, it will be possible to affect only very limited brain areas, particularly those of the "limbic" striatum that numerous studies have shown to be involved in reward, alcohol and drug abuse (Nestler, 1992).

References

Arnauld, E., Arsaut, J. and Demotes-Mainard, J. (1991) Differential plasticity of the dopa minergic D_2 receptor mRNA isoforms under haloperidol treatment, as evidenced by in situ hybridization in rat anterior pituitary. *Neurosci. Lett.* 130: 12-16.

Bouthenet, M.L., Souil, E., Martres, M.P., Sokoloff, P., Giros, B. and Schwartz, J-C. (1991) Localization of dopamine D_3 receptor mRNA in the rat brain using in situ hybridization histochemistry: comparison with dopamine D_2 receptor mRNA. *Brain Res.*

564: 203-219.

Bunzow, J.R., Van Tol, H.H.M., Grandy, D.K., Albert, P., Salon, J., Christie, McD., Machida, C.A., Neve, K.A. and Civelli, O. (1988) Cloning and expression of a rat D_2 dopamine receptor cDNA. *Nature* 336: 783-787.

Caine, S.B. and Koob, G.F. (1993) Modulation of cocaine self-administration in the rat through D_3 dopamine receptors. *Science* 260: 1814-1816.

Dal Toso, R., Sommer, B., Ewert, M., Herb, A., Pritchett, D.B., Bach, A., Shivers, B.D. and Seeburg, P.H. (1989) The dopamine D_2 receptor: two molecular forms generated by alternative splicing. *EMBO .J* 8: 4025-4034.

Dearry, A., Gingrich, J.A., Falardeau, P., Fremeau, R.T., Bates, M.D. and Caron, M.G. (1990) Molecular cloning and expression of the gene for a human D_1 dopamine receptor. *Nature* 347: 72-76.

Giros, B., Sokoloff, P., Martres, M.P., Riou, J.F., Emorine, L.J. and Schwartz, J-C. (1989) Alternative splicing directs the expression of two D_2 dopamine receptor isoforms. *Nature* 342: 923-926.

Giros, B., Martres, M.P., Sokoloff, P. and Schwartz, J.-C. (1990) cDNA cloning of the human dopaminergic D_3 receptor and chromosome identification. *CR. Acad. Sci. Paris III* 311: 501-508.

Giros, B., Martres, M.P., Pilon, C., Sokoloff, P. and Schwartz, J.-C. (1991) Shorter variants of the D_3 dopamine receptor produced through various patterns of alternative splicing. *Biochem. Biophys. Res. Commun.* 176: 1584-1592.

Grandy, D.K., Marchionni, M.A., Makan, H., Stofko, R.E., Alfano, M., Prothingham, L., Fischer, J.B., Burke-Howie, K.J., Bunzow, J.R., Server, A.C. and Civelli, O. (1989) Cloning of the cDNA and gene for a human D_2 dopamine receptor. *Proc. Natl. Acad. Sci. USA* 84: 9762-9766.

Grandy, D.K., Zhang, Y., Bouvier, C., Zhou, Q., Johnson, R.A., Allen, L., Buck, K., Bunzow, J.R., Salon, J. and Civelli, O. (1991) Multiple human D_5 dopamine receptor genes: a functional receptor and two pseudogenes. *Proc. Natl. Acad. Sci. USA* 88: 9175-9179.

Kebabian, J.W. and Calne, D.B. (1979) Multiple receptors for dopamine. *Nature* 277: 93-96.

Lannfelt, L., Sokoloff, P., Martres, M.P., Pilon, C., Giros, B., Jönsson, E., Sedvall, G. and Schwartz, J-C. (1992) Amine acid substitution in the dopamine D_3 receptor as a useful polymorphism for investigating psychiatric disorders. *Psychiatric Genetics* 2: 249-256.

Le Moine, C., Normand, E., Guitteny, A.F., Rouques, B., Teoule, R. and Bloch, B. (1990) Dopamine receptor gene expression by enkephalin neurons in rat forebrain. *Proc. Natl. Acad. Sci. USA* 87: 230-234.

Levesque, D., Diaz, J., Pilon, C., Martres, M.P., Giros, B., Souil, E., Schott, D., Morgat, J.L., Schwartz, J-C. and Sokoloff, P. (1992) Identification, characterization and localization of the dopamine D_3 receptor in rat brain using [^{3}H]7-hydroxy dipropylaminotretralin. *Proc. Natl. Acad. Sci. USA* 89: 8155-8159.

Mansour, A., Meador-Woodruff, J.H., Bunzow, J.R., Civelli, O., Akil, H. and Watson, S.J. (1990) Localization of dopamine D_2 receptor mRNA and D_1 and D_2 receptor binding in the rat brain and pituitary. An in situ hybridization receptor autoradiographic analysis. *J. Neurosci.* 10: 2587-2601.

Martres, M.P., Sokoloff, P., Giros, B. and Schwartz, J.-C. (1992) Effects of dopaminergic transmission interruption on the D_2 receptor isoforms in various cerebral tissues. *J. Neurochem.* 58: 673-679.

Monsma, F.J., McVittie, L.D., Gerfen, C.R., Mahan, L.C. and Sibley, D.R. (1989) Multiple D_2 dopamine receptors produced by alternative RNA splicing. *Nature* 342: 926-929.

Nestler, E. (1992) Molecular mechanisms of drug addiction. *J. Neurosci.* 12: 2439-2450.

Piomelli, D., Pilon, C., Giros, B., Sokoloff, P., Martres, M.P. and Schwartz, J.-C. (1991) Dopamine activation of the arachidonic acid cascade via a modulatory mechanism as a basis for D_1/D_2 receptor synergism. *Nature* 353: 164-167.

Schwartz, J.-C., Giros, B., Martres, M.P. and Sokoloff, P. (1992) The dopamine receptor family: molecular biology and pharmacology. In: Robbins, T. (ed.): *Seminars in the Neurosciences* Vol. 4, W.B. Saunders Company, London, pp. 99-104.

Schwartz, J.-C., Levesque, D. Martres, M.P. and Sokoloff, P. (1993) Dopamine D_3 receptor, basic and clinical aspects. *Clin. Neuropharm.* 16: 295-314.

Sokoloff, P., Giros, B., Martres, M.P., Bouthenet, M.L. and Schwartz, J.-C. (1990) Molecular cloning and characterization of a novel dopamine receptor (D_3) as a target for neuroleptics. *Nature* 347: 146-151.

Sokoloff, P., Andrieux, M., Besançon, R., Pilon, C., Martres, M.P., Giros, B. and Schwartz, J.-C. (1992) Pharmacology of human D_3 dopamine receptor expressed in a mammalian cell line : comparison with D_2 receptor. *Eur. J. Pharmacol. Mol. Pharmacol. Sect.* 255: 331-337.

Sokoloff, P., Martres, M.P., Giros, B., Levesque, D., Diaz, J., Pilon, C., Griffon, N. and Schwartz, J.-C. (1993) The dopamine D_3 receptor. In: Niznik, H.B. (ed.): *Dopamine receptor function and pharmacology*. Marcel Dekker Inc., New York, pp.165-188.

Spano, P.F., Govoni, S. and Trabucchi, M. (1978) Studies on the pharmacological properties of dopamine receptors in various areas of the central nervous system. *Adv. Biochem. Psychopharmacol.* 19: 155-165.

Sunahara, R.K., Niznik, H.B., Weiner, D.M., Stormann, T.M., Brann, M.R., Kennedy, J.L., Gelernter, J.E., Rozmahel, R., Yang, Y., Israel, Y., Seeman, P. and O'Dowd, B.F. (1990) Human dopamine D_1 receptor encoded by an intronless gene on chromosome 5. *Nature* 347: 80-83.

Sunahara, R.K., Guan, H.C., O'Dowd, B.F., Seeman, P., Laurier, L.G., Ng, G., George, S.R., Torchia, J., Van Tol, H.H.M. and Niznik, H.B. (1991) Cloning of the gene for a human dopamine D_5 receptor with higher affinity for dopamine than D_1. *Nature* 350: 614-619.

Tiberi, M., Jarvie, K.R., Silvia, C., Falardeau, P., Gingrich, J.A., Godinot, N., Bertrand, L., Yang-Feng, T.L., Fremeau, R.T. Jr and Caron, M.G. (1991) Cloning, molecular characterization, and chromosomal assignment of a gene encoding a second D_1 dopamine receptor subtype: differential expression pattern in rat brain compared with the D_{1A} receptor. *Proc. Natl. Acad. Sci. USA* 88: 7491-7495.

Vallar, L., Muca, C., Magni, M., Albert, P., Bunzow, J., Meldolesi, J. and Civelli, O. (1990) Differential coupling of dopaminergic D_2 receptors expressed in different cell types. *J. Biol. Chem.* 265: 10320-10326.

Van Tol, H.H.M., Bunzow, J.R., Guan, H.C., Sunahara, R.K., Seeman, P., Niznik, H.B. and Civelli, O. (1991) Cloning of the gene for a human dopamine D_4 receptor with high affinity for the antipsychotic clozapine. *Nature* 350: 610-614.

Van Tol, H.H.M., Wu, C.M., Guan, H.C., Ohara, K., Bunzow, J.R., Civelli, O., Kennedy, J., Seeman, P., Niznik, H.B. and Janovic, V. (1992) Multiple dopamine D_4 receptor variants in the human population. *Nature* 358: 149-152.

Zahm, D.B. and Brog, J.S. (1992) On the significance of subterritories in the "accumbens" part of the rat ventral striatum. *Neurosci.* 50: 751-767.

Zhou, Q.Z., Grandy, D.K., Thambi, L., Kushner, J.A., Van Tol, H.H.M., Cone, R., Pribnow, D., Salon, J., Bunzow, J.R. and Civelli, O. (1990) Cloning and expression of human and rat D_1 dopamine receptors. *Nature* 347: 76-80.

Toward a Molecular Basis of Alcohol Use and Abuse
ed. by B. Jansson, H. Jörnvall, U. Rydberg, L. Terenius & B. L. Vallee
© 1994 Birkhäuser Verlag Basel/Switzerland

Molecular pharmacology of serotonin receptors

Paul R. Hartig

Synaptic Pharmaceutical Corporation, 215 College Road, Paramus, NJ 07652

Summary

The serotonin system has long been thought to play a role at several steps in the cycle of alcohol abuse. Initial motivation may be triggered by anxiety, which may exhibit a serotonergic component (5-HT$_{1A}$ receptor). Alcohol can potentiate the opening of 5-HT$_3$ receptor ion channels, and agents which elevate serotonergic tone, including serotonergic agonists, uptake inhibitors and releasers, have shown promise in assisting with recovery from alcoholism. In this review, recent advances in serotonin receptor research are presented, with a special emphasis on the impact and interpretation of molecular biological data. Genetic and pharmacological concepts of receptor subtypes are reviewed and related to a new classification system for the 14 currently recognized subtypes of serotonin receptors. The current and likely future impact on drug design of the molecular approach to serotonin receptors is discussed. Finally, the question of why there are so many serotonin receptor subtypes is examined, along with possible roles of multiple G protein and second messenger pathways, and their effect on conserved domains of these receptor proteins.

Introduction

Serotonin and alcohol

Serotonin has been thought to play a role in many aspects of alcohol addiction, from motivation to mechanisms of intoxication to recovery from abuse. In most studies a general picture has emerged that increased central 5-HT decreases alcohol consumption while lowered 5-HT increases alcohol drinking (Sellers *et al.*, 1992), although some animal studies provide just the opposite picture (Korpi *et al.*, 1992). As reviewed by Sellers *et al.* (1992), Alcohol-Preferring and High-Alcohol-Drinking rat lines exhibit lowered 5-HT function, while decreased 5-HT in the nucleus accumbens and other brain regions encourages alcohol drinking in animal models. The serotonin metabolite 5-HIAA is decreased in the cerebrospinal fluid of many alcohol abusers, and in Alcohol-Preferring rats. Serotonin uptake blockers reduce alcohol consumption in both animal and human studies, and reduce the desire for alcohol. In addition, platelet serotonin uptake as well as general adenylate cyclase activity is reduced in families with a history of alcohol abuse. The known effects of uptake blockers on reduced food intake and enhanced satiety, as well as anxiolytic actions at the 5-HT$_{1A}$ receptor may indicate a nonspecific contribution to these alcohol effects. In addition to the 5-HT$_{1A}$ receptor, 5-HT$_2$ receptors have been implicated in alcohol consumption due to certain studies with ritanserin (Naranjo, 1994;

Sellers *et al.*, 1992), and opening of the 5-HT$_3$ receptor ion channel is potentiated by alcohol, which may indicate a role in the mechanism of alcohol intoxication (Sellers *et al.*, 1992).

Discussion

Serotonin receptor subtype diversity

At this writing, fourteen serotonin receptor subtypes have been identified, representing thirteen different G protein-coupled receptors, and one ligand-gated ion channel (5-HT$_3$). Clones have been reported for all but one of these subtypes (5-HT$_4$). This represents the largest proven diversity of G protein-coupled receptor subtypes responsive to any single transmitter. Undiscovered subunits of the 5-HT$_3$ receptor may well exist (Maricq *et al.*, 1991), as indicated by their diverse properties in different tissues (Richardson and Engel, 1986), which may also expand the number of 5-HT$_3$ receptor subtypes in the future. In this review, the current status of serotonin receptor characterization will be discussed, with comments on the use of cloned receptors in serotonergic drug development and alcoholism research.

Pharmacological subtypes, molecular subtypes and species homologues

In the discussion of receptor subtypes it is very useful to clearly distinguish between three different concepts of subtypes. These concepts are reviewed in more depth elsewhere (Hartig and Gluchowski, 1994). The term "pharmacological subtype" will be used to describe a particular set of binding or response data describing apparent binding affinities (K$_i$ or K$_d$) or tissue response parameters (pA$_2$, pK$_D$, pK$_A$) in a model system (for example, the 5-HT$_{1D}$ receptor was originally defined as a [^{3}H]5-HT binding site in bovine caudate (Heuring and Peroutka, 1987)). The term "molecular subtype" will be used to describe a single protein (in a particular animal species), and its associated cDNA or genomic clone. For G protein-coupled receptors, almost all molecular subtypes are encoded by different genes, so the one protein/one gene rule is generally applicable. Certain exceptions do exist, such as the long and short splice variants of the D2 dopamine receptor (Monsma *et al.*, 1989), but each serotonin receptor appears to be represented by a different gene. The term "species homologue" has been in use for some time to describe the "equivalent" receptor protein (and its associated gene) in a different species. In the neurotransmitter receptor field, all receptor clones identified thus far appear to have counterparts in all mammalian species examined. In other words, the genome of every mammal contains (so far) the same set of receptor subtype genes. The large differences in brain complexity and cognitive capabilities among mammals do not appear to derive from differences in the number of receptor subtypes encoded by their genomes.

Once the entire genomes of man, rats and other mammals have been mapped and characterized, it should be a rather straightforward task to assign to each human gene its animal equivalent (species homologue) in all other species, because high degrees of nucleotide and amino acid conservation are generally observed for species homologues obtained from different mammalian species. If you move far enough down the evolutionary ladder (far enough back in evolutionary time) then the complement of genes does, of course, change between the species. It has been interesting to learn just how ancient serotonin and other receptor systems are. For example, molecular subtypes first identified by cloning in the common fruit fly, drosophila (e.g. dro1 (Witz *et al.*, 1990)), have now recently been used to discover a homologous new serotonin receptor subtype (5-HT$_7$) in man (Bard *et al.*, 1993) and other species (Plassat *et al.*, 1993).

5-HT$_{1D}$ Receptors: subtypes and species homologues

In general, species homologues of receptors exhibit very similar amino acid sequences and pharmacological properties, leading to their assignment to the same pharmacological subtype. Different molecular subtypes generally exhibit much lower amino acid sequence and less pharmacological similarity, and are generally classified as different pharmacological subtypes based on their ligand binding properties. Cloning of the 5-HT$_{1D}$ receptor subfamily has provided noteworthy exceptions to both of these general rules, as discussed in Hartig *et al.* (1992) and reviewed briefly below.

The human 5-HT$_{1D\alpha}$ and 5-HT$_{1D\beta}$ receptor clones were both isolated by homology to the dog RDC4 receptor orphan clone (Weinshank *et al.*, 1992) and found to exhibit a degree of amino acid homology (77% transmembrane amino acid identity) that is typical of subfamilies of receptor subtypes (e.g. the three alpha-1 adrenergic receptors or the dopamine D1 and D5 receptors). For this reason, we expected to find related but different ligand binding properties when these clones were transfected and the receptors evaluated with compounds from several chemical classes. Instead, log-log plots of apparent K$_i$ values for 17 different compounds assayed at these two receptors yielded a correlation coefficient of r = 0.96 (Weinshank *et al.*, 1992), which is higher than appears to have been found between any other pair of cloned G protein-coupled receptors (except, perhaps, for the m2 and m4 muscarinic receptors (Dörje *et al.*, 1991)) and better than the correlation usually found when comparing receptor binding data on the same preparation between two different laboratories. Furthermore, it is equivalent to or better than the similarity in pharmacological properties usually found between species homologues of a receptor. Thus, the cloning and characterization of two 5-HT$_{1D}$ receptors demonstrated that a single pharmacological subtype could be displayed by different molecular subtypes. Further examination has revealed some compounds with notable separations in binding affinities at the two 5-HT$_{1D}$ receptors (e.g. ketanserin: Weinshank *et al.*, 1991), and ongoing research in several companies is making progress towards the design of compounds which discriminate between these subtypes. In this and other cases, we expect that the differences in amino acid sequence that do exist between these 5-HT$_{1D}$ receptor subtypes will be exploited by drug designers, eventually leading to highly selective drugs for each molecular subtype. However, early in the characterization of receptor clones a single (historical) pharmacological subtype (e.g. 5-HT$_{1D}$) may appear to be represented by two separate molecular subtypes of receptors (human 5-HT$_{1D\alpha}$ and human 5-HT$_{1D\beta}$).

Another important exception is found in the pharmacological properties of species homologues of the 5-HT$_{1D\beta}$ receptor. Although the 5-HT$_{1D\alpha}$ receptor has shown similar pharmacological properties in all species examined, the rat and human homologues of the 5-HT$_{1D\beta}$ receptor show quite different ligand binding profiles. For example, the expressed human and rat 5-HT$_{1D\beta}$ clones only exhibit a correlation coefficient of $r = 0.18$ in a log-log comparison of K$_i$ values for 11 compounds (Oksenberg *et al.*, 1992). This large difference in pharmacological properties for species homologues of the 5-HT$_{1D\beta}$ receptor led to designation of this molecular subtype as a 5-HT$_{1B}$ pharmacological subtype in the rat and mouse but a 5-HT$_{1D}$ pharmacological subtype in the human, dog, guinea pig and other species (Waeber *et al.*, 1989). The 5-HT$_{1B}$ receptor

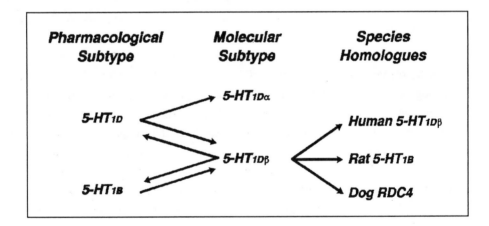

Fig. 1. Relationships between subtypes and species homologues of the 5-HT$_{1D}$ receptor subfamily.

is now understood to be a pharmacological subtype with no analogue in humans (Hartig *et al.*, 1992). A summary of the relationships between these subtypes and species homologues of 5-HT$_{1D}$ receptors is shown in Fig. 1.

Why have these two species homologues (human 5-HT$_{1D\beta}$, rat 5-HT$_{1B}$) diverged into two receptor subtypes with such different pharmacological properties? First, it must be kept in mind that these receptors evolved to recognize and respond to serotonin, not to the products of 20th century pharmaceutical science. Thus, their pharmacological differences may not be of any interest to the organism, and may simply reflect the extension of drug compounds into regions of the receptor that are not important for serotonin binding. An alternative, although highly speculative, idea is based on the fact that serotonin receptors exhibit high affinity for ergot and related alkaloids. It is conceivable that a particular alkaloid or other plant or fungal toxin may have been a toxic threat to the rat via an action on a serotonin receptor. In such a case the rat

SEROTONIN RECEPTORS

	$5\text{-}HT_1$	FAMILY			
Molecular Subtype	$5\text{-}HT_{1A}$	$5\text{-}HT_{1D\alpha}$	$5\text{-}HT_{1D\beta}$	$5\text{-}HT_{1E}$	$5\text{-}HT_{1F}$
Pharmacological Subtype	$5\text{-}HT_{1A}$	$5\text{-}HT_{1D}$	$5\text{-}HT_{1B}$	$5\text{-}HT_{1E}$	
Second Messenger	↓cAMP	↓cAMP	↓cAMP	↓cAMP	↓cAMP

	$5\text{-}HT_2$	FAMILY	
Molecular Subtype	$5\text{-}HT_{2A}$	$5\text{-}HT_{2B}$	$5\text{-}HT_{2C}$
Pharmacological Subtype	$5\text{-}HT_2$	old $5\text{-}HT_{2F}$	old $5\text{-}HT_{1C}$
Second Messenger	↑PI	↑PI	↑PI

		OTHERS				
Molecular Subtype	$5\text{-}HT_3$		$5\text{-}HT_{5A}$	$5\text{-}HT_{5B}$	$5\text{-}HT_6$	$5\text{-}HT_7$
Pharmacological Subtype		$5\text{-}HT_4$				
Second Messenger	ligand-gated ion channel	↑cAMP	?	?	↑cAMP	↑cAMP

Fig. 2. Classification scheme for serotonin receptors (Hoyer *et al.*, 1993a,b)

receptor would experience evolutionary pressure to not bind these toxins and thus may have developed a different pharmacological binding profile. Interestingly, a similar phenomenon has been proposed for dietary driven co-evolution of the rat P450 gene family in which the evolution of plant toxins (phytoalexins) appears to have caused a parallel rapid evolution and diversification of cytochrome P450 genes in order to detoxify these poisons (Gonzalez and

Nebert, 1990). Such a toxin-driven evolutionary pressure could conceivably account for the current differences in 5-HT$_{1D\beta}$ pharmacology in different species.

Serotonin receptor classification

Investigation of the molecular properties of 5-HT$_{1D}$ and other receptors has taught us that no single criterion is a reliable guide to the understanding and classification of all receptor subtypes. Recognizing this fact, and in an attempt to integrate past serotonin receptor classification schemes into a modern nomenclature, a revised serotonin receptor classification which incorporates structural (deduced amino acid and nucleotide sequence), operational (pharmacological profile) and transductional (second messenger coupling) components was recently proposed (Hoyer *et al.*, 1993a). A summary of these 14 known subtypes of serotonin receptors, including two subfamilies based on sequence homology and second messenger coupling, is provided in Figure 2. An extensive review of their pharmacological and physiological properties has recently been written (Hoyer *et al.*, 1993b), and should prove quite useful to serotonin researchers.

Why are there so many serotonin receptor subtypes?

The question of why 14 different subtypes of serotonin receptor are maintained in the genome is obviously of central importance both to neuroscientists and to drug designers. The question is especially intriguing for the case of receptor subfamilies such as the 5-HT$_1$ subfamily, where 5 different subtypes all appear to couple to the same second messenger response (inhibition of adenylate cyclase activity). Although no final answer is yet available to this question, several explanations appear to be ruled out by the current data and several others remain active possibilities, as discussed below.

Through much of the early 1980's, the question of which receptor subtypes were presynaptic and which were postsynaptic occupied many researchers. As suggested by earlier work, and now definitively proven by molecular biology, it seems that most G protein-coupled receptors can function as both pre- and postsynaptic receptors. The serotonin receptors also follow this rule. Receptor ligand autoradiography studies and *in situ* hybridization studies for receptor RNA have both demonstrated that all five 5-HT$_1$ receptors are expressed in multiple raphé nuclei (Bruinvels *et al.*, 1994). In addition, numerous *in situ* hybridization studies have shown that all of these subtypes are present in mammalian brain outside of the raphé nuclei and are thus postsynaptic (Gluchowski *et al.*, 1993). Thus, specialization of 5-HT$_1$ receptors for pre- or postsynaptic roles cannot account for the diversity of subtypes.

A plausible explanantion for subtypes would be that each is tuned to respond to a different concentration of serotonin. Based on our current knowledge of [^{3}H]5-HT binding affinities for the cloned human subtypes, this does not appear to be the case. All five cloned human serotonin 5-HT_1 receptors display serotonin binding affinities in the range of 1.9-9.7 nM, a range that hardly seems adequate for the maintenance of five separate genes. It is still possible, however, that the *in vivo* response midpoints may differ considerably from the high affinity serotonin binding state found in ligand binding studies. Response midpoints (EC_{50} values) appear to be highly sensitive to the experimental design (Thomsen *et al.*, 1988). Since it is not yet clear how to properly evaluate the *in vivo* response midpoints for each human serotonin receptor clone, this conclusion may need to be revised in the future.

Receptors of the 5-HT_1 subfamily do exhibit different regional distributions in brain and other tissues. It does not seem that separate 5-HT_1 receptor subtypes would be needed just to achieve different distributions in the brain if all of them have the same functional properties. It may, however, be advantageous to express a 5-HT_1-like receptor at different developmental times in different tissues of an organism. This would presumably be controlled primarily by tissue-specific promoter regions of the gene (primarily 5' upstream to the protein coding region). If multiple genes arose in order to achieve expression in a tissue and developmental time-specific order, it is reasonable that their coding regions would slowly change over evolutionary time leading to the different receptor subtypes we currently study. Our present knowledge of the untranslated regions of these genes, and of their role in control of gene expression, is too primitive to adequately evaluate this proposal.

A current, reasonable hypothesis for the diversity of 5-HT_1 receptor subtypes arises from the diversity of second messenger activities that single receptors have been known to trigger (also known as receptor promiscuity). Serotonin 5-HT_1 receptors have been shown to couple to 2 or 3 different second messenger systems in studies with the 5-HT_{1A} (reviewed in Hoyer and Boddeke, 1993) , 5-HT_{1D} (Zgombick *et al.*, 1993), and 5-HT_{1F} (Adham *et al.*, 1994) receptors. Certain of these activites are most likely artifacts of the high densities of receptor expression which can be achieved with cloned, transfected receptors (Hoyer and Boddeke, 1993). Others, such as intracellular Ca^{++} and phosphoinositide hydrolysis, may be downstream expressions of multiple signal points in the same pathway, while still others may represent true diversity of coupling also expressed *in vivo*. At present, it has not been possible to definitively separate these possibilities, so the question of diverse coupling will continue to be actively studied. An intriguing possibility, raised by a 5-HT_{1A} receptor study (Karschin *et al.*, 1991), is that G_i coupled receptors may directly couple to ion channels without an intervening second messenger. This possibility has not been actively investigated for most cloned 5-HT_1 receptors and could extensively reform our understanding of the direct biological actions of this subfamily.

Finally, an important aspect of this diverse second messenger coupling is the possibility that different 5-HT_1 receptors evolved to recognize different G protein isoforms. Preliminary data supportive of this view has been obtained from the 5-HT$_{1D}$ receptors. When the 5-HT$_{1D\alpha}$ receptor was transfected into different cell lines, those lines containing the G$_{i\alpha3}$ isotype exhibited strong coupling to inhibition of cAMP production but did not couple well to the 5-HT$_{1D\beta}$ receptor. Other lines containing other isotypes did couple to the 5-HT$_{1D\beta}$ receptor and the 5-HT$_{1D\alpha}$ receptor failed to couple well in cell lines containing G$_{i\alpha1}$ or G$_{i\alpha2}$ (Branchek *et al.*, 1992). This data is circumstantial evidence in favor of different G protein selectivity of the two 5-HT$_{1D}$ receptors. Definitive proof awaits gene or protein knockout studies, which are underway.

If the suggestion that the 5-HT$_{1D\alpha}$ and 5-HT$_{1D\beta}$ receptors are designed to couple to different G$_i$ isoforms is true, then one would expect to see a different amino acid sequence in the regions of these receptors that are known to be key domains for coupling to G proteins. Chimeric receptor and site-directed mutagenesis studies have demonstrated that at least three domains are critical determinants of the interaction between the receptor and its G protein: the amino end and carboxy ends of the third intracellular loop and the amino end of the carboxy tail (Dohlman *et al.*, 1991). If the 5-HT$_{1D\alpha}$ and 5-HT$_{1D\beta}$ receptors are specialized to recognize different G proteins, then we would expect these domains to differ strongly between these two 5-HT$_{1D}$ subtypes but to be very similar between species homologues of the same subtype. This is exactly the case, as illustrated by the 20 amino acids which comprise the amino end of the carboxy tail of these receptors. The human 5-HT$_{1D\beta}$ and rat 5-HT$_{1D\beta}$ (5-HT$_{1B}$) receptors are species homologues which are identical at all 20 of these amino acid positions. This is indicative of a protein region with some important functional role because random evolutionary change since the separation of rat and human lines some 80 million years ago should have caused 80% of these amino acids to change if they had no functional significance (based on the Neutral Theory of evolution, see discussion in Hartig *et al.*, 1990). The fact that such a high degree of conservation is observed suggests that this domain has a very important and operationally constrained function. In contrast, the human 5-HT$_{1D\alpha}$ and human 5-HT$_{1D\beta}$ receptors differ at 11 of these 20 positions (at 55% of positions) compared to only 37% amino acid differences overall (Weinshank *et al.*, 1992). Thus, this carboxy terminal domain is a hyper-variant region of the 5-HT$_{1D}$ receptors, implying evolutionary selection for a specific role. Therefore, the amino acid sequence information from the G protein coupling domains of the 5-HT$_{1D}$ receptors is quite supportive of the theory that these two receptors evolved to selectively couple to different G proteins.

Future directions

The rapid expansion of our knowledge of serotonin receptor subtypes will certainly stimulate a burst of research into the physiological and pathophysiological roles of these newly discovered subtypes. Much of the literature of the past will need to be reinterpreted in light of our new understanding that drugs and other pharmacological agents used in past studies were activating a much broader array of receptor subtypes than we had known at the time. As new, highly subtype-selective reagents with known reactivities against each species homologue of a receptor become available, a much deeper molecular understanding of the roles of each receptor subtype will form. In the alcohol addiction field, and other serotonin-related fields, new therapeutics will emerge with increased efficacy and decreased side effects from our new ability to target drugs to individual human serotonin receptor subtypes.

References

Adham, N., Borden, L., Schechter, L.E., Cochran, T., Vaysse, P., Weinshank, R.L. and Branchek, T.A. (1994) Cell-specific coupling of the cloned human 5-HT$_{1F}$ receptor to multiple signal transduction pathways, *N.-S. Arch. Pharmacol.*, in press.

Branchek, T., Adham, N., Zgombick, J., Schechter, L., Hartig, P., Gustafson, E. and Weinshank, R. (1992) Differential host preference in functional coupling of cloned 5-HT$_{1D}$ receptor subtypes in heterologous expression systems, Abstracts from the *Second International Symposium on Serotonin: From Cell Biology to Pharmacology and Therapeutics.* p.10

Bruinvels, A.T., Branchek, T.A., Gustafson, E.L., Durkin, M.M., Landwehrmeyer, B., Mengod, G., Hoyer, D. and Palacios, J.M. (1994) Localisation of 5-HT$_{1B}$, 5-HT$_{1D\alpha}$, 5-HT$_{1E}$, and 5-HT$_{1F}$ messenger RNA in rodent and primate brain. *Neuropharm.*, in press.

Dörje, F., Wess, J., Lambrecht, G., Tacke, R., Mutschler, E. and Brann, M.R. (1991) Antagonist binding profiles of five cloned human muscarinic receptor subtypes. *J. Pharmacol. Exp. Ther.* 256: 722-733.

Dohlman, H.G., Thorner, J., Caron, M.G. and Lefkowitz, R.J. (1991) Model systems for the study of seven-transmembrane-segment receptors. *Ann. Rev. Biochem.* 60: 653-688.

Gluchowski, C., Branchek, T.A., Weinshank, R.L. and Hartig, P.R. (1993) Molecular/Cell biology of G protein coupled CNS receptors. *Ann. Rep. Med. Chem.* 28: 29-38.

Gonzalez, F.J. and Nebert, D.W. (1990) Evolution of the P450 gene superfamily. *Trends Genetics* 6: 182-186.

Hartig, P.R., Kao, H.-T., Macchi, M., Adham, N., Zgombick, J., Weinshank, R. and Branchek, T. (1990) The molecular biology of serotonin receptors. *Neuropsychopharm.* 3: 335-347.

Hartig, P.R. and Gluchowski, C. (1994) Monoamine receptor clones: Targets for new drug design. *J. Med. Chem.*, in preparation.

Hartig, P.R., Branchek, T.A. and Weinshank, R.L. (1992) A subfamily of 5-HT$_{1D}$ receptor genes. *Trends Pharmacol. Sci.* 13: 152-159.

Heuring, R.E. and Peroutka, S.J. (1987) Characterization of a novel [^{3}H]5-Hydroxytryptamine binding site subtype in bovine brain membranes. *J. Neurosci.* 7: 894-903.

Hoyer, D., Hartig, P. and Humphrey, P.P.A. (1993a) A proposed new nomenclature for 5-HT receptors. *Trends Pharmacol. Sci.* 14: 233-236.

Hoyer, D., Fozard, J.R., Saxena, P.R., Mylecharane, E.J., Clarke, D.E., Martin, G.R. and Humphrey, P.P.A. (1993b) A new classification of receptors for 5-hydroxytryptamine (serotonin). *Pharmacol. Rev.*, in press.

Hoyer, D. and Boddeke, H. (1993) Partial agonists, full agonists, antagonists: dilemmas of definition. *Trends Pharmacol. Sci.* 14: 270-275.

Karschin, A., Ho, B.Y., Labarca, C., Elroy-Stein, O., Moss, B., Davidson, N. and Lester, H.A. (1991) Heterologously expressed serotonin 1A receptors couple to muscarinic K^+ channels in heart. *Proc. Natl. Acad. Sci. USA* 88: 5694-5698.

Korpi, E.R., Paivarinta, P., Abi-Dargham, A., Honkanen, A., Laruelle, M. Tuominen, K. and Hilakivi, L.A. (1992) Binding of serotonergic ligands to brain membranes of alcohol-preferring AA and alcohol-avoiding ANA rats. *Alcohol* 9: 369-374.

Maricq, A.V., Peterson, A.S., Brake, A.J., Myers, R.M. and Julius, D. (1991) Primary structure and functional expression of the 5-HT_3 receptor, a serotonin-gated ion channel. *Science* 254: 432-437.

Monsma, F.J., McVittie, L.D., Gerfen, C.R., Mahan, L.C. and Sibley, D.R. (1989) Multiple D_2 dopamine receptors produced by alternative RNA splicing. *Nature* 342: 926-929.

Naranjo, C. (1994) Effects of serotonin-altering medications on desire, consumption and effects of alcohol - treatment implications. *Nobel Symposium*, this volume.

Oksenberg, D., Marsters, S.A., O'Dowd, B.F., Jin, H., Havlik, S., Peroutka, S.J. and Ashkenazi, A. (1992) A single amino acid difference confers major pharmacological variation between human and rodent 5-HT_{1B} receptors. *Nature* 360: 161-163.

Plassat, J.-L., Amlaiky, N. and Hen, R. (1993) Molecular cloning of a mammalian serotonin receptor that activates adenylate cyclase. *Mol. Pharmacol.* 44: 229-236.

Richardson, B.P. and Engel, G. (1986) The pharmacology and function of 5-HT_3 receptors. *Trends Neurosci.* 9: 424-428.

Sellers, E.M., Higgins, G.A. and Sobell, M.B. (1992) 5-HT and alcohol abuse. *Trends Pharmacol. Sci.* 13: 69-75.

Thomsen, W.J., Jacquez, J.A. and Neubig, R.R. (1988) Inhibition of adenylate cyclase is mediated by the high affinity conformation of the α2-adrenergic receptor. *Mol. Pharmacol.* 34: 814-822.

Waeber, C., Dietl, M.M., Hoyer, D. and Palacios, J.M. (1989) 5-HT_1 Receptors in the vertebrate brain *N.-S. Arch. Pharmacol.* 340: 486-494.

Weinshank, R.L., Branchek, T.B. and Hartig, P.R. (1991) International Patent Application Number WO 91/17174 14-11-1991.

Weinshank, R.L., Zgombick, J.M., Macchi, M.J., Branchek, T.A. and Hartig, P.R. (1992) Human serotonin 1D receptor is encoded by a subfamily of two distinct genes: 5-$HT_{1D\alpha}$ and 5-$HT_{1D\beta}$. *Proc. Natl. Acad. Sci. USA* 89: 3630-3634.

Zgombick, J.M., Borden, L.A., Cochran, T.L., Kucharewicz, S.A., Weinshank, R.L. and Branchek, T.A. (1993) Dual coupling of the cloned human 5-$HT_{1D\alpha}$ and 5-$HT_{1D\beta}$ receptors to inhibition of adenylate cyclase and elevation in intracellular calcium concentrations via pertussis toxin-sensitive guanine nucleotide binding proteins. *Mol. Pharmacol.* 44: 575-582.

Toward a Molecular Basis of Alcohol Use and Abuse
ed. by B. Jansson, H. Jörnvall, U. Rydberg, L. Terenius & B. L. Vallee
© 1994 Birkhäuser Verlag Basel/Switzerland

Alcohol, the reward system and dependence

G.F. Koob, S. Rassnick, S. Heinrichs and F. Weiss

The Scripps Research Institute, Dept. of Neuropharmacology, 10666 N. Torrey Pines Road, La Jolla, California 92037, USA

Summary

Evidence is presented to show that multiple neurotransmitter systems of the brain reward systems including GABA, glutamate, dopamine, serotonin and opioid peptides are involved in alcohol reinforcement. Dependence is associated with changes in many of these same systems, but also with changes in other neurotransmitters, such as brain corticotropin releasing factor. A midbrain forebrain circuitry that involves parts of the *nucleus accumbens* and amygdala is hypothesized to be the focus for the neuropharmacology of alcohol reinforcement.

An animal model of alcohol reinforcement

A characteristic of many definitions of addiction and alcoholism is the compulsion to take the drug with a loss of control in limiting intake. This compulsion to take alcohol is thought to derive primarily from the factors that underlie its reinforcing actions and include the subjective sensations of tension reduction ("anxiolytic" actions) and euphoric effects. There are also reinforcing actions of alcohol that derive from the negative reinforcement associated with the development of dependence where subjects ingest alcohol to avoid the negative consequences of abstinence. The relative contribution of these various components of the reinforcing actions of alcohol is unknown; however, recent advances in animal models hold promise for elucidating the neuropharmacological substrates for these reinforcing actions and thus ultimately should provide important information about the etiology and maintenance of dependence.

Drug self-administration procedures have been an effective tool for the study of drug-seeking behavior and reinforcement particularly for opiate and stimulant drugs (Koob and Goeders, 1988). Studies involving intravenous self-administration of opiates and psychomotor stimulants have identified specific neural substrates that mediate the reinforcing actions of these drugs and suggest that these neural systems constitute part of the central nervous system reward systems that evolved for mediating motivated behavior and reinforcement in general (Koob and Bloom, 1988; Koob, 1992). The mesocorticolimbic dopamine system and its connections in the region of the *nucleus accumbens* appear to be critical for the reinforcing actions of indirect sympathomimetics such as cocaine and amphetamine, while opiate receptors in these same neural systems appear to have an important role in the reinforcing actions of opiate drugs (for recent review see Koob, 1992).

However, comparatively little is known about the neuropharmacological substrates of alcohol reinforcement. Factors limiting progress in understanding the neural substrates of alcohol reinforcement have centered on problems in developing spontaneous, voluntary oral alcohol self-administration at intoxicating doses and the lack of precise pharmacological knowledge of the actions of ethanol with the appropriate tools to interact with such actions. Recent developments in animal models for alcohol self-administration and evidence of multiple substrates for the action of alcohol have provided new impetus to the search for the neuropharmacological basis of alcohol reinforcement.

There are many procedures developed for successfully initiating oral ethanol intake including alcohol acclimatization, alcohol taste adulteration, and schedule-induced polydipsia, all of which involve some form of habituation to the aversive taste effects of alcohol (Samson, 1986). However, few of these models have utilized a motivated response to obtain the alcohol and few have conclusively demonstrated that the animals will drink for the pharmacological effects of alcohol (Weiss and Koob, 1991). Rats actually prefer alcohol at concentrations of up to 6%, but the amount consumed over 24 hours is typically below the rats' metabolic capacity.

A reliable operant oral self-administration model suited for the study of the neuropharmacological substrates of alcohol reinforcement has been pioneered by Samson and colleagues (1985; 1986). Here, a sweet solution substitution procedure is used where rats are trained to obtain water or alcohol reinforcers by responding on one of two levers in an operant chamber. This procedure involves training to respond first for a sweet solution (sucrose or saccharine) followed by a gradual fading in of an alcohol solution and a gradual fading out of the water solution. Using this procedure, unselected Wistar rats (Charles River, Kingston, New York) reliably self-administer alcohol in limited-access (30-60 minutes) sessions and obtain reliable blood alcohol levels in the range of 40-60 mg% (Weiss et al., 1990; Rassnick et al., 1993c).

The neuropharmacology of alcohol reinforcement in nondependent rats
Using this procedure, a number of studies have begun to explore the neuropharmacological mechanisms involved in alcohol reinforcement. The lack of a defined alcohol receptor and the

ability of alcohol to penetrate and perturb various biochemical and cellular processes has limited the tools available for elucidating its neuropharmacological mechanism. However, recent exciting data have been generated largely with systemic administration of neurotransmitter receptor antagonists and in some cases intracerebral administration of antagonists. Neurotransmitters implicated in the reinforcing actions of alcohol include gamma amino butyric acid (GABA), dopamine, serotonin, glutamate, and opioid peptides.

GABA has long been hypothesized to have a role in the intoxicating effects of alcohol based on the effectiveness of GABAergic antagonists to reverse the behavioral effects of alcohol and the effectiveness of GABAmimetic drugs to increase alcohol's actions. GABA antagonists decrease the ataxia produced by alcohol, the anesthesia produced by alcohol and the anticonflict effects of alcohol (Liljequist and Engel, 1984; Koob et al., 1988). At a neurochemical level, alcohol at physiological concentrations (10-50 mM) potentiates the stimulation by GABA of CL-transport in synaptosomes from the cerebral cortex and cerebellum (Sudzak et al., 1986). More recently, systemic administration of the partial inverse benzodiazepine agonist Ro 15-4513, which has been shown to reverse some of the sedative action of alcohol (Sudzak et al., 1986), produced dose-dependent reductions in alcohol self-administration (Samson et al., 1987; Rassnick et al., 1993c). The effective doses of Ro 15-4513 are significantly lower than those that produce anxiogenic-like effects and those that decrease responding for other reinforcers such as saccharine (Fig. 1; Rassnick et al., 1993c; Britton and Koob, 1986).

Several lines of evidence suggest a role for the mesocorticolimbic dopamine pathway in the acute reinforcing actions of alcohol. Alcohol dose-dependently increases the firing rate of ventral tegmental (VTA) dopamine neurons in vivo and in vitro over a range of behaviorally relevant concentrations, indicating that the drug activates the mesocorticolimbic dopamine system (Gessa et al., 1985). Consistent with this physiological data, low doses of systemically-administered alcohol as well as local intracerebral application of alcohol have been shown to increase dopamine release from the *nucleus accumbens* in a calcium-dependent manner (Engel et al., 1992). The dopaminergic hypothesis of ethanol reward has received further strong support by recent findings that self-administered alcohol stimulates dopamine release in the *nucleus accumbens* (Weiss et al., 1993), and that rats will self-administer ethanol directly into the ventral tegmental cell body region of the mesocorticolimbic dopamine system (Gatto et al., 1990).

106

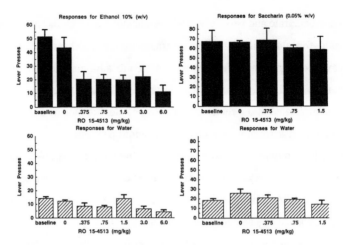

Fig. 1. Left: effects of Ro 15-4513 on responses for ethanol or water in the free choice operant task. Responses on a FR-1 schedule resulted in the delivery of response-contingent ethanol (10% w/v) or water reinforcement. Values shown here represent the mean (±SEM) number of lever presses for ethanol (10% w/v) or water during 30-min sessions. Right: effects of Ro 15-4513 on responses for saccharin or water in a free choice saccharin self-administration task. Responses at one of two levers on a FR-1 schedule produced contingent access to saccharin (0.05%) or water reinforcement. Values shown here represent the mean (±SEM) number of lever presses for saccharin or water during 30-min sessions. (Taken with permission from Rassnick et al., 1993c.)

A role of dopamine in alcohol reinforcement has been corroborated in behavioral work with oral alcohol self-administration. Pharmacological manipulation of dopamine transmission by systemic or intracerebral administration of dopamine agonists and antagonists alters self-administration in a direction generally consistent with a role of dopamine in alcohol reinforcement (Weiss et al., 1990; Rassnick et al., 1993ab; Samson et al., 1991). However, dopamine antagonists injected into the *nucleus accumbens* region not only decrease alcohol self-administration but also decrease responding for saccharin (Koob, unpublished results). Also, virtually complete 6-hydroxydopamine denervation of the *nucleus accumbens* failed to alter voluntary responding for alcohol (Rassnick et al.,1993c). Combined with the pharmacological data above, these results suggest that while mesocorticolimbic dopamine transmission may be associated with important aspects of ethanol reinforcement, it is not critical in this regard, and other neurochemical systems may participate in mediating reinforcing actions of ethanol. In fact, the view is emerging that multiple neurotransmitters collectively "orchestrate" the reward profile of alcohol (Engel et al., 1992).

Recent electrophysiological and neurochemical evidence suggests that alcohol in a physiological dose range may antagonize glutamate (Hoffman et al., 1989; Lovinger et al.,

1989). In a series of studies, a derivative of taurine called calcium homotaurine (acamprosate) had been shown to dose-dependently decrease alcohol drinking and self-administration in both nondependent and dependent rats (Boismare *et al.*, 1984; Le Magnen *et al.*, 1987; Rassnick *et al.*, 1992a). This compound also has efficacy in blocking relapse in detoxified alcoholics during the course of a daily treatment regimen in double-blind controlled studies (Lhuintre *et al.*, 1985). The exact neuropharmacological mechanism for acamprosate is unknown but there is mounting evidence that it can have anti-glutamate effects via an action on calcium channels (Zieglgansberger and Zeise, 1992). In addition, microinjection of a glutamate antagonist into the *nucleus accumbens* significantly decreases lever pressing for alcohol in nondependent rats (Rassnick *et al.*, 1993b). Taken together, these studies suggest that modulation of the NMDA glutamate receptor complex may also play a role in the reinforcing actions of alcohol.

Neuropharmacological manipulations of central nervous system serotonin have been shown to alter alcohol drinking in the rat. Increasing the synaptic availability of serotonin with precursor loading (5-hydroxytryptophan), by blockade of serotonin reuptake, or by central injection of serotonin itself all reduce voluntary intake of alcohol (Amit *et al.*, 1984; Rockman *et al.*, 1979; Daoust *et al.*, 1984; Murphy *et al.*, 1988; Zabik, 1989; McBride *et al.*, 1988). In more recent work with pharmacological agents selective for serotonin (5-HT) receptor subtypes, voluntary alcohol intake in rats was reduced by the specific 5-HT1A agonist, 8-OH-DPAT; the specific 5-HT3 receptor antagonists GR 38032 (ondansetron), zacopride, ICS 205-930, and MDL 72222 reduced voluntary alcohol intake in rats (Sellars *et al.*, 1992; Knapp *et al.*, 1992) and humans (ondansetron) (Toneatto *et al.*, 1991). Consistent with a serotonergic role in ethanol abuse, in several double-blind, placebo-controlled clinical studies, serotonin reuptake inhibitors have produced mild to moderate decreases in alcohol consumption in humans (Naranjo *et al.*, 1984; Amit *et al.*, 1984).

Opioid peptides have been implicated in alcohol reinforcement by numerous reports that the opiate antagonists naloxone and naltrexone reduce alcohol self-administration in several animal models (Myers *et al.*, 1986; Volpicelli *et al.*, 1986; Hubbell *et al.*, 1991). The inhibitory effects of opiate antagonists on ethanol intake may involve general, appetitive or consummatory aspects of behavior because these agents dose-dependently decrease consumption of sweet solutions or water as well as ethanol in operant, free-choice tests (Weiss *et al.*, 1990; Samson and Doyle, 1985). Although the lack of specificity of opiate antagonists for alcohol appears to rule out a selective mediation of alcohol reward by the opioid system, it is possible that antagonism of specific opiate receptor (sub)types (e.g., Hyytiä, 1993) in specific brain regions might reveal more selective effects. Also, intravenous alcohol self-administration by rhesus monkeys is suppressed by naloxone, a finding that suggests an attenuation of alcohol reinforcement independent of nonspecific taste factors (Altshuler *et al.*, 1980). Consistent with a role for opioid peptides in alcohol reinforcement, a double-blind, placebo-controlled clinical trial shows that naltrexone significantly reduces alcohol consumption, frequency of relapse, and "craving" for alcohol in humans (Volpicelli *et al.*, 1992). Thus, alcohol interactions with opioid neurotransmission may contribute to certain aspects of alcohol reinforcement that may be of particular importance to the dependence development.

Neuropharmacology of alcohol dependence

Alcohol dependence in humans is accompanied by a characteristic withdrawal syndrome upon cessation of chronic alcohol intake or an alcohol binge. This withdrawal is characterized during the first 1-2 days by physical symptoms such as tremor, mild disorientation, hallucinations, and a major sympathetic hyperactivity including increases in blood pressure, heart rate and body temperature. Seizures similar to those of grand mal epilepsy can also occur during this period. Severe withdrawal in later stages may include a syndrome called delirium tremens, which is characterized by marked tremor, anxiety, insomnia and autonomic hyperactivity. Individuals can become totally disoriented, with vivid hallucinations and irrational, psychotic-like behavior. Benzodiazepines are safe and effective treatment for alcohol withdrawal.

Alcohol withdrawal in animals is characterized by similar signs of central nervous system hyperexcitability that results in both physical and motivational signs of dependence. Motivational measures have included disruption of operant behavior (Denoble and Begleiter, 1976), increased vocalization upon handling, increased responsiveness in the acoustic startle test (Rassnick *et al.*, 1992b), and an increased sensitivity in the elevated plus maze test (i.e., less time spent exploring the open, exposed arms (Fig. 2). This test is very sensitive to "anxiolytic" and "anxiogenic" drugs.) (Baldwin *et al.*, 1991).

The neuropharmacology of the physical signs of alcohol withdrawal has pointed to an important functional role for central nervous system GABA. GABA agonists decrease the audiogenic seizures associated with alcohol withdrawal in animals (Cooper *et al.*, 1979). GABA antagonists increase many of the symptoms of alcohol withdrawal (Goldstein,1973), and the partial inverse benzodiazepine agonist Ro 15-4513 increases the incidence of seizures during alcohol withdrawal (Lister and Karanian, 1987). Neurochemically, chronic alcohol treatment has been shown to significantly decrease the potentiation by alcohol of GABA-stimulated chloride uptake, and this decrease appears to correspond to the time course of alcohol dependence (Morrow *et al.*, 1988).

Motivational measures of alcohol withdrawal have suggested a possible role for central nervous system corticotropin releasing factor (CRF) in alcohol dependence (Baldwin *et al.*, 1991). CRF, the major hypothalamic releasing factor responsible for pituitary stimulation of adrenocorticotropin hormone, is thought to have an extrahypothalamic role in mediating behavioral responses to stressors (Britton and Koob, 1987). Alcohol injected acutely can reverse the "anxiogenic-like" effects of intracerebroventricular administration of CRF (Britton and

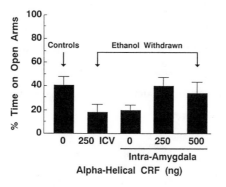

Fig. 2. Effects of microinfusion of alpha-helical CRF into the central amygdala nucleus and ICV administered alpha-helical CRF in the elevated plus maze during ethanol withdrawal. The data are expressed as mean ($\pm$SEM) of percent time exploring the open arms and are plotted as a function of alpha-helical CRF dose. (Taken with permission from Rassnick *et al.*, 1993d.)

Koob, 1986), and rats withdrawn from chronic alcohol show a stress-like response on the elevated plus maze which is reversed by intracerebroventricular administration of the CRF antagonist, alpha-helical CRF (Baldwin *et al.*, 1991). Much smaller amounts of the CRF antagonist, when injected into the amygdala, were effective in reversing the stress-like effects of alcohol on the plus maze (Rassnick *et al.*, 1993d). These results suggest that CRF in the central nervous system and perhaps particularly in the amygdala may have a role in the more motivational effects of alcohol withdrawal.

110

Other neurotransmitters implicated in alcohol dependence include norepinephrine, serotonin and glutamate. During alcohol dependence there is evidence of hyperactivity of the locus coeruleus norepinephrine system (Rogers *et al.*, 1980). Behavioral pharmacological studies have implicated serotonin (5-HT) in alcohol withdrawal in that 5-HT1A antagonists also can reverse the anxiogenic effects of alcohol in the elevated plus maze (Lal *et al.*, 1991). A possible role for glutamate in alcohol dependence comes from studies showing that noncompetitive glutamate antagonists can reverse the effects, signs and symptoms of alcohol withdrawal (Grant *et al.*, 1990).

In summary, the acute reinforcing actions of alcohol appear to be mediated by multiple neurochemical systems that include GABA, glutamate, dopamine, serotonin and opioid peptides (Fig. 3). The relative contributions of these neurotransmitters may change with the dose of alcohol and the switch from nondependence to dependence. During the development of dependence, compensatory adaptive processes in these same neurochemical systems may develop as has been proposed for other drugs of abuse, e.g., "within systems adaptations" (Koob and Bloom, 1988). Alternatively, changes in other neurochemical systems may be recruited during the dependence process to produce such compensatory adaptations, e.g., "between system adaptations." The results discussed above regarding changes in CRF function in the central nervous system may fit this theoretical construct.

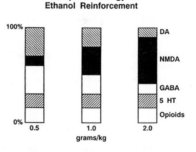

Fig. 3. Diagram depicting the hypothesized relative contribution of different neurotransmitter systems of the basal forebrain in the reinforcing actions of alcohol. Note that the relative contribution is hypothesized to change with dose. Based on a similar diagram developed for the alcohol drug discrimination stimulus by Dr. Kathleen Grant, Bowman Gray, Wake Forest University, North Carolina.

The exact neurobiological substrates for these reinforcing effects of alcohol are largely unknown at this time. However, one might speculate that the same midbrain forebrain system involving the region of the *nucleus accumbens* and its connections that is implicated in the effects of other drugs of abuse is also involved in the reinforcing actions of alcohol (Koob, 1992). An intriguing hypothesis for future work is that the "extended amygdala" which includes subregions of the *nucleus accumbens* and parts of the bed nucleus of the stria terminalis and amygdala (Heimer and Alheid, 1991) may be particularly important for the actions of alcohol.

Acknowledgements

This is manuscript number 8321-NP of The Scripps Research Institute. This research was supported in part by grants provided by NIH/NIAAA AA08459 and AA063420 and the Alcohol Beverage Medical Research Foundation. I would like to thank TSRI Word Processing Center and Diane Braca for help in preparation of this manuscript.

References

Altshuler, H.L., Phillips, P.E. and Feinhandler, D.A. (1980) Alterations of ethanol self-administration by naltrexone. *Life Sci.* 26: 679-688.

Amit, Z., Sutherland, E.A., Gill, K. and Ögren, S-O. (1984) Zimelidine: A review of its effects on ethanol consumption. *Neurosci. Biobehav. Rev.* 8: 35-54.

Baldwin, H.A., Rassnick, S., Rivier, J., Koob, G.F. and Britton, K.T. (1991) CRF antagonist reverses the "anxiogenic" response to ethanol withdrawal in the rat. *Psychopharmacology* 103: 227-232.

Boismare, F., Daoust, M., Moore, N., Saligaut, C., Lhuintre, J.P., Chretien, P. and Durlach, J. (1984) A homotaurine derivative reduces the voluntary intake of ethanol by rats: Are cerebral GABA receptors involved? *Pharmacol. Biochem. Behav.* 21: 787-789.

Britton, K.T. and Koob, G.F. (1986) Alcohol reverses the proconflict effect of corticotropin releasing factor. *Reg. Pept.* 16: 315-320.

Britton, K.T. and Koob, G.F. (1987) Behavioral effects of corticotropin releasing factor. In: Schwatzberg, A. and Nemeroff, C. (eds.): *HPA Physiology and Pathophysiology*, Raven Press, New York, pp. 55-66.

Cooper, B.R., Viik, K., Ferris, R.M. and White, H.L. (1979) Antagonism of the enhanced susceptibility to audiogenic seizures during alcohol withdrawal in the rat by gamma-aminobutyric acid (GABA) and GABA-mimetics. *J. Pharmacol. Exp. Ther.* 209: 396-408.

Daoust, M., Saligant, C., Chadeland, M., Chretien, P., Moore, N. and Boismare, F. (1984) Attenuation of antidepressant drugs of alcohol intake in rats. *Alcohol* 1: 379-383.

Denoble, U. and Begleiter, H. (1976) Response suppression on a mixed schedule of reinforcement during alcohol withdrawal. *Pharmacol. Biochem. Behav.* 5: 227-229.

Engel, J.A., Enerback, C., Fahlke, C., Hulthe, P., Hard, E., Johannessen, K., Svensson, L. and Söderpalm, B. (1992) Serotonergic and dopaminergic involvement in ethanol intake. In: Naranjo, C.A. and Sellars, E.M. (eds): *Novel Pharmacological Interventions for Alcoholism*, Springer, Inc., New York, pp. 68-82.

112

Gatto, G.J., Murphy, J.M., McBride, W.J., Lumeng, L. and Li, T.-K. (1990) Intracranial self-administration of ethanol into the ventral tegmental area of alcohol-preferring (P) rats. *Alcohol Clin. Exp. Res.* 14: 291.

Gessa, G.L., Montoni, F., Collu, M., Vargiu, L. and Mereu, G. (1985) Low doses of ethanol activate dopaminergic neurons in the ventral tegmental area. *Brain Res.* 348: 201-203.

Goldstein, D. (1973) Alcohol withdrawal reaction in mice: Effects of drugs that modify neurotransmission. *J. Pharmacol. Exp. Ther.* 186: 1-8.

Grant, K.A., Valverius, P., Hudspith, M. and Tabakoff, B. (1990) Ethanol withdrawal seizures and the NMDA receptor complex. *Eur. J. Pharmacol.* 176: 289-296.

Heimer, L. and Alheid, G. (1991) Piecing together the puzzle of basal forebrain anatomy. In: Napier, T.C., Kalivas, P. and Hanin, I. (eds.): *The Basal Forebrain: Anatomy to Function*, Plenum Press, New York, pp. 1-42.

Hoffman, P.L., Rabe, C., Moses, F. and Tabakoff, B. (1989) N-methyl-D-aspartate receptors and ethanol: Inhibition of calcium flux and cyclic GMP production. *J. Neurochem.* 52: 1937-1940.

Hubbell, C.L., Marlin, S.H., Spitalnic, S.J., Abelson, M.L., Wild, K.D. and Reid, L.D. (1991) Opioidergic, serotonergic and dopaminergic manipulations and rats' intake of a sweetened alcoholic beverage. *Alcohol* 8: 355-367.

Hyytiä, P. (1993) Involvement of μ-opioid receptors in alcohol drinking by alcohol-preferring AA rats. *Pharmacol. Biochem. Behav.* 45: 697-701.

Knapp, D.J., Pohorecky, L.A. and Zacopride, L.A. (1992) Zacopride, a 5-HT3 receptor antagonist, reduces voluntary ethanol consumption in rats. *Pharmacol. Biochem. Behav.* 41: 847-850.

Koob, G.F. (1992) Drugs of abuse: Anatomy, pharmacology, and function of reward pathways. *Trends Pharmacol. Sci.* 13: 177-184.

Koob, G.F. and Bloom, F.E. (1988) Cellular and molecular mechanisms of drug dependence. *Science* 242: 715-723.

Koob, G.F. and Goeders, N. (1988) Neuroanatomical substrates of drug self-administration. In: Liebman, J.M. and Cooper, S.J. (eds.): *Neuropharmacological Basis of Reward*, 6th Edition, Oxford University Press, Oxford, pp. 214-263.

Koob, G.F., Mendelson, W.B., Schafer, J., Wall, T.L., Thatcher-Britton, K. and Bloom, F.E. (1988) Picrotoxin receptor ligand blocks anti-punishment effects of alcohol. *Alcohol* 5: 437-443.

Lal, H., Prather, P.L. and Rezazadeh, S.M. (1991) Anxiogenic behavior in rats during acute and protracted ethanol withdrawal: Reversal by buspirone. *Alcohol* 8: 467-471.

Le Magnen, J., Tran, G., Durlack, J. and Martin, C. (1987) Dose-dependent suppression of the high alcohol intake of chronically intoxicated rats by CA-acetyl homotaurinate. *Alcohol* 4: 97-192.

Lhuintre, J.P., Daoust, M., Moore, N.D, Chretien, P., Saligaut, C., Tran, G., Boismare, F. and Hillemand, B. (1985) Ability of calcium bis acetyl homotaurine, a GABA agonist, to prevent relapse in weaned alocholics. *Lancet* 1: 1014-1016.

Liljequist, S. and Engel, J.A. (1984) The effects of GABA and benzodiazepine receptor antagonists on the anti-conflict actions of diazepam or ethanol. *Pharmacol. Biochem. Behav.* 21: 521-525.

Lister, R.G.., and Karanian, J.W. (1987) RO15-4513 induces seizures in DBA/2 mice undergoing alcohol withdrawal. *Alcohol* 4(5): 409-411.

Lovinger, D.M., White, G. and Weight, F.F. (1989) Ethanol inhibits NMDA-activated ion currents in hippocampal neurons. *Science* 243: 1721-1724.

McBride, W.J., Murphy, J.M., Lumeng, L. and Li, T-K (1988) Effects of Ro-15-4513, fluoxetine and desipramine on the intake of ethanol, water and food by the alcohol-preferring (P) and non-preferring (NP) lines of rats. *Pharmacol. Biochem. Behav.* 30: 1045-1050.

Morrow, A.L., Suzdak, P.D., Karanian, J.W. and Paul, S.M. (1988) Chronic ethanol administration alters gamma-aminobutyric acid, pentobarbital and ethanol-mediated ^{36}Cl⁻ uptake in cerebral cortical synaptoneurosomes. *J. Pharmacol. Exp. Ther.* 246: 158-164.

Murphy, J.M., Waller, M.B., Gatto, G.J., McBride, W.J., Lumeng, L. and Li, T-K (1988) Effects of fluoxetine on the intragastric self-administration of ethanol in the alcohol-preferring P line of rats. *Alcohol* 5: 283-286.

Myers, R., Borg, S. and Mossberg, R. (1986) Antagonism by naltrexone of voluntary alcohol selection in the chronically drinking macaque monkey. *Alcohol* 7: 383-388.

Naranjo, C.A., Sellers, E.M., Roach, C.A., Woodley, D.V., Sanchez-Craig, M. and Sykora, K. (1984) Zimelidine-induced variations in alcohol intake by nondepressed heavy drinkers. *Clin. Pharmacol. Ther.* 35: 374-381.

Rassnick, S., D'Amico, L., Pulvirenti, L., Zieglgansberger, W. and Koob, G.F. (1992a) GABA and *nucleus accumbens* glutamate neurotransmission modulate ethanol self-administration in rats. *Ann. NY Acad. Sci.* 654: 502-505.

Rassnick, S., Koob, G.F. and Geyer, M.A. (1992b) Responding to acoustic startle during chronic ethanol intoxication and withdrawal. *Psychopharmacology* 106: 351-358.

Rassnick, S., Pulvirenti, L. and Koob, G.F. (1993a) SDZ-205,152, a novel dopamine receptor agonist, reduces oral ethanol self-administratoin in rats. *Alcohol* 10: 127-132.

Rassnick, S., Pulvirenti, L. and Koob, G.F. (1993b) Oral ethanol self-administration in rats is reduced by the administration of dopamine and glutamate receptor antagonists into the *nucleus accumbens*. *Psychopharmacology* 109: 92-98.

Rassnick, S., D'Amico, L., Riley, E. and Koob, G.F. (1993c) GABA antagonist and benzodiazepine partial inverse agonist reduce motivated responding for ethanol. *Alcohol Clin. Exp. Res.* 17: 124-130.

Rassnick, S., Heinrichs, S.C., Britton, K.T. and Koob, G.F. (1993d) Microinjection of a corticotropin-releasing factor antagonist into the central nucleus of the amygdala reverses anxiogenic-like effects of ethanol withdrawal. *Brain Res.* 605: 25-32.

Rockman, G.E., Amit, Z., Carr, G., Brown, Z.W. and Ogren, S.O. (1979) Attenuation of ethanol by 5-hydroxytryptamine blockade in laboratory rats. I. Involvement of brain 5-hydroxytryptamine in the mediation of positive reinforcing properties of ethanol. *Arch. Int. Pharmacodyn. Ther.* 241: 245-259.

Rogers, J., Siggins, G.R., Schulman, J.A. and Bloom, F.E. (1980) Psychological correlates of ethanol intoxication, tolerance, and dependence in rat cerebellar Purkinje cells. *Brain Res.* 196: 183-198.

Samson, H.H. (1986) Initiation of ethanol reinforcement using a sucrose-substitution procedure in food- and water-sated rats. *Alcohol Clin. Exp. Res.* 10: 436-442.

Samson, H.H. and Doyle, T.F. (1985) Oral ethanol self-administration in the rat: Effects of naloxone. *Pharmacol. Biochem. Behav.* 22: 91-99.

Samson, H.H., Tolliver, G.A., Pfeffer, A.O., Sadeghi, K.G. and Mills, F.G. (1987) Oral ethanol reinforcement in the rat: Effect of the partial inverse benzodiazepine agonist R015-4513. *Pharmacol. Biochem. Behav.* 27: 517-519.

114

Samson, H.H., Tolliver, G.A., Haraguchi, M. and Kalivas, P.W. (1991) Effects of d-amphetamine injected into the n. accumbens on ethanol reinforced behavior. *Brain Res. Bull.* 27: 267-271.

Sellars, E.M., Higgins, G.A. and Sobell, M.B. (1992) 5-HT and alcohol abuse trends. *Pharmacol. Sci.* 13: 69-75.

Sudzak, P.D., Glowa, J.R., Crawley, J.W., Schwartz, R.D., Skolnick, P. and Paul, S.M. (1986) A selective imidazobenzodiazepine antagonist of ethanol in the rat. *Science* 234: 1243-1247.

Toneatto, T., Romach, M.K., Sobell, M.K., Sobell, L.C., Somer, G.R. and Sellers, E.M. (1991) Ondansetron, a 5-HT3 antagonist, reduces alcohol consumption in alcohol abusers. *Alcohol Clin. Exp. Res.* 15: 382.

Volpicelli, J.R., Alterman, A.I., Hayashida, M. and O'Brien, C.P. (1992) Naltrexone in the treatment of alcohol dependence. *Arch. Gen. Psychiatry* 49: 876-880.

Volpicelli, R., Davis, M.A. and Olgin, J.E. (1986) Naltrexone blocks the post-shock increse of ethanol consumption. *Life Sci.* 38: 841-847.

Weiss, F. and Koob, G.F. (1991) The neuropharmacology of ethanol self-administration. In: Meyer, R.F., Koob, G.F., Lewis, M. and Paul, S. (eds.): *Ethanol Reinforcement*, Birkhauser, Boston, pp. 125-162.

Weiss, F., Lorang, M.T., Bloom, F.E. and Koob, G.F. (1993) Ethanol self-administration stimulates dopamine release in the rat *nucleus accumbens*: Genetic and motivational determinants. *J. Pharmacol. Exp. Ther.,* in press.

Weiss, F., Mitchiner, M., Bloom, F.E. and Koob, G.F. (1990) Free-choice responding for ethanol versus water in Alcohol-Preferring (P) and unselected Wistar rats is differentially altered by naloxone, bromocriptine and methysergide. *Psychopharmacology* 101: 178-186.

Zabik, J.E. (1989) Use of serotonin-active drugs in alcohol preference studies. *Recent Dev Alcohol* 7: 211-223.

Zieglgansberger, W. and Zeise, M.L. (1992) Calcium-diacetyl-homotaurinate which prevents relapse in weaned alcoholics decreases the action of excitatory amino acids in neocortical neurons of the rat *in vitro*. In: Naranjo, C.A. and Sellers, E.M. (eds.): *Pharmacological Interventions for Alcoholism*, Springer, Heidelberg, New York, pp. 337-341.

Toward a Molecular Basis of Alcohol Use and Abuse
ed. by B. Jansson, H. Jörnvall, U. Rydberg, L. Terenius & B. L. Vallee
© 1994 Birkhäuser Verlag Basel/Switzerland

Clinical aspects on molecular probes, markers and metabolism

Ulf Rydberg

Department of Clinical Neuroscience, Section on Clinical Alcohol and Drug Addiction Research, Karolinska Hospital, Karolinska Institutet, S-171 76 Stockholm, Sweden

Summary

Though today techniques utilizing molecular methods are emphasized, we must not neglect to put the molecular data into a clinical context. Extrapolations should be handled with great care. The knowledge on alcohol has many levels, and facts are often taken out of context. For a proper report of a clinical state or situation, validity lies both in the precision of clinical descriptions and in the accuracy of the molecular parameter. Alcohol as a medical risk factor is often disregarded, and there is often a communication gap.

Introduction

"Probes" can be described as "instruments or chemical reagents to analyze structure and function". Regarding "markers" we usually differentiate between "trait markers" and "state markers". Trait markers have to do with those possible differences in genetic properties that might influence the outcome of the contact between the substance "ethanol" and the living organism. Although there is usually a dose-effect relationship between the ethanol exposure and disease, there are exceptions. At a certain level of exposure only a fraction of the exposed individuals develop the disease. Liver cirrhosis, Wernicke's encephalopathy and the fetal alcohol syndrome are examples where high ethanol exposure causes disease in a fraction of the exposed persons, but where knowledge about the differences in vulnerability at the molecular level is fragmentary. HLA genes, transketolase, alcohol dehydrogenase and the cytochrome P450 system are some candidate markers to explain in part the differences in vulnerability to ethanol.

It has been known since about 1870 that alcoholism occurs in families. But, it has taken more than a century to differentiate between single genetic factors and more complex variables, such as psychological and sociological features. Moving from a descriptive, phenomenological level to a definition of a gene locus is a tremendous step, considering the perhaps 70-80,000 genes of the human genome. General guidelines for effective work strategies and orientation must be adapted to the field of alcohol studies.

Knowledge regarding the molecular and genetic basis for risks of alcohol dependence has advanced. Studies of the dopamine D2 receptor (Uhl *et al.*, 1993), although not yet conclusive, have helped to reveal genetic complexity. Studies of genetic markers and other risk factors are sought continously. Cytochromes P450 are important liver factors, which may contribute to individual differences in the outcome of therapy with different drugs.

Positron emission tomography has been used in studies of alcohol dependent persons to test the possibility of an uneven distribution of relevant ligands in brain (Litton *et al.*, 1993). Generally the clinical description of the patients must be precise, in order fully to utilize the gains with this analytical technique. Information will generally be optimized by a detailed clinical description of patients.

There are many possibilities to develop "state markers". Ethanol is believed to influence hundreds of chemical reactions in the human body. Some effects are related to intoxication itself, others to hangover (Rydberg 1977). Still others reflect a prolonged and high consumption (Nilssen *et al.*, 1992, Nyström *et al.*, 1992). New state markers such as carbohydrate-deficient transferrin (CDT) must be characterized not only in relation to specificity and sensitivity but also regarding cost-effectiveness to elucidate clinical usefulness (Stibler 1991).

Alcohol - a risk factor

There are many levels of research and many levels of discussion, which sometimes are in striking contrast to common sense. Since alcohol is known to all in a colloquial manner, many believe that they also have a great deal of scientific knowledge about alcohol and the medical implications of alcohol use and abuse. Probably, there are few areas in which extrapolations have been made on as erroneous a basis as on alcohol. Minor details are often seized out of context and used for generalizations regarding the status of other organs, the influence in other age groups etc., resulting in conclusions which are both logically and scientifically improbable. The "J-curve" is one example from which too wide extrapolations have been made in recent years. This relationship may be valid in some age groups and in some populations, but it is not justified, however, to extrapolate from these limited studies to entire populations.

In "riskology" one differentiates between risk measurement, risk assessment, risk management and risk perception. In the alcohol area, risk perception is commonly minimal even by the medical profession.

The interdisciplinary approach

How can we link alcohol studies at the molecular level to that of clinical research and good clinical practice? A quote from Wallgren and Barry (1971) may be appropriate:

"In the alcohol area, certain topics often become popular. Research on important topics has often been retarded by a strong influence of fads and fashions for certain topics, such as biochemical actions on the liver. A balanced effort is the most effective way for the accumulation of knowledge". A diagram may help to illustrate the different levels of abstraction necessary to clarify the effects of alcohol. (Fig. 1).

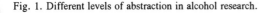

man	in	the	total	environment

man in the laboratory ??????????
??????????
animal experiments ?????????????
isolated organs ??????????????
??????????????
cell cultures ??????????????????
???????????????????
molecular level ????????????????????????

Fig. 1. Different levels of abstraction in alcohol research.

Alcohol studies may be viewed from the level of society to that of molecules. We might oscillate between different levels of abstraction, but it is not immediately evident that we can extrapolate from one level to the other. By increasingly focusing on the molecular level, there is a risk of neglecting the macro perspective.

From an operational description of the risk of developing dependence, we derive the formula: Risk of dependence $= F(f_1; f_2; f_3; f; .... f_n)$. where f constitute single factors. The risk of dependence is a function of multiple factors, and only recently has it been possible to measure some of them individually, but it is not yet known how to balance them against each other. Should they be added, multiplied or should only fractions of them be used? (Rydberg 1974). It

118

is evident that a true multifactorial view on dependence is emerging, where we see that the genetic, molecular, group and societal levels all are important.

During recent years a shift in fashion of topics for alcohol research is noticable, in two opposite and perhaps polarized directions, towards both the societal and molecular levels. At the societal level, epidemiological and statistical research urge strict descriptions of samples and populations to guarantee representation. Moreover, there are rapidly increasing cost interests such as Employment Assistance Programs and cost-effectiveness of preventive measures. At the molecular level, interest has shifted to ultrasophisticated studies using molecular biology, keeping in mind biochemical individuality.

From molecular development to a trade mark pharmaceutical - a pathway with obstacles
There is important news regarding drugs influencing alcohol drinking behaviour (Naranjo *et al.*, 1992, Volpicelli *et al.*, 1992). Many studies hitherto focused on serotonin, but other approaches are to be expected. The heterogeniety of the systems involved may predict differences in effect.

Broader knowledge of the molecular actions of ethanol may lead to a more specific approach to developing targeted drugs. Hitherto, the large pharmaceutical companies have had limited tools to tailor drugs for the treatment of alcohol and drug dependence or other alcohol-related disabilities. One must also realize that the path from a number of targeted substances and molecules in a "phase I" setting is long and bestrewed with thorns. Only a minute number of substances suggested for further development reach tests in animals, in healthy subjects and, still fewer, reach the state of "phase IV" tests in adequate patient groups. In the information process, these limitations must be kept in mind.

Molecular biology in its context
Torsten Wiesel, Nobel Laureate in 1985, stated (1992) that "Physicians/scientists who think about both clinical and basic research often have insights that escape researchers whose attention is confined to one of those domain's alone. Combining clinical and basic research can bring about faster application of laboratory ideas and testing of bedside hypotheses, crucial for those who suffer from diseases with near-hopeless prognoses. Unfortunately, the past several decades have witnessed a decline in medical scientists interested in integrating basic science and medicine..... There is, as a consequence, an overwhelming national need for superbly trained

physician/scientists as we enter an age of molecular medicine - an era that requires coupling laboratory experimentation with careful bedside observation."

In a recent monograph by Hyman and Nestler (1993), a chapter states that ".....when posed as a rigid dichotomy, the nature versus nurture question leads to false conclusions and confusion, and also that identification of specific genetic abnormalities by which environmental factors influence the expression of those genes should revolutionize diagnostic, treatment and preventive aspects of psychiatric practice.....understanding of the brain's neurotransmitter-receptor systems, intracellular signal transduction pathways, and genetic regulatory mechanisms might lead to such a promising future for psychiatry".

A problem of communication

Alcohol is not the kind of risk factor favored by media. Disasters and risks where one can blame an expert are of course more popular than alcohol, an instant in which the afflicted person himself is responsible (British Medical Association, 1987, Sjöberg, 1991). Empirical studies on alcohol as a risk factor have shown that of 25 investigated risks, "alcohol" is the one with the highest discrepancy, 1.01 units on a grade scale of 4, between objective risk and subjectively perceived risk (Sjöberg, 1991). This implies that the information and communication problems are maximal. Also mathematically, (Rydberg *et al.*, 1993) alcohol as a risk factor has often been looked upon in an unprofessional manner.

Conclusion

Thus for a better knowledge on probes, markers and metabolism in this area, we urgently need a precise molecular approach on alcohol problems, but carefully combined with a respect for the complexity of the living organism, its psychological and social properties and its subtle relationship to the environment.

Acknowledgements

Thanks are due to Professor Anders Rane, MD, PhD and to Dr. Johan Franck, MD, PhD for valuable comments, and to Mr. Staffan Lindberg, MA for skilful help with the figure.

120

References

British Medical Association: Alcohol. (1987) In: *Living With Risk: A Report From the BMA Professional and Scientific Division.* Wiley Publ., Chichester, England, pp. 58-61.

Hyman, S.E. and Nestler E.J. (1993) The Molecular Foundations of Psychiatry. American Psychiatric Press, Washington D C, p. 239.

Litton, J.-E., Neiman J., Pauli S., Farde L., Hindmarsh, T., Halldin, C. and Sedvall, G. (1993) PET analysis of 11Cflumazenil binding to benzodiazepine receptors in chronic alcohol-dependent men and healthy controls. *Psychiatry Res.: Neuroimaging* 50: 1-13.

Naranjo, C. and Bremner, K.E. (1992) Evaluation of serotonin uptake inhibitors in alcoholics: a review. In Naranjo C.A. and Sellers E.M. (eds).: *Novel Pharmacological Interventions for Alcoholism.* Springer-Verlag, New York, pp. 105-117.

Nilssen, O., Huseby, N.E., Höyer, G., Brenn, T., Schirmer, H. and Förde, O. H. (1992) New alcohol markers - how useful are they in population studies: The Svalbard study 1988-89. *Alcohol Clin. Exp. Res.* 16: 82-86.

Nyström, M., Peräsalo J. and Salaspuro, M. (1992) Carbohydrate-deficient transferrin (CDT) in serum as a possible indicator of heavy drinking in young university students. *Alcohol Clin. Exp. Res.* 16: 93-97.

Rydberg, U. (1974) Alkoholens omsättning och medicinska verkningar. Kap. 2 i: Alkohol-politiska utredningens huvudbetänkande. Alkoholpolitik, del 1, Bakgrund. SOU 90:23-57.

Rydberg, U. (1977) Experimentally induced alcoholic hangover. In: Ideström, C.-M. (ed.): *Recent advances in the study of alcoholism.* Excerpta Medica International Congress Series 407: 32-40.

Rydberg, U., Damström-Thakker K. and Skerfving, S. (1993) The toxicity of ethanol. In: Costa è Silva, J.A. and Nadelson, C. (eds.): *International Review of Psychiatry,* Vol. I, American Psychiatric Press, Washington DC, pp. 563-600.

Sjöberg, L. (1991) Alkoholens risker, upplevda och verkliga. *Nordisk Alkohol-Tidskrift* 8:253-67.

Stibler, H. (1991) Carbohydrate-deficient transferrin in serum: a new marker of potentially harmful alcohol consumption reviewed. *Clin. Chem.* 37: 2029-2037.

Uhl, G.R., Blum, K. and Smith, S.S. (1993) Substance abuse vulnerability and D2 receptor genes. *Trends Neurosci.* 16: 83-88.

Volpicelli, J.R., Alterman, A.I., Hayashida, M. and O'Brien, C.P (1992) Naltrexone in the treatment of alcohol dependence. *Arch. Gen. Psychiatry* 49: 876-880.

Wallgren, H. and Barry, H. III (1971) Actions of Alcohol, Vol. I and II. Elsevier, Amsterdam - London - New York, p. 871.

Wiesel, T. (1992) The Rockefeller Hospital and the future of clinical research. *Search, The Rockefeller University Magazine* 2: 2.

Toward a Molecular Basis of Alcohol Use and Abuse
ed. by B. Jansson, H. Jörnvall, U. Rydberg, L. Terenius & B. L. Vallee
© 1994 Birkhäuser Verlag Basel/Switzerland

Familial alcoholism: Family, twin adoption and high risk studies

Joachim Knop

Department of Psychiatry, Gentofte University Hospital, DK-2900 Hellerup, and Institute of Preventive Medicine, Kommunehospitalet, DK-1399 Copenhagen, Denmark

Summary

The nature-nurture question in the etiology of alcoholism is discussed. The research results from twin and adoption studies indicate a considerable genetic (=biological) component in the etiology of alcoholism. A longitudinal high risk study of alcoholism is presented. The sons of alcoholic men and matched controls have been followed prospectively since before birth. The main results from previous phases of the study and a recent 30-year follow-up assessment are presented.

Introduction

Almost every medical discipline is represented in contemporary alcohol research, but still little is known about etiology. The nature-nurture dialogue is still going on due to methodological difficulties in separating genetic and environmental factors. In this contibution, some new epidemiological research strategies will be presented: family, twin, adoption, and high risk studies in alcoholism.

Family studies

For thousands of years it has been a well-established observation that alcohol runs in families. Throughout this century several epidemiological studies have found a familial accumulation of alcoholism. Cotton (1979) has reviewed 39 family studies of alcoholism over the past 40 years. The survey covered 6241 alcoholics and 4083 non-alcoholic probands. The incidence of alcoholism was found to be 31% among relatives of alcoholics compared to 5% among relatives of non-alcoholics. The concept of subgrouping alcoholism into a familial and non-familial form has developed from these family studies (Frances *et al.*, 1980; Cloninger *et al.*, 1981).

However, the running-in-families phenomenon does not necessarily imply inheritance. Several environmental stressors have been proposed as significant etiological factors of drinking problems later in life. It has been a predominant hypothesis that both biological, psychological and social factors contribute to the etiology of alcoholism (Meyer and Babor, 1989). The interaction of genetic and environmental factors are difficult to separate in traditional family study designs. But modern epidemiology has developed new research strategies as valuable contributions to the ongoing nature-nurture question. In the following, twin, adoption and high risk studies will be described.

Twin studies

The classical design in twin studies includes comparison of concordance rates between monozygote (MZ) and dizygote (DZ) twin pairs.

The first twin study focusing on alcoholism was performed in Sweden by Lennart Kaij (1960). He examined 174 male twin pairs of which at least one in the pair was registered as an alcoholic. The concordance rate among MZ-pairs was 55% compared to 30% among DZ-pairs. This tendency increased when milder cases of alcohol abuse were excluded. Another important finding was that intellectual deterioration was more closely correlated to zygocity than was severity of alcohol abuse.

Several twin studies have been applied to alcohol-related problems (Hrubec and Omenn, 1981; Jonsson and Nilsson, 1968; Koehlin, 1972; Murray et al., 1966; Vesell et al., 1971). Issues like concordance rates of alcoholism, alcohol-induced psychoses, cirrhosis, EEG-response and metabolism have been studied. The results from these studies are not unambiguous - some point to a considerable genetic component, while others conclude that environmental factors may play a predominant role in etiology. In a recent paper, Marshall and Murray (1989) have reviewed the twin literature of alcoholism. They call attention to several methodological problems in traditional twin research: Apart from sharing genes, most MZ twins also share environment, have lower birth weights and have higher rates of birth complications. The study of disconcordant MZ-pairs seems to be a suitable alternative strategy.

The overall conclusion of twin studies of alcoholism is that genetic factors contribute to some extent to the etiology, even if the results are not unambiguous.

Half sibling study

Schuckit et al. (1972) examined 151 half siblings of 61 alcoholic parents. 65% of the alcoholic half siblings had an alcoholic biological parent compared to 20% of the non-alcoholic half siblings. This result also reflects a considerable genetic component in the etiology of alcoholism.

Adoption studies

Almost 50 years ago, Roe *et al.* (1945) applied the adoption method on alcohol research. The design is very simple: Study adopted-away offspring of biological parents with severe alcoholism, but reared by foster-parents without alcohol problems.

Cadoret and Gath (1978) conducted a follow-up study of 1646 adopted-away children, born between 1939 and 1955. A small subgroup of six adoptees with alcoholic biological fathers and 78 adoptees without parental alcoholism were examined. Three of the six index subjects and only one of the 78 controls were (blindly) diagnosed as alcoholics.

Bohman (1978) examined 89 male adopted-away children whose biological fathers figured in the Swedish Alcohol Register. 40% of the offspring were registered as alcoholics themselves, in contrast to 13% of 723 adoptees without alcoholism in the biological parents.

A Danish adoption study

Throughout the 1970s a Danish adoption study of alcoholism was conducted, and its three phases will be described in detail:

Phase one

The material comprises all adoption cases (approx. 5000) in Copenhagen betwen 1927 and 1947. All biological parents were screened in the Danish Psychiatric Register with regard to psychiatric discharge diagnoses (according to the ICD-system). A total of 133 men (average age 30 years) were interviewed by a psychiatrist. They were divided into two groups: 55 adopted-away sons with a biological father who had been admitted to a psychiatric ward due to severe alcoholism, and 78 matched controls (no registered parental alcoholism). Only two variables distinguished the two groups significantly: Alcoholism (18% vs. 5%) and divorce rate (27% vs. 9%) were more pronounced in the index group. All other characteristics, including social, educational and psychopathological factors, were almost equally distributed in the two groups (Goodwin *et al.*, 1973).

Phase two

A subgroup of the adopted-away sons of the alcoholic fathers had biological brothers who were raised by their biological parents. They were also interviewed by a psychiatrist, using the same interview instrument as in phase one. Interviews were also conducted with matched non-adopted-away control subjects to avoid bias. The main finding of the phase two-study was that sons raised by their alcoholic parents had the same increased risk of becoming alcoholics themselves as their adopted-away biological brothers (Goodwin *et al.*, 1974). The main conclusion from the phase one and two studies is that sons of alcoholics have a 3- to 4-fold higher risk of developing alcoholism than do sons of non-alcoholics. The increased risk seems to be partly independent of the environment in which they were raised, indicating a genetic/biological component in the etiology of alcoholism.

Phase three

The same design was applied to the adopted-away daughters of biological alcoholic parents. The sample included 49 index and 48 control subjects (mean age 35 years). The rate of alcoholism and serious drinking problems were approx. 4% in both groups. More than 90% of the women in both groups were abstainers or very light drinkers. The results are not conclusive, and social factors may influence the drinking pattern more than the situation among males (Goodwin *et al.*, 1977a,b).

The next step in genetic-epidemiological alcohol research is to identify specific premorbid risk factors in a prospective context. A suitable method is to study subjects with a well-defined and increased risk for developing alcoholism. As described above, parental alcoholism seems to be the strongest predictor among other relevant factors.

A longitudinal study of sons of alcoholic fathers

The subjects were selected from a cohort of children born between 1959 and 1961. The cohort included 8949 consecutive deliveries at the maternity department of Rigshospitalet in Copenhagen, all examined intensively in a perinatal study (Zachau-Christiansen and Ross, 1975). From the obstetrical records we identified 8440 fathers. In 1979 we screened these fathers in the Danish (nationwide) Psychiatric Register for psychiatric diagnoses. In this manner, a total of 448 fathers were identified as alcoholics. Their 255 sons were selected for the study as the high risk (HR) group. Thirty-two were excluded due to perinatal death and emigration, leaving 223 HR sons available for the study. They were matched pairwise in accordance with social characteristics, and for each HR pair one low risk (LR) subject (n = 105) was selected from the remaining pool of sons without registered alcoholism in the parents.

Data

From the previous 30 years a series of data sources (prospective of nature) have been available for the study:

1959-61   pregnancy, delivery and postnatal data

1960-62   examination at one year of age

1966-78   school physician records

school psychologist records

school teacher questionnaire

1980-81   premorbid 20 year assessment

1984-85   postal questionnaire, seeking our refusals

1990-92   30 year follow-up assessment

1993   screening in Danish Death Register

1993   screening in Danish Psychiatric Register

1994   screening in Danish Criminality Register

1994   screening in Danish Military Service Register

Results

Pregnancy and delivery

The obstetrical and pediatric examinations were aggregated into six composite scales: pregnancy complications; delivery complications; neonatal physical status; physical status at the age of 1 year; motor development at the age of 1 year.

The HR and LR group did not differ significantly on any of the scales.

Schooling

We obtained data about the subject's school career from structured questionnaires completed by teachers. The child's intellectual, behavioral and social functioning was elicited. The HR boys' school carrier has been significantly more difficult than the LR boys (disciplinary problems, repeating a grade). Most striking is that 51% of HR boys vs. 34% of LR boys have been referred to a school psychologist.

Seven a priori scales focusing on alcohol-related antecedents were constructed. Only two scales distinguished the HR and LR groups significantly: Impulsivity/restlessness (p < 0.05) and verbal deficiency scale (p < 0.05).

School medical records have been collected for 83% of the HR and 86% of the LR cases. These records are based on annual examinations throughout the schooling period. They have been reviewed by a pediatrician, who coded mental and CNS-disorders according to the ICD-8 diagnostic criteria, the diagnostic system used in Danish pediatric practice at that time. The HR group had significantly more language and learning disorders than the LR group (p = 0.011). Among CNS-disorders we only found a marginally significant excess of cranial fractures and concussions in the HR group (p = 0.067). Results from the schooling are described by Knop *et al.* (1985; 1993).

Twenty years assessment

At age 19-20 years, the subjects participated in an extensive assessment. The aim was to collect premorbid data and to test the validity of hypotheses and selected variables by comparing HR and LR subjects. A total of 204 subjects completed the entire assessment. An outline of the results from the premorbid assessment is described below.

Among 12 neuropsychological tests, only three separated the two risk groups significantly: The Halstead Category Test, where HR subjects made more errors compared to LR subjects (p = 0.018). In the vocabular subtest of the WAIS battery the HR subjects had poorer scores than the LR group (p = 0.042). The HR subjects scored poorer points in the Porteus Maze test (p = 0.05) (Drejer *et al.*, 1980).

Every subject was examined with EEG. A subset of 44 HR and 28 LR subjects participated in an alcohol session, in which EEG was recorded after alcohol ingestion (0.5 g of ethanol per kg body weight). Blood alcohol concentration was measured simulatneously during the EEG-session. The HR subjects reacted with a significantly greater reduction of alpha energy 120 min after alcohol administration compared to LR subjects (Pollock *et al.*, 1983). No differences were found regarding blood alcohol concentration between the two risk groups.

Self-ratings of intoxication and objective measures of reaction on alcohol administration demonstrated a significantly lower sensitivity to alcohol among HR subjects than LR controls (Pollock *et al.*, 1986).

Visually evoked potentials (VEP) were measured during this session. The results indicate that right-handed HR subjects showed more symmetry in a positive component (latency, 242 ms) than

LR subjects. After alcohol ingestion, the occipital P 100 component latency demonstrated lateralized differencies between the two risk groups, providing speculations about ethanol's "normalizing" effect on brain function among predisposed individuals (Pollock *et al.*, 1988).

Blood samples were obtained from all four subjects for measurement of monoamine oxidase (MAO) activity in blood platelets, which has been proposed as a biochemical marker in alcoholism. However, we did not find any differences in MAO-activity beween the two risk groups (Potkin *et al.*, unpublished).

Every subject underwent an extensive psychopathological evaluation covering 28 diagnostic categories and personality traits. This evaluation did not demonstrate significant differences between the HR and LR groups. The prevalence of marked psychopathology was infrequent in both groups, as could be expected in a sample of healthy young men (Schulsinger *et al.*, 1986).

The 20 year follow-up assessment was intended to be premorbid. Therefore evaluation of their drinking practice was an essential part of the assessment. The mean alcohol consumption was equally distributed in the two risk groups (mean 17.2 drinks during last week), an amount close to the mean consumption in the Danish population of males at the same age (Vilstrup and Nielsen, 1981). No subject was found to alcohol-dependent, indicating that the assessment was indeed premorbid. Drug experiences were relatively infrequent without significant group differences.

Thirty year follow-up assessment

Recently, the study group has completed the data collection for the subjects, now about 30 years old. The overall aims of this follow-up have been:

- to compare the two risk groups on a series of dimensions including drinking behavior, rate of alcohol/drug dependence, concomitant physical health problems and degree of psychosocial functioning;
- to determine whether the differences between the HR and LR groups at age 20 years continue to distinguish the risk groups; and
- to determine whether measures from the perinatal and schooling period and 20 years assessment successfully predict current measures of dependence and abuse (according to DSM-III-R criteria) at age 30 years.

A trained psychiatrist, who was blind concerning the subjects' risk status, conducted home interviews with the subjects. It was essential to select variables compatible to the data collected at previous stages of the study. The interview instruments included:

- Psychiatric Diagnostic Interview-revised (PDI-R)
- Structured Clinical Interview for DSM-R-III, personality disorders (SCID-II)
- Tridimensional Personality Questionnaire (TPQ)
- Millon Clinical Multiaxial Inventory (MCMI)
- Neuropsychological test battery
- Psychosocial history
- Medical history
- Family history
- Alcohol and drug history
- Michigan Alcohol Screening Test (MAST)

In addition, we obtained access to nationwide registers:
- Danish Psychiatric Register
- Danish Death Register
- Danish Criminality Register
- Danish Military Service Register

Only a few overall results from the 30 year follow-up will be presented here, since the data collection terminated recently.

A total of 241 home interviews have been conducted (161 HR and 80 LR subjects). Due to emigration, death and disappearance the number of available subjects was reduced from 328 (20 year assessment) to 305, resulting in a participation rate of approx. 80%.

Dependence (according to DMS-III-R criteria) is significantly more frequent in the HR subjects, while abuse is almost equally distributed in the two risk groups (Table I).

Apart from alcohol, cannabis seems to be the preferred drug among our subjects (in particular the HR group). Concerning other psychopathology, the risk groups did not differ significantly in frequency of anxiety disorders, mood disorders, schizophrenia, antisocial personality disorder and other personality disorders. Only anxiety disorders are close to reaching significance (HR 13%, LR 5%, $p = 0.055$).

Table I. Substance use disorders (DSM-III-R) in HR and LR subjects at age 30 years (in percent)

DSM-III-R diagnoses	HR (n=161)	LR (n=80)
Alcohol or drug abuse	19	16
Alcohol or drug dependence	29	15*

*p=0.02

However, when the sample is grouped according to severity of substance use disorders (and not by risk status), some characteristic differences appear (Table II).

The lowest rate of psychopathology is found in the no abuse group. Most psychopathology is found in the dependence group compared with the abuse group. Most striking is the comorbid correlation between dependence and antisocial personality disorder (p = 0.001).

Table II. Psychiatric diagnoses in 241 subjects at age 30 years grouped according to substance use disorders (%)

DSM-III-R diagnoses	No abuse n=139	Abuse n=43	Dependence n=59
No diagnosis	84	72	31
Anxiety disorders	6	12	19
Mood disorders	5	14	17
Schizophrenia	0	0	7
Other Axis 1 diagnosis	1	0	3
Antisocial personality disorder	1	5	41*
Other personality disorder	3	2	12

The figures are not additive, since 16 subjects have more than one diagnosis.
*p<0.0001

Final remarks

An extensive series of data analyses will examine 1) differences between HR and LR subjects at different phases in their lives, 2) early and late predictors of current drinking categories, and 3) the extent that premorbid differences between HR and LR subjects successfully forecasted different levels of problem drinking at the age of 30 years.

The study group plans a final follow-up examination of the sample at the age of 40 years.

References

Bohman, M. (1978) Some genetic aspects of alcoholism and criminality. *Arch. Gen. Psychiatry* 35: 269-276.

Cadoret, R.J. and Gath, A. (1978) Inheritance of alcoholism in adoptees. *Br. J. Psychiatry* 132: 252-258.

Cloninger, C.R., Bohman, M. and Sigvardsson, S. (1981) Inheritance of alcohol abuse: Cross-fostering analysis of adopted men. *Arch. Gen. Psychiatry* 38: 861-868.

Cotton, N.S. (1979) The familial incidence of alcoholism. *J. Stud. Alcohol* 40: 89-116.

Dupont, A., Videbech, T. and Weeke, A. (1974) A cumulative national psychiatric register: Its structure and application. *Acta Psychiatr. Scand.* 50: 161-173.

Drejer, K., Theilgaard, A., Teasdale, T.W. and Goodwin, D.W. (1980) A prospective study of young men at high risk for alcoholism: Neuropsychological assessment. *Alcohol Clin. Exp. Res.* 9: 298-302.

Frances, R., Timm, S. and Bucky, S. (1980) Studies of familial and non familial alcoholism. I. Demographic Studies. *Arch. Gen. Psychiatry* 37: 564-569.

Goodwin, D.W., Schulsinger, F., Hermansen, L., Guze, S.B. and Winokur, G. (1973) Alcohol problems in adoptees rasied apart from alcohol biological parents. *Arch. Gen. Psychiatry* 28: 238-243.

Goodwin, D.W., Schulsinger, F., Møller, N., Hermansen, L., Winokur, G. and Guze, S.B. (1974) Drinking problems in adopted and non-adopted sons of alcoholics. *Arch. Gen. Pychiatry* 31: 164-169.

Goodwin, D.W., Schulsinger, F., Knop, J., Mednick, S.A. and Guze, S.B. (1977a) Psycho-pathology in adopted and non-adopted daughters of alcoholics. *Arch. Gen. Psychiatry* 34: 751-755.

Goodwin, D.W., Schulsinger, F., Knop, J., Mednick. S.A. and Guze, S.B. (1977b) Alcoholism and depression in adopted-out daughters of alcoholics. *Arch. Gen. Psychiatry* 34: 751-755.

Hrubec, Z. and Omenn, G.S. (1981) Evidence of genetic predisposition to alcoholic cirrhosis and psychosis: Twin concordance for alcoholism and its biological end points by zygocity among male veterans. *Alcohol Clin. Exp. Res.* 5: 207-215.

Jonsson, A. and Nilsson, T. (1968) Alkoholkonsumption hos monozygota och dizygota tvillingar. *Nord. Hyg. Tidskr.* 49: 21-25.

Kaij, L. (1960) *Alcoholism and twins.* Almqvist & Wiksell, Stockholm (dissertation).

Koehlin, J.C. (1972) An analysis of alcohol-related questionnaire items from the nationalmerit twin study. *Ann. NY Acad. Sci.* 197: 117-120.

Knop, J., Teasdale, T.W., Schulsinger, F. and Goodwin, D.W. (1985) A prospective study of young men at high risk for alcoholism: School behaviour and achievement. *J. Stud. Alcohol* 46: 273-278.

Knop, J., Goodwin, D.W., Jensen, P., Penick, E., Pollock, V., Gabrielli, W., Teasdale, T.W. and Mednick, S.A. (1993) A 30-year follow-up study of the sons of alcoholic men. *Acta Psychiatr. Scand.* Suppl. 370: 48-53.

Marshall, E.J. and Murray, R.M. (1989) The contribution of twin studies to alcohol research. In: Goedde, H.W. and Agarwal, D.P. (eds.): *Alcoholism: Biochemical and genetic aspects.* Pergamon Press, New York, pp. 277-289.

Meyer, R. and Babor, T.F. (1989) Explanatory models of alcoholism. In: Tasman, A., Hales, R. and Frances, A. (eds.): *Review of Psychiatry*, Vol. 8, American Psychiatric Press, Washington DC, pp. 273-292.

Murray, R.M., Clifford, C., Gurling, H.M.D., Tophan, A., Clow, A. and Bernadt, M. (1983) Current genetic and biological approaches to alcoholism. *Psychiatr. Dev.* 2: 179-192.

Partanen, J., Bruun, K. and Markkanen, T. (1966) Inheritance of drinking behaviour: *A study of intelligence, personality and use of alcohol in adult twins.* Finnish Foundation for Alcohol Studies, Helsinki (Publication No. 14).

Pollock, V.E., Volovka, J., Goodwin, D.W., Mednick, S.A., Gabrielli, W.F., Knop, J. and Schulsinger, F. (1983) The EEG after alcohol administration in men at high risk for alcoholism. *Arch. Gen. Psychiatry* 40: 857-861.

Pollock, V.E., Teasdale, T.W., Gabrielli, W.F. and Knop, J. (1986) Subjective and objective measures to alcohol in young men at high risk for alcoholism. *J. Stud. Alcohol* 47: 297-304.

Pollock, V.E., Volovka, J., Goodwin, D.W., Gabrielli, W.F., Mednick, S.A., Knop, J. and Schulsinger, F. (1988) Pattern reversal visual evoked potentials after alcohol administration among men at risk for alcoholism. *Psychiatry Rev.* 26: 191-202.

Potkin, S., Goodwin, D.W., Phelps, B., Gabrielli, W.F., Knop, J., Jansson, L., Wyatt, R., Mednick, S.A. and Schulsinger, F. Evaluation of platelet monoamine oxidase activity as a predictor of future alcoholims: A controlled high risk study in sons of alcoholics. (Unpublished manuscript)

Roe, A., Burks, B.S. and Mittelman, B. (1945) Adult adjustment of foster children of alcoholic and psychiatric parentage and the influence of the foster home. *Memoires Section Alc. Stud.* 3.

Schuckit, M.A., Goodwin, D.W. and Winokur, G. (1972) A study of alcoholism in half siblings. *Am. J. Psychiatry* 128: 122-125.

Schulsinger, F., Knop, J., Goodwin, D.W., Teasdale, T.W. and Mikkelsen, U. (1986) A prospective study of young men at high risk for alcoholism: Social and psychological characteristics. *Arch. Gen. Psychiatry* 43: 755-760.

Vesell, E.S., Page, J.G. and Passananti, G.T. (1971) Genetic and environmental factors affecting ethanol metabolism in men. *Clin. Pharmacol. Ther.* 12: 192-201.

Vilstrup, H. and Nielsen P.E. (1981) Alkoholforbrugets fordeling i den danske befolkning i 1979. *Ugeskr. Laeger* 143: 1047-1052 (Summary i English)

Zachau-Christiansen, B. and Ross, E.M. (1975) *Human development during first year.* Wiley & Sons, New York.

Toward a Molecular Basis of Alcohol Use and Abuse
ed. by B. Jansson, H. Jörnvall, U. Rydberg, L. Terenius & B. L. Vallee
© 1994 Birkhäuser Verlag Basel/Switzerland

Association strategies in substance abuse

George R. Uhl

Molecular Neurobiology Branch, Addiction Research Center, National Institutes of Drug Abuse, and Departments of Neurology and Neuroscience, Johns Hopkins University School of Medicine, Box 5180 Baltimore, MD 21224, USA

Summary

Classical genetic studies now provide fairly strong support to the idea that polygenic influences contribute to interindividual differences in vulnerability to drug abuse and dependence. Molecular genetic association studies have identified the D2 dopamine receptor gene as a candidate to contribute some of this genetic influence, although several cautions in utilizing association approaches must be recognized in interpreting their data. Clinical features of drug abuse that make association studies likely to play increasingly large roles in elucidating its underlying molecular genetics are discussed.

Human substance abuse vulnerability and genetic influences

Individuals are differentially vulnerable to substance abuse. Not everyone has an opportunity to use addictive substances, not everyone who has an opportunity to use an addictive substance does so and not everyone who uses an addictive substance becomes addicted. Both genetic and environmental conditions that differentially predispose individuals to drug-taking behavior and to the transition from drug taking behavior to established and maintained drug abuse might thus be found.

Processes involved in substance abuse are behaviorally complex; genetic mechanisms contributing to interindividual differences in vulnerability are likely to be equally complex. Genetic contributions to the initiation of drug use may differ from those that contribute to heavier drug use (Scheier and Newcomb, 1991) or drug dependence (Glantz and Pickens, 1992). Drug use vulnerability might also be modified by protective factors that could contribute to drug abstinence or failure to develop regular use patterns or drug dependence (Brooks *et al.*, 1989a; 1989b). Allelic variants of specific genes could mediate differential drug reinforcing properties, alter drug pharmacodynamics or pharmacokinetics, influence "sensation seeking" personality traits facilitating exposure to drugs, exacerbate drug toxicities or minimize "protective" factors such as hangovers. Individuals also differ in their specific drugs of choice.

Familial and population genetic studies are beginning to reveal possible genetic bases for some of the interindividual differences in vulnerability to substance abuse. Twin and adoption studies suggest that both genetic and environmental influences are likely to be involved in drug use or dependence (see below). Perhaps the most convincing evidence for the nature and degree of genetic bases for substance abuse comes from the convergence of results obtained through methodologically-distinct approaches.

Application of molecular genetic approaches to studies of substance abuse is relatively recent. Nevertheless, recent studies with polymorphic genetic markers at several candidate gene loci have initiated the search for specific genes whose allelic variants could contribute to genetic differences in substance abuse vulnerability (see below). Substantial collective work on markers at the dopamine D2 receptor gene locus (DRD2) has continued to suggest differences between populations of substance abusers and controls in work from several, but not all, groups (Uhl *et al.*, 1992; 1993; but see Gelernter *et al.*, 1993).

Focus on genetic influences should not obscure the significant environmental nature of many risk factors for substance abuse. Indeed, one major fruit of the labors of elucidating genetic influences on substance abuse behaviors will be that the environmental components of vulnerability will be more readily identified so that their independent and interactive components can be more easily assessed. Appropriate recognition of genetic and environmental vulnerabilities can lead to improved strategies for both treatment and prevention.

Table I. DRD2 TaqI A and B RFLP frequencies in drug abusers and controls

	%A1		%B1	
	Drug Abusers	Controls	Drug Abusers	Controls
Smith *et al.*, 1992, and O'Hara *et al.*, 1993	41% (96/237)	28% (45/160)	32% (76/237)	22% (35/160)
Noble *et al.*, 1993	51% (27/53)	16% (16/100)	38% (20/52)	13% (7/53)
Comings *et al.*, 1993	42% (45/106)	29% (221/763)		
Gelernter *et al.*, 1993*	45% (49/108)	36% (24/68)		
Summarized:	**44% (217/504)**	**28% (306/1091)**	**33% (95/289)**	**(20% (42/213)**

* Estimated from presented data, assuming Hardy-Weinberg equilibrium

Twin studies determine if within-pair similarity for substance abuse is greater in genetically identical monozygotic (MZ) twins than in less-genetically-similar fraternal dizygotic (DZ) twins (Table I). Careful studies of these relatively rare subjects have allowed the first tentative estimates of the magnitude of the genetic components of drug abuse vulnerability (e.g., Pickens *et al.*, 1991; Goldberg *et al.*, 1993).

Pickens and coworkers (1991) found a significantly greater concordance for substance abuse in monozygotic males than in dizygotic males. These workers then used population prevalence data to estimate components of liability variance, estimates of the proportion of genetic contribution to substance abuse. When substance abuse and/or dependence in men was considered, 31% of the variance could be attributed to genetic components. The corresponding figures for alcohol dependence were 60%; more prominent genetic components for individuals with substance dependence might have been determined if sufficient power had been available. More limited data from females failed to show a significant genetic influence on regular use of any illicit drug, although data provided a trend in this direction for substance abuse and/or dependence and for alcohol abuse (Pickens *et al.*, 1991).

Study of Vietnam Era twin pair registrants has also indicated genetic components to drug abuse, with significant heritability scores for abuse of hallucinogens, stimulants, opiates, sedatives, and marijuana (Goldberg *et al.*, 1993). Only cannabis users displayed shared environmental components. This large twin study thus provides the first evidence that vulnerability to each of these classes of addictive substances may display a genetic component, and the conclusion that vulnerability to no substance can be attributed to shared environmental features alone. Twin data now provides significant support for the idea that drug abuse vulnerability displays significant genetic components that may be more prominent in the more severe abusers.

Allelic association and dopaminergic genes in drug abuse
Significant data now supports the idea that virtually every abused drug, and ethanol, can induce behavioral reinforcing properties by altering function in brain dopamine circuits arising from the ventral midbrain (Di Chiaro *et al.*, 1988; Kuhar *et al.*, 1991). Genes important in the mesolimbic/mesocortical dopaminergic pathways, strong candidate genes for possible contributions to interindividual differences in substance abuse vulnerability, have been identified

in recent molecular cloning studies. The dopamine transporter that is the pharmacologically-defined cocaine receptor (Shimada *et al.*, 1991; Vandenbergh *et al.*, 1992), the synaptic vesicular monoamine transporter that may be involved in amphetamine action (Surratt *et al.*, 1993), and the dopamine synthetic enzymes tyrosine hydroxylase and dopa decarboxylase, and the D1-D5 G-linked postsynaptic dopamine receptors are each encoded by such candidate genes (See chapter by Schwartz, this volume).

Association studies compare genetic markers at specific gene sites in affected and unaffected individuals. Since association studies can study affected and unaffected individuals in the same cohort, they can be less susceptible to secular trends in substance abuse than classical family likage studies. Genetic markers at a gene locus are thus **associated** with a disease if they are present more often in unrelated individuals displaying the disease than in unaffected individuals.

For allelic association to be detected, chromosomal recombination between a polymorphic marker and a postulated gene contributing to substance abuse in the population must occur infrequently within the many generations separating the inheritance of the population sampled. A polymorphic marker such as a specific RFLP successfully used in association studies could thus lie very close to the functional gene defect contributing to substance abuse vulnerability. More often, associations are based on considerations of chromosomal recombination, especially linkage disequilibrium. The meiotic events forming the chromosomes for each human generation result in "crossing over" recombinant events splicing sequences from one chromosomal copy with those of the other chromosomal copy. The average rate of this process allows 1% of chromosomal loci separated by 1 million base pairs of DNA to recombine in each generation. However, this average rate of recombination can vary substantially across different chromosomal loci (see Uhl *et al.*, 1993). Linkage disequilibrium can thus allow a genetic marker used in an allelic association study to provide information about not only whether closely-adjacent DNA contains a functional gene defect, but also about the possibility that DNA many thousands of bases removed from the polymorphic genetic marker but in linkage disequilibrium with it could also contain a functional gene defect.

The D_2 dopamine receptor (DRD2) gene encodes a G-protein linked 7-transmembrane region receptor protein expressed abundantly in dopaminergic circuits important for behavioral reward (Uhl *et al.*, 1993). Genetic polymorphic markers have been identified at several DRD2 loci. A *Taq*I A restriction fragment length polymorphism (RFLP) is located in the 3' flanking region of the DRD2 gene. A *Taq*I B RFLP, lies more 5' (Hauge *et al.*, 1991). A *Taq*I C RFLP provides a polymorphic marker for a site lying between *Taq*I A and B (Bolos *et al.*, 1990; Parsian *et al.*, 1991b; Sarkar *et al.*, 1991).

Blum *et al.* (1990) provided the first evidence that the DRD2 gene might display population variants influencing susceptibility to alcoholism. These workers found a striking allelic association between the *Taq*I A1 RFLP form and alcoholism. However, these results were viewed with caution for several reasons (Cloninger *et al.*, 1991; Uhl *et al.*, 1992). Thousands of different genes are expressed in the human brain; the *a priori* probability of identifying a vulnerability-enhancing allele of one these genes was low. Genetic linkage of this marker in several families in which alcohol abuse appeared to pass from generation to generation in nearly-mendelizing fashion was not supported (Bolos *et al.*, 1990; Parsian *et al.*, 1991a). The *Taq*I A RFLPs used in this study were demonstrated to be inhomogeneously distributed in different human populations (Uhl *et al.*, 1991; O'Hara *et al.*, 1993), with high A1 allele frequencies in black, Asian and American Indian populations (Uhl *et al.*, 1991; O'Hara *et al.*, 1993; Gelernter *et al.*, 1993). Spurious associations not indicative of true causal links between DRD2 allelic status and substance abuse could thus result from sample stratification, disproportionate sampling of abusers or controls from population subgroups displaying atypical RFLP frequencies.

Despite these cautions, we and three other groups of investigators have examined DRD2 gene markers in drug abusers and control populations (Table I). Linkage disequilibrium has now been well documented at the dopamine D2 receptor gene locus, suggesting that *Taq*I A1 and B1 DRD2 RFLPs are interesting reporters for events in significant portions of the DRD2 gene locus in caucasians. More than 95% of the possible linkage disequilibrium (D'/D_{max}) is maintained between these loci (Hauge *et al.*, 1991; O'Hara *et al.*, 1993). *Taq*I A and B genotypes *could* thus reliably mark a structural or functional gene variant at the DRD2 locus that could be directly involved in altering behavior. Further, A1 and B1 markers appear more frequently in drug abusers than in control populations in each of four currently-available studies (Smith *et al.*, 1992; O'Hara *et al.*, 1993; Comings *et al.*, in press; Noble *et al.*, in press; Gelernter *et al.*, 1993). Meta analyses of these data suggest that differences between drug abuser and control populations are highly significant for both the A1 (four studies), and B1 (two studies) loci. These meta analyses suggest an odds ratio of drug abuse likelihood for individuals possessing an A1 allele of 2.4 and a 3.3-fold odds ratio for those having a B1 allele (p < 0.001 in both cases;

calculated per Smith *et al.*, 1992).

The most severe abusers of addictive substances may manifest higher A1 and B1 DRD2 gene marker frequencies (Commings *et al.*, 1991), while "control" comparison groups studied carefully to eliminate individuals with significant use of any addictive substance appear to display lower A1 and B1 frequencies than unscreened control populations (see Uhl *et al.*, 1993).

No data available to date derives from true population based sampling techniques. The theoretical possibility of false-positive error based on sample stratification thus remains. This possibility is especially prominent if control populations are not drawn from the same genetic pools from which substance abusers were drawn. Racial differences in the frequencies of DRD2 gene markers, in possible "founder" chromosome populations, and in the strength of linkage disequilibrium are all now reasonably well-documented. Ethnic differences between Caucasian groups can also be demonstrated in small samples drawn from the literature. Careful attention to control populations is thus essential to avoid spurious results.

Association studies for polymorphic markers at loci for other dopaminergic genes have also been suggested. None of these genes has been documented to display substantial linkage disequilibrium. With this caveat, neither markers at the dopamine transporter, synaptic vesicular transporter nor tyrosine hydroxylase loci have yielded positive allelic association results (Blum *et al.*, 1990; Noble *et al.*, 1992; Persico *et al.*, 1993; C. Surratt, A. Persico and GRU, in preparation). Polymorphic elements determined by EcoRI RFLPs at DRD1, BalI and MspI RFLPs at DRD3, and both HincII RFLPs and a variable number repetitive element at the DRD4 each provide plausible markers for substance abuse.

We know how addictive processes start: drug occupies a brain receptor. Processes directly modulating receptor function, however, have not been able to account for significant fractions of the biochemical bases of addiction (Nestler, 1992). Other candidate molecular mechanisms of information storage can be tentatively postulated as possibly involved in addiction (Uhl, in press). However, direct genetic approaches not dependent on biochemical hypotheses may be more likely to identify genes that could contribute to interindividual differences in substance abuse vulnerability than candidate gene approaches that assume greater knowledge of addiction process biochemistry than we may actually possess.

Classical linkage approaches to this problem involve use of polymorphic genetic markers, spaced throughout the genome, and assessment of the extent to which a particular marker form is co-inherited with the disorder. This approach suffers as genetic heterogeneity, for example, a single gene leading to substance abuse in only a subset of families, reduces its statistical power. In drug abuse, characterized by strong secular trends in the abused substance and poor mendelizing patterns, this approach is even more difficult.

Association genome scanning can utilize the rapidly growing number of polymorphic markers available at different chromosomal loci to select "hot spots" where marker frequencies differ between chromosomal substance abusing and control populations. When only single populations of abusers and controls are compared, stratification may account for frequency differences at many sites. However, if a marker provides such a "hot spot" in further comparisons of association performed in different abuser and nonabuser populations, then surrounding DNA can be sequenced to identify involved genes.

Each of these steps should challenge the state-of-the-art of current molecular behavioral genetics. Rapidly evolving methods now allow potential assessments of gene marker frequency differences in pooled DNA samples from abusers or controls aid the approaches feasibility. Rapid progress in understanding the human genome will simplify identification of the genes adjacent to identified "hot spots." Statistical approaches to better understanding population genetics and linkage disequilibrium will also aid in quantitative results. Perhaps most important for such studies are careful classical genetic studies to sharpen focus on questions concerning which features of the drug abusing phenotype are inherited. The initial results in this area of drug abuse molecular genetics can provide a framework for attacks on this complex problem with the concerted power of current state-of-the-art classical and molecular genetic tools.

Acknowledgements

The work in the authors' laboratory was supported by the intramural program of NIDA. Carol Sneeringer preparation with the manuscript.

References

Blum, K., Noble, E.P., Sheridan, P.J., Montgomery, A., Ritchie, T., Jagadeeswaran, P., Nogami, H., Briggs, A.H. and Cohn, J.B. (1990) Allelic association of human dopamine D2 receptor gene in alcoholism. *JAMA* 263: 2055-2060.

Bolos, A.M., Dean, M., Lucas-Derse, S., Ramsburg, M., Brown, G.L. and Goldman, D. (1990) Population and pedigree studies reveal a lack of association between the dopamine D2 receptor gene and alcoholism. *JAMA* 264: 3156-3160.

140

Brooks, J.S., Nomura, C. and Cohen, P. (1989 a) A network of influences on adolescent drug involvement: Neighborhood, school, peer, and family. *Genet. Soc. Gen. Mono.* 115: 125- 145.

Brooks, J.S., Nomura, C. and Cohen, P. (1989 b) Prenatal, perinatal, and early childhood risk factors and drug involvement in adolescence. *Genet. Soc. Gen. Mono.* 115: 223-241.

Cloninger, C.R. and Begleiter, H. (1990) *Genetics and biology of alcoholism*, Cold Spring Harbor Press, New York.

Cloninger, C.R. (1991) D2 dopamine receptor gene is associated but not linked with alcoholism. *JAMA* 266: 1833-1834.

Comings, D.E., Comings, B.G., Muhleman, D., Dietz, G., Shahbahrami, B., Tast, D., Knell, E., Kocsis, P., Baumgarten, R., Kovacs, B.W., Levy, D.L., Smith, M., Borison, R.L., Evans, D.D., Klein, D.N., MacMurray, J., Tosk, J.M., Sverd, J., Gysin, R. and Flanagan, S.D. (1991) The dopamine D2 receptor locus as a modifying gene in neuropsychiatric disorders. *JAMA* 266: 1793-1800.

Comings, D.E., MacMurray, J., Johnson, J.P., Muhleman, D., Ask, M.N., Ahn, C., Gysin, R. and Flanagan, S.D. (1993) The dopamine D_2 receptor gene: A genetic risk factor in polysubstance abuse. *Alcohol Drug Depend (in press)*. (This is not mentioned in the text)

Di Chiara, G. and Imperato, A. (1988) Drugs abused by humans preferentially increase synaptic dopamine concentrations in the mesolimbic system of freely moving rats. *Proc. Natl. Acad. Sci.* 85: 5274-5278.

Gelernter, J., Goldman, D. and Risch, N. (1993) The A1 allele at the D2 dopamine receptor gene and alcoholism: a reappraisal. *JAMA* 269: 1673-1677.

Gelernter, J., Kranzler, H. and Satel, S. (1993) No association between DRD2 alleles and cocaine abuse. *Fifty-Fifth Annual Scientific Meeting, College on Problems of Drug Dependence, Inc.*, poster.

Glantz, M. and Pickens, R.W. (1992) *Vulnerability to Drug Abuse*. American Psychological Association, Washington, D.C.

Goldberg, J., Lyons, M.J., Eisen, S.A., True, W.R. and Tsuang, M. (1993) Genetic influence on drug use: A preliminary analysis of 2674 Vietnam era veteran twins. *Behav. Genet. (abstract),* in press.

Hauge, X.Y., Grandy, D.K., Eubanks, J.H., Evans, G.A., Civelli, O. and Litt, M. (1991) Detection and characterization of additional DNA polymorphisms in the dopamine D2 receptor gene. *Genomics* 10: 527-530.

Kuhar, M.J., Ritz, M.C. and Boja, J.W. (1991) The dopamine hypothesis of the reinforcing properties of cocaine. *TINS* 14: 299-302.

Nestler, E.R. (1992) Molecular mechanisms of drug addiction. *J. Neurosci.* 12: 2439-2450.

Noble, E.P., Blum, K. and Khalsa H. (1992) Allelic association of the D2 dopamine receptor gene in cocaine dependence: Proceedings of the XVIIIth Congress of the Collegium Internationale NeuroPsychopharmacologium. *Clin. Neuropharm.* 15: 99b.

Noble, E.P., Blum, K., Khalsa, M.E., Ritchie, T., Montgomery, A., Wood, R.C., Fitch, R.J., Ozkaragoz, T., Sheridan, P.J., Anglin, M.D., Paredes, A., Treiman, L.J. and Sparkes, R.S. (1993) D_2 dopamine receptor gene alleles in treatment-seeking cocaine dependent Caucasian subjects. *Drug Alcohol Depend.*, in press.

O'Hara, B.F., Smith, S.S., Bird, G., Persico, A., Suarez, B., Cutting, G.R. and Uhl, G.R.(1993) Dopamine D2 receptor RFLPs, Haplotypes and their association with substance use in black and caucasian research volunteers. *Hum. Hered.* 43: 209-218.

Parsian, A., Todd, R.D., Devor, E.J., O'Malley, K.L., Suarez, B.K., Reich, T. and Cloninger CR. (1991a) Alcoholism and alleles of the human D2 dopamine receptor locus. *Arch. Gen. Psych.* 48: 655-663.

Parsian, A., Fisher, L., O'Malley, K.L. and Todd, R.D. (1991b) A new TaqI RFLP within intron 2 of human dopamine D2 receptor gene (DRD2). *Nucl. Acid Res.* 19: 6977.

Persico, A.M., O'Hara, B.F., Farmer, S., Gysin, R., Flanagan, S. and Uhl, G.R. (1993) Dopamine D2 receptor gene TaqI "A" locus map including A4 variant: Relevance for alcoholism and drug abuse. *Drug Alcohol Depend.* 31: 229-234.

Pickens, R.W., Svikis, D.S., McGue, M., Lykken, D.T., Heston, L.L. and Clayton, P.J. (1991) Heterogeneity in the inheritance of alcoholism: a study of male and female twins. *Arch. Gen. Psych.* 48: 19-28.

Sarkar, G. and Sommer, S.S. (1991) Haplotyping by double PCR amplification of specific alleles. *Biotechniques* 10: 436-440.

Scheier, L.M. and Newcomb, M.D. (1991) Differentiation of early adolescent predictors of drug use versus abuse: A developmental risk-factor model. *J. Subst. Abuse* 3: 277-299.

Shimada, S., Kitayama, S., Lin, C.L., Nanthakumar, E., Gregor, P., Patel, A., Kuhar, M.J. and Uhl, G.R. (1991) Cloning and expression of a cocaine-sensitive dopamine transporter cDNA. *Science* 254:576-578.

Smith, S.S, O'Hara, B.F, Persico, A.M, Gorelick, D.A, Newlin, D.B, Vlahov, D., Solomon, L., Pickens, R. and Uhl GR. (1992) Genetic vulnerability to drug abuse: the dopamine D2 receptor TaqI B1 RFLP is more frequent in polysubstance abusers. *Arch. Gen. Psych.* 49: 723-727.

Surratt, C.K., Persico, A.M., Yang, X.-D., Edgar, S.R., Bird, G.S., Hawkins, A.L., Griffin, C.A., Li, X., Jabs, E.W. and Uhl, G.R. (1993) A human synaptic vesicle monoamine transporter cDNA predicts posttranslational modifications, reveals chromosome 10 gene localization and identifies TaqI RFLPs. *FEBS Lett.* 318: 325-330.

Uhl, G.R, Persico, A.M. and Smith, S.S. (1992) Current excitement with D2 dopamine receptor gene alleles in substance abuse. *Arch. Gen. Psych.* 49: 157-160.

Uhl, G.R. (1993) Molecular and genetic studies of the targets of acute drug action, substrates for interindividual differences in vulnerability to substance abuse, and candidate mechanisms for addiction. In: Chiarello, E. (ed.): *NIDA Res Monogr*, in press.

Uhl, G.R., Blum, K. and Smith, S.S. (1993) Substance abuse vulnerability and D2 receptor genes. *TINS* 16: 83-88.

Vandenbergh, D.J., Persico, A.M. and Uhl, G.R. (1992) A human dopamine transportercDNA predicts reduced glycosylation, displays a novel repetitive element and provides racially-dimorphic TAQ I RFLPs. *Mol. Brain Res.* 15: 161-166.

Toward a Molecular Basis of Alcohol Use and Abuse
ed. by B. Jansson, H. Jörnvall, U. Rydberg, L. Terenius & B. L. Vallee
© 1994 Birkhäuser Verlag Basel/Switzerland

PET-determination of benzodiazepine receptor binding in studies on alcoholism

Lars Farde, Stefan Pauli, Jan-Eric Litton, Christer Halldin, Jack Neiman and Göran Sedvall

Department of Clinical Neuroscience, Karolinska Hospital, S-171 76 Stockholm, Sweden

Summary

Positron Emission Tomography (PET) and the radioligand [^{11}C]flumazenil were used to examine benzodiazepine (BZ) receptor binding in the human brain. In a first study of healthy males acute ingestion of alcohol did not alter total radioactivity uptake or specific [^{11}C]flumazenil binding in the neocortex or cerebellum. In a second study [^{11}C]flumazenil binding was determined in 5 healthy male controls and 5 chronic alcohol dependent men using a saturation procedure with two PET experiments. Mean values for BZ-receptor density and affinity were similar in the two groups but the B_{max} variance for the alcohol dependents was significantly larger ($p < 0.05$) for all regions. The present studies do not support the view that alcohol affects central BZ receptor binding in man.

Introduction

The effects of alcohol on cerebral blood flow (CBF), glucose metabolism and receptor binding have been studied by Positron Emission Tomography (PET). Most investigators have reported a decrease in CBF of alcohol-dependent patients (Lofti and Meyer 1989; Mathew and Wilson 1991). Decreased uptake of [^{18}F]fluorodeoxyglucose has been reported after acute administration of ethanol to healthy subjects and alcohol-dependent patients (Volkow *et al.*, 1990a; de Wit *et al.*, 1992; Frascella and Brown, 1992) and in studies on chronic patients (Kessler *et al.*, 1984; Sachs *et al.*, 1987; Volkow *et al.*, 1990b; Adams *et al.*, 1993). In a study by Wik and collaborators (1988) twelve healthy controls were compared to nine male alcohol-dependent patients who had abstained from alcohol and drugs for more than four weeks. [^{11}C]Glucose metabolism was reduced by 30% in the whole brain. Cortical areas were most affected but significant reductions were also seen in subcortical areas.

The findings of reduced CBF and glucose metabolism shows that alcohol consumption can induce measurable effects on physiological parameters of brain function. A more detailed biochemical understanding of such effects may be provided by PET-examination of neurotransmission. Various neurotransmission systems have been discussed as targets for ethanol

(Samson and Harris 1992; Cooper *et al.,* 1993). The GABA-benzodiazepine receptor complex has been suggested for several reasons. Benzodiazepine (BZ) receptor agonists have, like alcohol, anxiolytic, muscle relaxant, and sedative-hypnotic properties. On the basis of experimental studies in animals it has been suggested that abnormalities of BZ receptor function are related to the development of alcoholism and may be the result of prolonged heavy drinking (Samson and Harris, 1992). BZ receptor densities determined by [^{3}H]flunitrazepam binding *in vitro* has been reported to be reduced in brain tissue from alcoholic patients (Freud and Ballinger, 1988).

Several radioligands have been developed for PET-analysis of BZ-receptor binding. The most widely used is the antagonist [^{11}C]flumazenil (Mazière *et al.,* 1984; Persson *et al.,* 1985). With this ligand BZ receptor density can be determined in a saturation analysis with at least two PET experiments in each subject (Persson *et al.,* 1989). A recently developed ligand is [^{11}C]Ro15-4513 (Halldin *et al.,* 1992) a ligand of particular interest for research on alcohol effects (cf discussion). The aim of our first alcohol PET-project was to examine the effect of acute alcohol administration on the binding of central [^{11}C]flumazenil binding in healthy human subjects. The second project was to compare [^{11}C]flumazenil binding to that in detoxified alcohol-dependent patients. The results were reported in detail by Pauli *et al.* (1992) and Litton *et al.* (1993).

Material and Method

Healthy subjects

For the study on the effects of acute alcohol administration 4 men aged 26-35, body weight 70-86 kg, were recruited. As control subjects in the comparative study on alcohol-dependent patients 5 healthy men aged 22-28 were selected. The subjects were healthy according to history, physical examination, CT or MRI examination of the head and blood and urine chemistry. They denied intake of any medication for at least 8 weeks before the experiments. They had abstained from alcohol for at least 1 week before the experiment.

Alcohol-dependent subjects

Five male patients, aged 29-55, satisfied the criteria for alcohol dependence and uncomplicated alcohol withdrawal according to DSM IIIR. Laboratory investigations revealed moderate changes in serum transaminases, serum gamma glutamyl transpeptidase and mean corpuscular volume.

No patient had liver disease, mental disease, systemic illness dependent on medication, history of severe head trauma, stroke or epileptic seizures. The duration of alcohol abuse ranged from 6 to 20 years. The drinking period before last admission ranged from 30 days to several years with an estimated ethanol consumption of 2.7-3.3 g/kg body wt/day. The cumulative lifetime benzodiazepine exposure prior to admission was estimated from the records to be between 0 and 3900 mg of oxazepam equivalents. The patients entered the detoxification program at the Magnus Huss clinic. During the first 3-6 days they received decreasing doses of oxazepam. The total dose of oxazepam during detoxification was 110-640 mg (mean352 mg). The PET-investigation was performed 10-22 days after withdrawal of oxazepam.

Design

In the study on acute alcohol administration each subject participated in two PET experiments on the same day. The first was a placebo experiment performed at about 11 a.m. The second was performed at 2 p.m. after oral intake of 3.0 ml vodka per kg body weight (Absolute® vodka, 40% ethanol, AB Vin & Spritcentralen, Stockholm), corresponding to 1.0 g/kg alcohol (ethanol). The alcohol or placebo were given as three separate drinks 75, 65 and 55 min prior to the injection of [^{11}C]flumazenil.

In the study on alcohol-dependency each subject participated in two PET experiments performed on the same day. In the first [^{11}C]flumazenil of high specific radioactivity was injected (400-3400 Ci/mmol) and in the second [^{11}C]flumazenil of low specific radioactivity (1.0-1.2 Ci/mmol).

Chemistry

[^{11}C]Flumazenil was prepared by N-methylation of the desmethyl compound (Ro 15-5528) using [^{11}C]methyl iodide (Halldin *et al.*, 1988). The alcohol concentration (mM) in plasma was analyzed by head space gas chromatography (Jones and Schubert, 1989).

Imaging procedures

To allow accurate positioning of the brain and to transfer the positioning between MRI and PET an individual fibreglass helmet was made (Bergström *et al.*, 1981; Litton *et al.*, 1993). This system has a reproducibility error in the positioning between experiments of less than 2 mm

(Bergström *et al.*, 1981). MRI was performed on a 1.0 T unit (Siemens - Magnetom) using a standard spin-echo sequence. Regions of interest (ROI´s) were drawn in the MRI images and transferred to the PET images. The MRI images were evaluated with respect to morphological changes and signal changes indicating pathologic processes as described by Litton *et al.*, (1993). The instrument used was the Scanditronix PC2048-15B brain tomograph. The spatial resolution is 4.5 mm full width at half maximum (FWHM) (Litton *et al.*, 1990). A venous cannula was inserted into the right antecubital vein. An arterial cannula was inserted into the left brachial artery. [^{11}C]Flumazenil (150-300 MBq) was injected i.v. as a bolus. Each PET experiment comprised 22 sequential scans varying from 20 sec to 6 min in duration with a total acquisition time of 45-65 min.

Arterial blood was drawn and analyzed for total radioactivity and unchanged [^{11}C]flumazenil. During the first 4 min after radioligand injection, an automatic blood sampling system (ABSS) was used to measure the radioactivity in the arterial blood every second (Eriksson *et al.*, 1988). Manual blood samples (2 ml) were drawn at each scan during the PET experiment. Radioactivity in the manually drawn samples of whole blood and in plasma was measured in a well counter. Plasma from each arterial blood sample was aspirated and frozen at -80°C.

Calculations

The ABSS captured the initial peak of radioactivity in arterial blood after ligand injection. The start time for appearence of radioactivity in arterial blood was defined as the time at which the leading edge of the curve reached 10% of its maximum value.

The radioactivity (nCi/ml) in a ROI was obtained from reconstructed images of each scan, corrected for ^{11}C-decay (20.3 min) and plotted versus time. To obtain fitted uptake curves a sum of three exponentials was fitted to the time-activity data in a least square sense as described by (Pauli *et al.*, 1992).

The radioactivity in pons, a region with low density of BZ receptors, was used as an estimate of the free ligand concentration (C_f). Specific ligand binding (C_b) to BZ receptors in neocortex and cerebellum was calculated as the difference between the radioactivity (C_t) in these regions and the radioactivity in pons (C_f)

$$C_b = C_t - C_f$$

In the study on acute alcohol administration the difference between the alcohol and placebo experiment was formed for each subject. Student's t-test was used to test differences.

In the study on alcohol-dependent subjects the receptor density, B_{max}, and affinity, K_d, were calculated as described previously (Farde *et al.*, 1986; 1989; Persson *et al.*, 1989) The f-test was used to test the hypothesis of equal variances for the B_{max} and K_d values of the healthy volunteers and the alcohol-dependent patients. The test was made for all ROIs. Fisher's exact test was used to test if signs of atrophy were more prevailent among alcohol-dependent patients. The Student's t-test was used to test if the ventricular size differed between the two groups.

Results

Acute alcohol administration

The concentration of alcohol in plasma was 18.9 ± 5.6 mM (mean $\pm$SD, n=4) at the start of the PET experiment after alcohol administration. In all four subjects alcohol induced a similar change in the plasma radioactivity curves. The radioactivity peak was higher, narrower and occurred earlier than in the placebo experiments (Fig. 1). The mean start time in the alcohol experiments was 8 ± 3 sec (mean $\pm$SD; n=4) earlier than in the control experiments (p< 0.02).

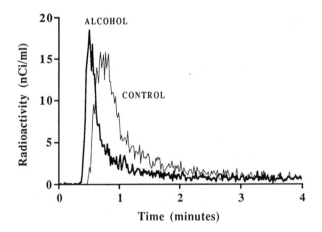

Fig. 1. Radioactivity in arterial blood (nCi/ml) of a healthy man after injection i.v. of 150 MBq [^{11}C]Flumazenil before and after alcohol administration.

The mean regional uptake of radioactivity in the brain after i.v. injection of [^{11}C]flumazenil is shown in Fig. 2. After alcohol there was a tendency towards lower uptake between 2 and 30

148

min after injection of [^{11}C]flumazenil in all regions examined. After 30 minutes the radioactivity was similar in the control and alcohol experiments.

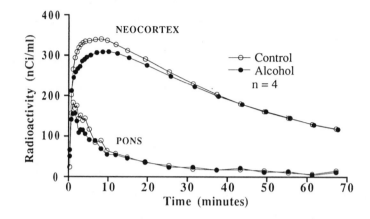

Fig. 2. Effect of alcohol on radioactivity in the neocortex and pons in four healthy men after i.v. administration of [^{11}C]flumazenil. Average curves and standard deviations are plotted.

At time of equilibrium, the binding ratio for neocortex, Cb/Cf, was 5.0 $\pm$ 0.9 and 5.2$\pm$ 1.3 (mean $\pm$SD; n=4) before and after alcohol administration respectively. The corresponding values for cerebellum were 2.9 $\pm$ 0.6 and 2.9 $\pm$ 0.5. Alcohol had no statistically significant influence on the binding ratios of [^{11}C]flumazenil. For further details, see Pauli *et al.* (1992).

Study on alcohol-dependent subjects

Examination by MRI indicated that the alcoholic patients had a tendency for more signs of atrophy (Fisher's exact test, p=0.08) and a preponderance for ventricular enlargement for both the frontal horn index (t-test, p<0.10) and the transverse ventricular diameter (t-test, p<0.10). The BZ-receptor density (B_{max}), affinity (K^d) and the ratio of specific binding to free radioligand concentration (C_b/C_f), are presented for frontal and cerebellar cortex in Table I. Values for the temporal and occipital cortex were similar to those for the frontal cortex (Litton *et al.*, 1993). There was no significant difference between the mean B_{max} values of the two groups.

Three alcohol-dependent patients had at least 20 % lower B_{max} values as compared to the controls in all regions examined. The other two patients had at least 25 % higher B_{max} values.

Table I. Individual B_{max}, K_d and Cb/Cf values in five healthy controls and in five alcohol-dependent subjects

	Frontal Cortex			Cerebellum		
Healthy controls age (years)	B_{max}	K_d	Cb/Cf	B_{max}	K_d	C_b/C_f
24	85	22	3.8	44	18	2.4
28	71	15	4.6	44	17	2.6
22	84	33	2.5	34	20	1.7
27	74	18	3.2	51	16	2.5
25	67	21	4.0	45	21	2.5
Mean	76	22	3.6	44	18	2.3
SD	7.8	6.8	0.8	5.9	2.1	0.4
Alcohol-dependent						
39	66	30	2.2	19	8	2.2
29	58	24	2.5	16	14	1.1
32	66	18	3.6	38	18	2.1
55	108	33	3.2	61	24	2.5
33	136	34	4.0	54	22	2.5
Mean	87	28	3.1	38	17	2.1
SD	34	6.7	0.7	20	6.3	0.6

This is reflected in the SD of the B_{max} values for the group of healthy volunteers which was 10% whereas the alcohol-dependent had a significantly ($p < 0.05$) greater SD (40%). For the K_d values no significant difference in SD was found between the groups.

Discussion

Acute effects of alcohol

Using PET the brain disposition of [^{11}C]flumazenil was studied in four healthy volunteers after ingestion of a high dose of alcohol or give plasma concentrations corresponding to alcohol intoxication. The results were consistent in all subjects. Alcohol did not increase either total radioactivity uptake or specifically bound BZ receptor related radioactivity in the brain. The results do not support the view that alcohol induced effects are related to altered BZ receptor density or affinity.

However, analysis of [^{11}C]flumazenil time-activity curves indicated some minor alterations. Alcohol had a small but significant effect on the time curve for radioactivity in arterial blood. After alcohol administration, the plasma radioactivity peak was higher, more narrow and occurred earlier than in the control experiments. These effects might be caused by an increased peripheral blood flow which causes an earlier start time and a lesser spread of the plasma radioactivity peak.

For all subjects there was a consistent delay of the time t_{max} for the maximal total radioactivity uptake in neocortex and cerebellum after alcohol as compared to that in the control experiment. This occurred in spite of the fact that the plasma peak in the alcohol experiments occurred earlier than in the control experiments. The delayed brain uptake may reflect the alcohol induced reduction in cerebral blood flow which has been reported in several studies (Mathew and Wilson, 1991).

Alcohol dependency

Central BZ receptor binding was determined in 5 healthy male volunteers and 5 male alcohol-dependent patients. The mean B_{max} values for the healthy volunteers was similar to the values found previously for [^{11}C]flumazenil (Persson *et al.*, 1989) and for [^{3}H]flumazenil binding to benzodiazepine receptors in human postmortem brain homogenates (Kopp *et al.*, 1990; Hall *et al.*, 1992).

BZ receptor densities determined by [^{3}H]flunitrazepam binding in vitro has been reported to be reduced in post mortem brain tissue of alcoholic patients as compared to normal controls (Freud and Ballinger, 1988). This study in vivo did not confirm that there is a general reduction in BZ-receptor binding in the brain of alcohol-dependent patients.

Three of five alcohol-dependent subjects had B_{max} values lower than the range of the control values whereas two of them had higher values. B_{max} variance for the alcoholics was found to be significantly higher for all ROIs compared to the variance for healthy volunteers ($p < 0.05$). The duration of life-time alcohol problems, hereditary background and type of alcoholism were similar in subjects with high and low BZ receptor binding. Alcoholics with higher B_{max} values had clinically more pronounced withdrawal symptoms and received higher doses of oxazepam during detoxification. It remains unclear whether the administration of oxazepam in 3-6 days could induce long-term upregulation in benzodiazepine receptor numbers in alcoholics. Animal experiments do not support the view that upregulation is likely (Gallager *et al.*, 1984). Further studies are required to determine whether or not the increased variance in receptor density is an indication of the existence of two different groups of alcoholics.

Ro 15-4513, a partial inverse agonist at the benzodiazepine (BZ) receptor site antagonizes the effects of ethanol in experimental studies in rats and has affinity to the GABA-complex with a6 subunits (Bonetti *et al.*, 1985; Fadda *et al.*, 1987; Bonetti *et al.*, 1989). We have labelled Ro 15-4513 with ^{11}C for PET (Halldin *et al.*, 1992). In autoradiographic studies *in vitro* [^{11}C]Ro 15-4513 bound specifically in the neocortex, the basal ganglia and the cerebellar cortex of the human brain (Hall *et al.*, 1992). Flumazenil and clonazepam inhibited the binding in cerebral regions, but a significant proportion of binding in the cerebellum was not inhibited by these agents. This proportion may represent α_6-containing BZ receptors. PET-examination of [^{11}C]Ro 15-4513 binding in Cynomolgus monkeys demonstrated high uptake of radioactivity in neocortex. The uptake of radioactivity was markedly displaced by high doses of unlabelled Ro 15-4513 or clonazepam. [^{11}C]Ro 15-4513 has recently been used to demonstrate specific BZ-receptor binding in the human brain (Inoue *et al.*, 1992). The antagonistic effect of Ro15-4513 on ethanol induced effects makes [^{11}C]Ro 15-4513 an interesting radioligand for future PET-studies on BZ-receptor binding in the brain of alcohol-dependent subjects.

Acknowledgements

The precursor for [^{11}C]flumazenil, Ro 15-1528, was kindly supplied by Dr. W. Hunkeler, Hoffman La Roche, Basle, Switzerland. The assistance by members of the Karolinska PET center is gratefully acknowledged. The study was supported by grants from the Bank of Sweden Tercentenary Foundation (0427), The Swedish Alcohol Research Fund (90/32), the Swedish Medical Research Council (03560), and the Karolinska Institute.

References

Adams, K., Gilman, S., Koeppe, R., Kluin, K., Brunberg, J., Dede, D., Berent, S. and Kroll, P. (1993) Neuropsychological deficits are correlated with frontal hypometabolism in Positron Emission Tomography studies of older alcoholic patients. *Alcohol Clin. Exp. Res.* 17: 205-210.

Bergström, M., Boëthius, J., Eriksson, L., Greitz, T., Ribbe, T. and Widén, L. (1981) Head fixation device for reproducible position alignment in transmission CT and positron emission tomography. *J. Comput. Assist. Tomogr.* 5: 136-141.

Bonetti, E., Buckard, W., Gabl, M., Hunkeler, W., Lorez, H-P., Martin, J., Moehler, H., Osterrieder, W., Pieri, L., Pole, P., Richards, J., Schaffner, R., Scherschlicht, R., Schoch, P. and Haefely, W. (1989) Ro 15-4513 partial inverse antagonism at BZR and interaction with ethanol. *Pharmacol. Biochem. Behav.* 31: 733-749.

Bonetti. E., Burkard, W., Gabl, M. and Möhler, H. (1985) The partial inverse benzodiazepine agonist Ro 15-4513 antagonizes acute ethanol effects in mice and rats. *Br. J. Pharmacol.* 86: 463.

Cooper, M., Metz, J., de Wit, H. and Mukherjee, J. (1993) From the cradle to the grave: Alcohol and its effects upon the brain. *J. Nucl. Med.* 34: 798-803.

de Wit, H., Metz, J. and Cooper, M. (1992) Studying psychoactive drugs with positron emission tomography: Relationships between mood and metabolic rate. In: Frascella, J. and Brown, R. (eds.): *Neurobiological approaches to brain behavior interaction*. U.S. Department of Health and Human Services, Rockville, MD, pp. 117-134.

Eriksson, L., Holte, S., Bohm, C., Kesselberg, M. and Hovander, B. (1988) Automated blood sampling systems for positron emission tomography. *IEEE Trans. Nucl. Sci.* 35: 703-707.

Fadda, F., Mosca, E., Colombo, G. and Gessa, G. (1987) Protection against ethanol mortality in rats by the imidazobenzodiazepine Ro 15-4513. *Eur. J. Pharmacol.* 136: 265-266.

Farde, L., Eriksson, L., Blomquist, G. and Halldin, C. (1989) Kinetic analysis of central [^{11}C]raclopride binding to D_2-dopamine receptors studied by PET - A comparison to the equilibrium analysis. *J. Cereb. Blood Flow Metab.* 9: 696-708.

Farde, L., Hall, H., Ehrin, E. and Sedvall, G. (1986) Quantitative analysis of D_2 dopamine receptor binding in the living human brain by PET. *Science* 231: 258-261.

Frascella, J. and Brown, R. (1992) Studying psychoactive drugs with positron emission tomography: Relationships between mood and metabolic rate. *NIDA Research Monogr.* 124: 117-134.

Freud, G. and Ballinger, W. (1988) Decrease of benzodiazepine receptors in frontal cortex of alcoholics. *Alcohol* 5: 275-282.

Gallager, D., Lakoski, J., Gonsalves, S. and Rauch, S. (1984) Chronic benzodiazepine treatment decreases postsynaptic GABA sensitivity. *Nature* 308: 74-77.

Hall, H., Litton, J., Halldin, C., Kopp, J. and Sedvall, G. (1992) Studies on the binding of [3H]flumazenil and [3H]sarmazenil in post-mortem human brain. *Human Psychopharmacol.* 7: 367-377.

Halldin, C., Högberg, T., Hall, H., Karlsson, P., Hagberg, C., Ström, P. and Farde, L. (1992) [11C]Ro 15-4513, a ligand for visualization of benzodiazepine receptor binding. Preparation, autoradiography and positron emission tomography. *Psychopharmacol.* 108: 16-22.

Halldin, C., Stone-Elander, S., Thorell, J.-O., Persson, A. and Sedvall, G. (1988) ^{11}C-labelling of Ro 15-1788 in two different positions, and also ^{11}C-labelling of its main metabolite Ro 15-3890, for PET studies of benzodiazepine receptors. *Appl. Radiati. Isot.* 39(9): 993-997.

Inoue, O., Suhara, T., Itoh, T., Kobayashi, K., Suzuki, K. and Tateno, Y. (1992) In vivo binding of [11C]Ro15-4513 in human brain measured with PET. *Neurosci. Lett.* 145: 133-136.

Jones, A. and Schubert, J. (1989) Computer-aided head space gas chromatography applied to blood-alcohol analysis: Importance of online process control. *J. Forensic Sci.* 34: 1116-1127.

Kessler, R., Parker, E., Clark, C., Martin, P., George, D. and Weingartner, H. (1984) Regional cerebral glucose metabolism in patient with alcoholic Korsakoff's syndrome. *Neuroscience* 10: 541.

Kopp, J., Hall, H., Persson, A. and Sedvall, G. (1990) Temperature dependence of [³H]Ro 15-1788 binding to benzodiazepine receptors in human brain homogenates. *J. Neurochem.* 55: 1310-1315.

Litton, J., Holte, S. and Eriksson, L. (1990) Evaluation of the Karolinska new positron camera system; the Scanditronix PC2048-15B. *IEEE Trans. Nucl. Sci.* 37: 743-748.

Litton, J., Neiman, J., Pauli, S., Farde, L., Hindmarsh, T., Halldin, C. and Sedvall, G. (1993) PET-analysis of [11C]flumazenil binding to benzodiazepine receptors in chronic alcohol-dependent men and healthy controls. *Psychiatry Res.* 50: 1-13.

Lofti, J. and Meyer, J. (1989) Cerebral hemodynamics and metabolic effects of chronic alcoholism. *Cerebrovasc. Brain Metab. Rev.* 1: 2-25.

Mathew, R. and Wilson, W. (1991) Substance abuse and cerebral blood flow. *Am. J. Psychiatry* 148: 292-305.

Mazière, M., Hantraye, P,. Guibert, B., Kaijima, M., Prenant, C., Sastre, J., Crouzel, M., Naquet, R. and Comar, D. (1984) RO 15 1788-[11] CA specific radioligand for an "*in vivo* " study of central bezodiazepine receptors, by positron emission tomography. *Clin. Neuropharmacol.* 7: 662-663.

Pauli, S., Liljequist, S., Farde, L., Swahn, C.G., Halldin, C., Litton, J.E. and Sedvall, G. (1992) PET analysis of alcohol interaction with the brain disposition of [^{11}C]flumazenil. *Psychopharmacol.* 107: 180-185.

Persson, A., Ehrin, E., Eriksson, L., Farde, L., Hedström, C-G., Litton, J-E., Mindus, P. and Sedvall, G. (1985) Imaging of ^{11}C-labelled Ro 15-1788 binding to benzodiazepine receptors in the human brain by positron emission tomography. *J. Psychiat. Res.* 19: 609-622.

Persson, A., Pauli, S., Halldin, C., Stone-Elander, S., Farde, L., Sjögren, I. and Sedvall, G. (1989) Saturation analysis of specific [^{11}C]Ro 15-1788 binding in the human neocortex using positron emission tomography. *Hum. Psychopharmacol.* 4: 21-31.

Sachs, H., Russel, J., Christman, D. and Cook, B. (1987) Alteration of regional cerebral glucose metabolic rate in non-Korsakoff chronic alcoholism. *Arch. Neurology* 44: 1242-1251.

Samson, H. and Harris, A. (1992) Neurobiology of alcohol abuse. *Trends Pharmacol. Sci.* 13: 206-211.

Volkow, N., Hitzemann, R., Wolf, A., Logan, J., Fowler, J., Christman, D., Dewey, S., Schlyer, D., Burr, G,. Vitkun, S. and Hirschhowitz, J. (1990a) Acute effects of ethanol on regional brain glucose metabolism and transport. *Psychiatry Res. Neuroimaging.* 35: 39-48.

Volkow, N., Hitzemann, R., Wolf, A., Logan, J., Fowler, J., Christman, D., Dewey, S., Schlyer, D., Burr, G., Vitkun, S. and Hirschhowitz, J. (1990b) Decreased brain metabolism in neurologically intact healthy alcoholics. *Am. J. Psychiatry* 149: 1016-1022.

Wik, G., Borg, S., Sjögren, I., Wiesel, F., Blomqvist, G., Borg, J., Greitz, T., Nybäck, H., Sedvall, G., Stone-Elander, S. and Widén, L. (1988) PET-determination of regional cerebral glucose metabolism in alcohol-dependent men and healthy controls using [^{11}C]glucose. *Acta Psychiatr. Scand.* 78: 234-241.

Toward a Molecular Basis of Alcohol Use and Abuse
ed. by B. Jansson, H. Jörnvall, U. Rydberg, L. Terenius & B. L. Vallee
© 1994 Birkhäuser Verlag Basel/Switzerland

Serotonin, violent behavior and alcohol

Markku Linnoila, Matti Virkkunen[2], Theodore George, Michael Eckardt, James D. Higley,
David Nielsen[1], and David Goldman[1]

*Laboratories of Clinical Studies and [1]Neurogenetics, Division of Intramural Clinical and Biological Research,
National Institute on Alcohol Abuse and Alcoholism, Bethesda, Maryland, USA, [2]Department of Psychiatry,
University of Helsinki, Helsinki, Finland.*

Summary

At the NIAAA intramural research program, in collaboration with investigators at the Department of Psychiatry,
University of Helsinki, we have mounted an extensive research program on early onset male alcoholism. A central
serotonergic deficit is common among these patients. This finding has led to behavioral, biochemical, physiological
and molecular genetic studies on the serotonin system in early onset, antisocial and violent male alcoholics and in
appropriate control populations. The results of the studies completed by the fall of 1993 are summarized in this
communication.

Introduction

Alcoholism is the most common mental disorder among men (Robbins *et al.*, 1984). Alcohol
abuse and alcoholism are strongly associated with violent crime, especially murder (Murdoch
et al., 1990). Moreover, alcoholics are at a high risk of attempting or committing suicide as
compared to the rest of the population (Roy and Linnoila, 1986). Thus, attempts to elucidate
relationships between alcoholism and violent behavior are potentially of importance for
improving public health.

Subtyping alcoholism

Similar to many other mental disorders, alcoholism is not only common but also heterogenous.
A subtype of particular interest for studying alcoholism and violence are early onset alcoholics
with antisocial behavioral traits. Tarter was among the first to draw attention to the observation
that early onset male alcoholism was often preceded by symptoms of hyperactivity and minimal
brain dysfunction (Tarter *et al.*, 1977). Cloninger *et al.* (1981), in a large-scale adoption study,
defined a male-limited or Type II form of alcoholism, which was characterized by high heritability from fathers to sons, early onset and antisocial behavioral traits. Later, von Knorring *et al.*

(1985) operationalized clinical criteria to permit classification of alcoholics into Types I and II.

Biochemical concomitants of alcoholic subtypes

In additional psychobiological studies, von Knorring *et al.* (1985) established that compared to healthy volunteers and Type I alcoholics, Type II alcoholics had low platelet monoamine oxidase (MAO) activity. Platelet MAO activity has been reported to show a low positive correlation with cerebrospinal fluid (CSF) 5-hydroxyindoleacetic acid (5-HIAA) concentration, and suggested to reflect "functional capacity" of the central serotonin system (Oreland, 1980). Low platelet MAO activity has also been reported to be associated with personality characteristics of poor impulse control and monotony avoidance (Schalling *et al.*, 1987).

In the late 1970's, before the description of the Type I and II subtyping of alcoholics, Ballenger et al. investigated CSF monoamine metabolite concentrations in short-term abstinent, relatively young male alcoholics (Ballenger *et al.*, 1979). They found that, compared to age and sex-matched controls who were admitted with a suspected neurological condition, but none was found, the alcoholics had significantly lower mean CSF 5-HIAA concentration. In our latest unpublished studies in Bethesda, which address specifically the issue of CSF 5-HIAA in early versus late onset male alcoholics, we find that early onset males have a low mean CSF 5-HIAA concentration. Moreover, late onset male alcoholics have higher CSF concentrations of anxiety-associated neuropeptides corticotropin releasing hormone (CRF) and diazepam binding inhibitor (DBI) than early onset male alcoholics (Roy *et al.*, 1991).

Serotonin receptor challenges in early vs. late onset male alcoholics

Experimental animals performing a drug discrimination task cannot reliably distinguish the serotonin 1C and 2 receptor agonist trifluoromethylpiperazine (TFMPP) from ethanol (Schechter, 1988). Based on this finding and the Ballenger *et al.* (1979). CSF study, we decided to investigate effects of meta-chlorophenylpiperazine (mCPP), which is pharmacologically and chemically very similar to TFMPP in male alcoholics. We divided our sample into Types I and II according to the von Knorring criteria (von Knorring *et al.*, 1985) and collected an additional sample of age and sex-matched healthy volunteers. At the time of the study, the alcoholics had been abstinent for a minimum of three weeks. All subjects received a placebo infusion followed by an mCPP infusion. They were rated for desire to drink by a psychiatrist blind to the subtype and the time of starting the mCPP infusion. Visual analogue scales were filled out by the subjects. Blood was sampled through an intravenous catheter continuously during the procedure. Compared to the healthy volunteers, alcoholics regardless of subgrouping, had blunted adrenocorticotropin (ACTH) responses to mCPP. Type II alcoholics reported significantly more strong urges to drink alcohol or "craving" during the mCPP infusion than Type I alcoholics or healthy volunteers. Type I alcoholics, on the other hand, reported more anxiety in response to mCPP than Type II alcoholics or healthy volunteers. Interestingly, our unpublished data show no difference between alcoholics, regardless of subgrouping, and healthy volunteers in behavioral and neuroendocrine responsiveness to low dose clomipramine infusions. Studies to more precisely define the role of serotonin 1C and 2 receptors in alcoholism are clearly warranted.

Studies on alcoholic violent offenders in Finland

Finns are a large, relatively homogenous, genetic isolate, because the country was inhabited by a small founder population more than 1,000 years ago. Now the population numbers about 5 million. Because of the great differences in language between the Finns and their neighbors, there has been relatively little genetic mixing between them. It is our assumption that the genetic vulnerabilities conducive of the development of alcoholism would be more homogenous in a relative genetic isolate population than in a more mixed population.

We have further narrowed the phenotype by investigating impulsive, violent alcoholics, who based on their behaviors are postulated to represent an extreme group of Type II alcoholics (Virkkunen *et al.*, 1993a). At the biochemical level, our probands are characterized by a very low CSF 5-HIAA concentration of less than 50 pmol/ml. For the past three years, we have been collecting pedigrees of such probands and a male population sample for genetic association and linkage studies. All subjects have been administered the Structured Clinical Interview for DSMIIIR (SCID) I and II (Spitzer *et al.*, 1990) interviews by a research psychiatrist and they have also filled out the Karolinska Scales of Personality (Schalling *et al.*, 1987) and the Michigan Alcoholism Screening Test (Selzer, 1971). Children and adolescents between the ages of 8 and 16 have been administered the Diagnostic Interview for Children and Adolescents

(DICA) (Reich *et al.*, 1990). All subjects have provided a blood sample. Their leukocytes have been harvested and immortalized by transformation with the Epstein-Barr virus. Serotonin receptor, transporter and synthesizing enzyme genes are currently being systematically analyzed in this sample.

In a parallel series of physiological studies on the alcoholic, violent offenders described below we are elucidating pathophysiological phenomena associated with violent behavior under the influence of alcohol.

CSF 5-HIAA as a biological marker
"Neuroanatomy" of 5-HIAA

Quantification of CSF 5-HIAA has been alleged to yield less than useful information, because CSF 5-HIAA was thought to reflect nonspecifically general serotonergic activity within the central nervous sytem. The first piece of information challenging this view comes from a Dutch study on brain-injured victims of traffic accidents. Patients with primarily frontal, closed brain injuries had lower mean CSF 5-HIAA concentrations than patients with brain injuries of similar severity but primarily affecting other parts of the central nervous system (van Woerkom *et al.*, 1977). In a postmortem study, Stanley et al. found a strong positive correlation between lumbar CSF and frontal cortex 5-HIAA concentrations (Stanley *et al.*, 1985). Thus, CSF 5-HIAA concentration may closely parallel serotonin turnover rate in the frontal cortex. If this finding is corroborated, then it becomes easier to understand how CSF 5-HIAA concentration may be positively correlated with impulse control, which has been demonstrated to be a function under the control of frontal cortical structures (Miller, 1992).

Seasonality

In a small scale, cross-sectional study, Brewerton *et al.* (1988) found an apparent seasonal rhythm of CSF 5-HIAA and homovanillic acid (HVA) concentrations in healthy volunteers. The concentrations were highest in the summer and lowest in the late winter-early spring. Such a seasonal rhythm was not found in either American (Roy *et al.*, 1991a) or Finnish (Virkkunen *et al.*, 1993b) alcoholics. It was also lacking in a recent large scale Swedish study (Blennow *et al.*, 1993), which used largely preoperative patients free of mental disorders as subjects and an unknown fraction of them was studied during the summer. Longitudinal studies by Träskman-Benz *et al.* (1993) on suicidal depressed patients have not found significant seasonal variation either. Clearly, more studies are needed to address this issue.

Heritability

Oxenstierna *et al.* (1986), examined CSF monoamine metabolite concentrations in mono and dizygotic twins and found a higher intrapair correlation for 5HIAA in the monozygotic twins. They did not, however, find significant heritability of CSF 5HIAA. In a large scale study on rhesus monkeys, designed to investigate genetic and environmental influences on CSF monoamine metabolities, Higley et al. found significant maternal and paternal genetic contributions to CSF 5HIAA concentration. Rather drastic environmental manipulations such as rearing in peer groups or by an unrelated female monkey did not significantly influence CSF 5HIAA concentrations. Thus, it appears that, at least in the nonhuman primate, CSF 5HIAA may be a useful phenotypic marker (Higley *et al.*, 1993).

Physiological concomitants of low CSF 5HIAA

In rodents the serotonergic input to the suprachiasmatic nucleus is important for the entrainment of endogenous circadian rhythms by light (Morin and Blanchard, 1991). We have previously reported that many alcoholic, impulsive, violent offenders complained of insomnia during their stay on the forensic psychiatry ward (Roy *et al.*, 1986). Thus, we postulated that the insomnia was the consequence of desynchronized diurnal sleep-activity rhythm and investigated this issue by asking the subjects to wear physical activity monitors. We found that offenders with intermittent explosive disorder, who had a low mean CSF 5HIAA concentration, had very disturbed sleep-activity rhythm. Offenders with antisocial personality disorder, who also had low mean CSF 5HIAA concentrations, showed greater physical activity throughout the ten day monitoring period than healthy volunteers. Thus, a low CSF 5HIAA concentration is not per se conducive of a diurnal sleep-activity rhythm disturbance but within certain diagnostic groups, it is associated with such a disturbance (Virkkunen *et al.*, 1993b).

Low blood glucose concentrations after an oral glucose challenge in alcoholic violent offenders with intermittent explosive disorder were reported by Virkkunen et al. in the 1970s

(Roy *et al.*, 1986). The hypoglycemic tendency coexists with low CSF 5HIAA more often than expected by chance but the blood glucose nadir after an oral glucose challenge and CSF 5HIAA do not correlate with each other (Virkkunen *et al.*, 1993b). The lack of correlation may be the result of the blood glucose nadir and circadian rhythm disturbances reflecting primarily hypothalamic and CSF 5HIAA primarily frontal serotonin functions.

Behavior correlates of low CSF 5HIAA

The early literature on CSF 5HIAA in humans emphasized an association between the violent nature of the suicide attempts and interpersonal aggression and low CSF 5HIAA concentrations. A series of studies by us has highlighted the role of serotonin in impulse control (Virkkunen and Linnoila, 1993). Indeed, it appears that low CSF 5HIAA is primarily associated with increased irritability and impaired impulse control in violent alcoholics with antisocial personality disorder or intermittent explosive disorder (Virkkunen *et al.*, 1993a). Among healthy volunteers low CSF 5HIAA is associated with increased acting out hostility (Roy *et al.*, 1988). The consequences of the impaired impulse control associated with low CSF 5HIAA are clearly different depending on the diagnostic group under study.

We performed discriminant function analyses between impulsive and nonimpulsive alcoholic offenders and alcoholic violent offenders and healthy volunteers in our most recent unpublished studies. These analyses demonstrated that a low CSF 5HIAA was clearly a concomitant of impaired impulse control and CSF free testosterone explained most of the variance for interpersonal aggressiveness or violence (Virkkunen *et al.*, 1993b).

CSF biochemistries and predicting future behavior

If a low CSF 5HIAA and a low blood glucose nadir are biochemical traits with strong behavioral correlates then they could be expected to predict future behaviors to some extent. This reasoning has been supported for low CSF 5HIAA predicting the risk of suicide attempts in patients with unipolar depression (Träskman *et al.*, 1981) and in alcoholic, violent offenders (Virkkunen *et al.*, 1989b). The combination of low CSF 5HIAA and a low blood glucose nadir after an oral glucose challenge were somewhat predictive of recidivism of violent crimes after release from prison (Virkkunen *et al.*, 1989a). The ages of the subjects may be a critical variable contributing to the prediction of recidivist criminality by alcoholic, violent offenders, because patients exhibiting antisocial behaviors are believed to "burn out" in their 40s. It remains to be seen whether the "burning out" phenomenon is more strongly associated with reduced CSF free testosterone or increased CSF 5HIAA concentrations or a combination of the two with age. A higher mortality of the more impulsive alcoholic offenders could also be conducive of an apparent "burning out" over time.

Molecular biology of tryptophan hydroxylase in Finnish violent alcoholics

Recently, we have discovered a tryptophan hydroxylase polymorphism which appears to be associated with low CSF 5HIAA and increased lifetime risk of suicide attempts among alcoholic, impulsive violent offenders (Nielsen *et al.*, 1993).

Future directions

We are breeding nonhuman primates for extremes of CSF 5HIAA concentrations and manipulating their development to investigate the role of frontal serotonin turnover in specific vulnerabilities towards abnormal behaviors including alcohol abuse. As informative families of the Finnish alcoholic violent offenders become available we will be performing genetic linkage analyses. Lastly, we are developing serotonergic ligands for functional brain imaging studies.

References

Ballenger, J., Goodwin, F., Major, L. and Brown, G. (1979) Alcohol and central serotonin metabolism in man. *Arch. Gen. Psychiatry* 36: 224-227.

Blennow, K., Wallin, A., Gottfries, C.-G., Karlsson, I., Mansson, J.-E., Skoog, I., Wikkelso, C. and Svennerholm, L. (1993) Cerebrospinal fluid monoamine metabolites in 114 healthy individuals 18-88 years of age. *Eur. Neuropsychopharmacol.* 3: 55-62.

Brewerton, R.D., Berrittini, W.H., Nurnberger, J.I. and Linnoila, M. (1988) Analysis of seasonal fluctuations of CSF monoamine metabolites and neuropeptides in normal controls: findings with 5-HIAA and HVA. *Psychiat. Res.* 23: 257-265.

Cloninger, C.R., Bohman, M. and Sigvardsson, S. (1981) Inheritance of alcohol abuse: cross-fostering analysis of adopted men. *Arch. Gen. Psychiatry* 38: 861-868.

Higley, J.D., Thompson, W.T., Champoux, M., Goldman, D., Hasert, M.F., Kraemer, G.W., Scanlan, J.M., Suomi, S.J. and Linnoila, M. (1993) Paternal and maternal genetic and environmental contributions to CSF monoamine metabolite concentrations in rhesus monkeys (macaca mulatta). *Arch. Gen. Psychiatry* 50: 615-623.

162

Miller, L.A. (1992) Impulsivity, risk-taking, and the ability to synthesize fragmented information after frontal lobectomy. *Neuropsychol.* 30: 69-79.

Morin, L.P. and Blanchard, J. (1991) Depletion of brain serotonin by 5, 7 DHT modifies hamster circadian rhythm response to light. *Brain Res.* 566: 173-185.

Murdoch, D., Pihl, R.O. and Ross, D. (1990) Alcohol and crimes of violence: Present issues. *Int. J. Addict.* 25: 1065-1081.

Nielsen, D.A., Goldman, D., Virkkunen, M., Tokola, R., Rawlings, R. and Linnoila, M. (1993) Low 5-HIAA and a suicide attempts associate with a polymorphism of the tryptophan hydroxylase gene. *Arch. Gen. Psychiatry*, in press.

Oreland L. (1980) Monoamine oxidase activity and affective illness. *Acta Psychiat. Scand.* 280: 41-47.

Oxenstierna, G., Edman, G., Iselius, L., Oreland, L., Ross, S.B. and Sedvall, G. (1986) Concentrations of monoamine metabolites in the cerebrospinal fluid of twins and unrelated individuals: a genetic study. *J. Psychiat. Res.* 20: 19-29.

Reich, W., Welner, Z. and Herjanic, B. (1990) DICA-R, Multi-Health Systems Inc, North Tonawanda, NY.

Robbins, L.N., Helzer, J.E., Weissman, M.M., Orvaschel, H., Gruenberg, E., Burke, J.D. and Regier, D.A. (1984) Lifetime prevalence of specific psychiatric disorders in three sites. *Arch. Gen. Psychiatry* 41: 949-958.

Roy, A., Adinoff, B., DeJong, J. and Linnoila, M. (1991) Cerebrospinal fluid variables among alcoholics lack seasonal variation. *Acta Psychiat. Scand.* 84: 579-582.

Roy, A., Adinoff, B. and Linnoila, M. (1988) Acting out hostility in normal volunteers: negative correlation with CSF 5-HIAA levels. *Psychiat. Res.* 24: 187-194.

Roy, A., DeJong, J., Lamparski, D., Adinoff, B., George, D.T., Moore, V., Garnett, D., Kerich, M. and Linnoila, M. (1991) Mental disorders among alcoholics: Relationship to age of onset and cerebrospinal fluid neuropeptides. *Arch. Gen. Psychiatry* 48: 423-427.

Roy, A. and Linnoila, M. (1986) Alcoholism and suicide. *Suicide and Life-Threatening Behavior* 16: 162-191.

Roy, A., Virkkunen, M., Guthrie, S. and Linnoila, M. (1986) Indices of serotonin and glucose metabolism in violent offenders, arsonists and alcoholics. In: Mann, J.J. and Stanley, M. (eds.): *Psychobiology of Suicidal Behavior*. Annals of the New York Academy of Sciences. Vol. 487, N.Y., pp. 202-220.

Schalling, D., Åsberg, M., Edman, G. and Oreland, L. (1987) Markers for vulnerability to psychopathology: Temperament traits associated with platelet MAO activity. *Acta Psychiatr. Scand.* 76: 72-182.

Schechter, M.D. (1988) Use of TFMPP stimulus properties as a model of 5-HTIB receptor activation. *Pharmacol. Biochem. Behav.* 31: 53-57.

Selzer, M.L. (1971) The Michigan Alcoholism Screening Test: The quest for a new diagnostic instrument. *Am. J. Psychiatry* 127: 1653-1658.

Spitzer, R.L., Williams, J.B.W., Gibbon, M. and First, M.B. (1990) *Structured Clinical Interview for DSM IIIR*. American Psychiatric Press Inc., Washington, D.C.

Stanley, M., Träskman-Benz, L. and Dorovini-zis, K. (1985) Correlations between aminergic metabolites simultaneously obtained from human CSF and brain. *Life Sci* 37: 1279-1286.

Tarter, R., McBride, H., Buonpane, N. and Schneider, D. (1977) Differentiation of alcoholics according to childhood history of minimal brain dysfunction, family history and drinking pattern. *Arch. Gen. Psychiatry* 34: 761-768.

Träskman, L., Åsberg, M., Bertilsson, L. and Sjöstrand, L. (1981) Monoamine metabolites in CSF and suicidal behavior. *Arch. Gen. Psychiatry* 38: 631-636.

Träskman-Benz, L., Alling, C., Alsen, M., Regnell, G., Simonsson, P.and Ohman, R. (1993) The role of monoamines in suicidal behavior. *Acta Psychiatr. Scand.* 371: 45-47.

van Woerkom, T.C.A.M., Teelken, A.W. and Minderhound, J.M. (1977) Difference in neuro-transmitter metabolism in frontotemporal lobe contusion and diffuse cerebral contusion. *Lancet* i: 812-813.

Virkkunen, M., DeJong, J., Bartko, J., Goodwin, F.D., Linnoila, M. (1989a) Relationship of psychobiological variables to recidivism in violent offenders and impulsive fire setters: a follow up study. *Arch. Gen. Psychiatry* 46: 600-603.

Virkkunen, M., DeJong, J., Bartko, J. and Linnoila, M. (1989b) Psychobiological concomitants of history of suicide attempts among violent offenders and impulsive fire setters. *Arch. Gen. Psychiatry* 46: 604-606.

Virkkunen, M. and Linnoila, M. (1993) Brain serotonin, Type II alcoholism and impulsive violence. *J. Stud. Alcohol* 11: 163-169.

Virkkunen, M., Kallio, E., Rawlings, R., Tokola, R., Poland, R., Guidotti, A., Nemeroff, C., Bissette, G., Kalogeras, K., Karonen, S. and Linnoila, M. (1993a) Personality profiles and state aggressiveness in Finnish violent offenders, impulsive fire setters, and healthy volunteers. *Arch. Gen. Psychiatry*, in press.

Virkkunen, M., Rawlings, R., Tokola, R., Poland, R., Guidotti, A., Nemeroff, C., Bissette, G., Kalogeras, K., Karonen, S.L. and Linnoila, M. (1993b) CSF biochemistries, glucose metabolism, and diurnal activity rhythms, in violent offenders, impulsive fire setters, and healthy volunteers. *Arch. Gen. Psychiatry*, in press.

von Knorring, A.L., Bohman, M., von Knorring, L. and Oreland, L. (1985) Platelet MAO activity as a biological marker in subgroups of alcoholism. *Acta Psychiatr. Scand.* 72: 51-58.

Toward a Molecular Basis of Alcohol Use and Abuse
ed. by B. Jansson, H. Jörnvall, U. Rydberg, L. Terenius & B. L. Vallee
© 1994 Birkhäuser Verlag Basel/Switzerland

Neuropeptides and alcohol addiction in monkeys

J.M. van Ree, M. Kornet[1], C. Goosen[1]

Department of Pharmacology, Rudolf Magnus Institute, Utrecht University, Vondellaan 6, 3521 GD Utrecht, The Netherlands and [1]TNO-Primate Center, Rijswijk, The Netherlands

Summary

Neuropeptides have been implicated in experimental drug addiction. Desglycinamide (Arg^8) vasopressin (DGAVP) attenuates heroin and cocaine intake during initiation of drug selfadministration in rats. β-Endorphin is self-administered in rats and a role of endogenous opioids in cocaine reward has been proposed. The present studies deal with voluntary alcohol consumption in monkeys under free choice conditions. Monkeys initiated alcohol drinking within a few days and after a stable drinking pattern was acquired increased their ethanol consumption during a short period following interruption of the alcohol supply (relapse). The alcohol drinking behavior seems under the control of reinforcement principles. DGAVP reduced the acquisition of alcohol drinking in the majority of treated monkeys. Initiation of alcohol drinking induced modifications in neuroendocrine homeostasis e.g. an increased plasma β-endorphin. Both the opioid antagonist naltrexone and the opioid agonist morphine dose-dependently decreased alcohol intake during continuous supply and after imposed abstinence. The monkeys were more sensitive to both drugs after imposed abstinence. The effects are interpreted in the context of the endorphin compensation hypothesis of addictive behavior. It is suggested that endorphins may be particularly implicated in craving for addictive drugs and in relapse of addictive behavior.

Introduction

For a growing number of peptide hormones it has become clear that they exert a direct influence on the central nervous system in addition to their classical peripheral activities on endocrine organs (Van Ree *et al.*, 1978; De Wied and Jolles, 1982). Detailed structure activity studies have shown that different parts of the hormone molecule may contain the information for peripheral and central effects and for the various influences on brain function. The principle that peptide hormones may be the precursors for a number of small peptides with different activities on the nervous system, is known as the neuropeptide concept (De Wied, 1969).

Neuropeptides derived from the neurohypophyseal hormone vasopressin and the precursor molecule pro-opiomelanocortin (POMC) have been implicated in experimental drug addiction (Van Ree, 1983). Initially, experiments have been focused on the influence of desglycinamide (Arg^8) vasopressin (DGAVP) on intravenous heroin selfadministration in rats. DGAVP is practically devoid of vasopressin's endocrine effects, but still exhibits certain effects of vasopressin on brain processes (Van Ree *et al.*, 1978). It was found that subcutaneous treatment with DGAVP decreased heroin intake during the initiation of heroin selfadministration (Van Ree and De Wied, 1977a). Subsequent studies have shown that the peptide is also effective following oral or intracerebroventricular administration (for ref. see Van Ree, 1983, 1987). DGAVP decreased ventral tegmental brain selfstimulation and the initiation of opiate selfadministration directly into the ventral tegmental area. The peptide also decreased cocaine selfadministration behavior during the initial phase of this behavior (Van Ree *et al.*, 1988). Intracerebroventricular administration of antiserum against vasopressin enhanced heroin and cocaine intake during the initiation of drug selfadministration, suggesting that vasopressin plays a physiological role in this behavior (Van Ree and De Wied, 1977b; De Vry *et al.*, 1988). From these experiments it was concluded that DGAVP attenuates the positive reinforcing efficacy of drugs, probably through interaction with mesolimbic reward pathways (Wise, 1978). The effects of DGAVP are only present under certain conditions i.e. during the initial phase of drug selfadministration and not during maintenance, and when reward control over behavior is changed. On basis the of these findings, two clinical trials (double blind and placebo controlled) have been carried out in which DGAVP was given to mild to moderate heroin addicts during the initial phase of a methadon detoxification programme (Van Beek-Verbeek *et al.*, 1983; Fraenkel *et al.*, 1983). Both heroin and cocaine use was found to be decreased by the peptide treatment, suggesting that drug seeking behavior in general was attenuated.

Pro-opiomelanocortin is the precursor molecule of adrenocorticotrophic hormone (ACTH) and β-lipotropine (β-LPH). β-LPH is in turn a precursor molecule for the potent opiate-mimicking peptide, β-endorphin. Soon after the discovery of endorphins, these peptides have been implicated in drug addiction. Experiments on β-endorphin intracerebroventricular selfadministration in rats have demonstrated that β-endorphin acts as an endogenous positive reinforcer of behavior, and thus might exert intrinsic control over behavior, including selfadministration of addictive drugs (Van Ree *et al.*, 1979). Evidence that endogenous opioids are involved in reward processes has been provided by the effects of opiate agonists and antagonists on brain selfstimulation (for ref. see Schaefer, 1988). Thus, endogenous opioids may be involved in selfadministration of opiate and non-opiate drugs. Results from animal studies in which the effect of opioid antagonism on cocaine selfadministration was examined suggest a role of endogenous opioids specifically in the initiation of intravenous cocaine selfadministration, and less so in maintenance (Carroll *et al.*, 1986, De Vry *et al.*, 1989; Ettenberg *et al.*, 1982). The opioid antagonist naltrexone induced a rightward shift of the dose response curve for cocaine reward during the initiation of cocaine selfadministration (De Vry *et al.*, 1989). In addition, β-endorphin-immunoreactivity was decreased in the anterior part of the limbic system during initiation of cocaine or heroin selfadministration in rats (Sweep *et al.*, 1989). Interestingly, this effect was present when the desire for the drug was high, and not when the animals had just selfadministered the drug, which might implicate β-endorphin in craving.

In the present experiments, the influence of DGAVP and the involvement of endogenous opioids were examined in another model of addictive behavior i.e. alcohol drinking in monkeys. The relationship between neurohypophyseal neuropeptides and alcohol has been studied only with respect to the development and maintenance of tolerance and physical dependence (Crabbe and Rigter, 1980; Hoffman and Tabakoff, 1984) and under forced ingestion conditions (Mucha and Kalant, 1979). Much more research has been performed with respect to endogenous opioids and alcohol. It was demonstrated that alcohol could interact with endogenous opioid activity e.g. by stimulating the release of β-endorphin from the pituitary (Gianoulakis *et al.*, 1989). Changes in β-endorphin levels in the CSF of alcoholics have been reported (Genazzani *et al.*, 1982). Studies with opioid agonists and antagonists have revealed that opioid modulation can significantly affect alcohol consumption in animals (Altshuler *et al.*, 1980; Sinclair *et al.*, 1973a; Hubbell and Reid, 1990). An endorphin compensation hypothesis has been presented, in which a predisposed or acquired deficiency of endorphinergic activity is supposed to be compensated by alcohol ingestion (Blum, 1983; Erickson, 1990; Volpicelli *et al.*, 1986).

Material and Methods

The experiments were performed at the Primate Center TNO-Rijswijk, The Netherlands. Subjects were male adult rhesus monkeys (Macacca mulatta), who were 2 to 13 years of age and weighed between 3 and 12 kg. They were housed in a separate cage in the same room, that was temperature (24^0C) and humidity (60%) controlled and illuminated from 08.00 to 18.00 h. In the DGAVP experiment the monkeys were with four together daily between 10.00 and 12.30 h in a large play cage. Diet was composed of monkey chow (200 g per day at 09.00 h), and fruit, vegetables and bread in the afternoon.

Each subject had access to three identical drinking cylinders attached at the side of the home cage behind an opaque board, with only the three drinking nipples protruding into its cage. During the experimental period one cylinder provided tap water, the other two provided ethanol/water solutions in different concentrations (a low and a high concentration). Position of the drinking cylinders was changed daily. Liquids were available ad lib. For further details see Kornet et al., 1990a,b; 1991a,b; 1992a,b.

Results

Alcohol drinking was initiated in all subjects within a few days without any specific induction procedure. After drinking behavior was established, the relationship between drinking behavior and ethanol concentrations was studied by changing the ethanol concentrations (supplied concentrations: 2 and 4; 4 and 8; 8 and 16; 16 and 32 percent v/v) every 14 days. When ethanol concentrations in the solutions increased, consumption of ethanol solutions decreased, consumption of drinking water increased and of total water decreased. Net ethanol intake remained (2-6 ml.kg^{-1} per day) relatively constant across supplied concentrations, although the two lowest concentrations offered resulted in a somewhat lower intake. When a similar experiment was performed in monkeys who had experience with alcohol drinking for 6 months, it was found that net ethanol intake progressively increased as a function of increasing concentrations (Kornet, Goosen and Van Ree, unpublished). It was concluded that the observed alcohol drinking resulted primarily from a central reinforcement of ethanol intake.

Monkeys who have had free access to water and ethanol solutions concurrently for about one year, were subjected to interrupted alcohol supply for 1, 2 or 7 days (Kornet et al., 1990b). The previously acquired ethanol consuming behavior appeared very resistant to extinction, since ethanol consumption was immediately resumed after renewed access. Overt physical withdrawal distress was not observed. In the first two hours after renewed access to the alcohol solutions, an increased alcohol consumption was observed, which was higher when interruption lasted longer. After an interruption of 7 days, the increased ethanol intake was also present during the subsequent night. The increased alcohol consumption after interruption may resemble the 'catch up' phenomenon as observed in human alcoholics.

The effect of DGAVP was studied on the acquisition of voluntary, free choice alcohol drinking in naive monkeys, that had access to either 1 and 2% or to 4 and 8% ethanol/water solutions in addition to drinking water (Kornet et al., 1991a). Half of the monkeys were injected twice daily with 50 μg.kg⁻¹ of DGAVP for 14 days, the other half with placebo. DGAVP decreased net ethanol intake in the animals drinking 1 and 2% solutions, while in the placebo treated animals net alcohol intake increased over time. The effect of DGAVP was time dependent and long lasting. When acquisition was studied with the 4 and 8% solutions, the animals showed a different acquisition pattern as compared to the lower concentrations, both in the placebo and DGAVP treated monkeys. It was concluded that DGAVP inhibits acquisition of alcohol drinking in the majority of monkeys under free choice conditions.

During the DGAVP experiment, blood was taken from the monkeys before alcohol supply and after two and four weeks of alcohol drinking of the low concentrations (1 and 2%) and an endocrine profile was assessed (Kornet et al., 1992b). After two weeks of alcohol drinking significant increases were found in β-endorphin and ACTH. After four weeks prolactin was increased, cortisol decreased and particularly β-endorphin remained significantly increased. Testosterone was not significantly changed. No differences in endocrine response existed between DGAVP and placebo treated subjects, although the increase in prolactin and testosterone was less pronounced in the DGAVP treated monkey. No relationship was found between basal hormonal levels and subsequent ethanol intake. However, two placebo treated subjects that showed the highest increase in ethanol intake over time reacted differently, by reducing β-endorphin and ACTH levels over time, showing the largest decrease in cortisol and hardly any prolactin response.

In two separate sets of experiments the effect of graded doses of naltrexone and morphine on alcohol consumption of monkeys who had been drinking ethanol/water solutions for more than one year, was assessed under two conditions a) during continuous supply of drinking water and two ethanol/water solutions (16 and 32%) and b) after two days of alcohol abstinence (Kornet et al., 1991b; 1992a). Naltrexone was intramuscularly administered in doses of 0.02, 0.06, 0.17, 0.5, 1.0 and 1.5 mg.kg⁻¹ and morphine in doses of 0.03, 0.06, 0.17, 0.50 and 1.5 mg.kg⁻¹. A dose dependent decrease of alcohol intake was observed both with naltrexone and morphine during the first 17 h after injection under the conditions of continuous supply and after two days of alcohol abstinence, while consumption from the bottle with drinking water was not changed in that period, except a decrease of water intake in the morphine treated animals, but only under the condition of continuous supply. After imposed abstinence the monkeys were more sensitive to naltrexone and morphine with respect to the drug-induced decrease of alcohol intake as compared to the condition of continuous supply (a significant effect after a dose of 0.17 versus 0.5 mg.kg^{-1}). The effect of naltrexone and morphine was present during the first hours after injection. The reduction lasted for the subsequent night, in particular after morphine treatment, under the condition of continuous supply, but not after imposed abstinence. The results indicate that a rather low dose of both the opioid antagonist naltrexone and the opiate agonist morphine reduces the alcohol consumption in monkeys.

Discussion

All rhesus monkeys in the presented studies initiated alcohol drinking within a few days by no manipulation other than making ethanol/water solutions available in addition to drinking water. Alcohol drinking was maintained as long as alcohol remained available. The results of the experiments in which the ethanol concentrations were changed and the alcohol supply was interrupted, indicate that the alcohol drinking behavior in the monkeys is under the control of reinforcement principles in particular in experienced monkeys. This accords well with other findings that low doses of ethanol are positively reinforcing (Bain and Kornetsky, 1989). The increase in ethanol intake after imposed interruption of alcohol supply has also been reported in rats and humans (Sinclair *et al.*, 1973b; Burish *et al.*, 1981) and seems to reflect a reinstatement of the previously acquired ethanol-reinforced behavior with a temporary increased motivation for ethanol. This behavior induced by imposed interruption might be mediated by the same mechanism(s) as the relapses in alcohol addicts (Barnes, 1988). Several hypotheses on relapse - the major problem in addiction - have been formulated i.e. the influence of incentive motivation processes (Goldberg *et al.*, 1990), drug-opposite conditioned responses (O'Brien, 1986) and internal deficiency, induced by frequent drug ingestion e.g. endorphin deficiency (Genazzani *et al.*, 1982). The presented model for relapse in monkeys could contribute to further analyse this phenomenon and to test methods for prevention of relapse.

The neuropeptide DGAVP reduced the acquisition of alcohol drinking in the majority of treated monkeys. This effect is in line with studies in rats, showing that initiation of heroin and cocaine selfadministration is decreased by the peptide (Van Ree and De Wied, 1977a; De Vry *et al.*, 1988). That DGAVP exerts a similar effect for different classes of addictive drugs and that electrical brain selfstimulation is also attenuated by DGAVP (Dorsa and Van Ree, 1979), indicate that this peptide modulates brain reward and thereby decreases the positive reinforcing effects of addictive drugs. But, the peptide is only effective in situations in which the reinforcement control over behavior is developing or changed (Van Ree, 1987). Concerning the endocrine profile during initiation of alcohol drinking in monkeys, of particular interest is that plasma β-endorphin remained significantly increased over time, which may be an early stage of ethanol-induced modification in neuroendocrine homeostasis. The decrease of plasma cortisol over time may represent an impaired pituitary-adrenal function, as has been reported for human alcoholics (Marks, 1979). Two monkeys that developed high increase in ethanol intake over time seemed to deviate from the others in endocrine reactions. This may have some relevance for human studies in the prediction of susceptibility for alcohol addiction (Schuckit et al., 1988).

That rather low doses of naltrexone and morphine decreased alcohol consumption in monkeys, supports the idea that endogenous opioid systems are implicated in alcohol addiction. That both an opioid antagonist and agonist produced the same effect is not easy to understand, and is not in line with reported data on the influence of these drugs on alcohol consumption in rats (Hubbell and Reid, 1990). The findings in the monkey can best be explained in the context of the so called 'endorphin-compensation' hypothesis that presumes that alcohol drinking is reinforcing by increasing activity at opioid receptors (Volpicelli et al., 1990). Predictions for this hypothesis are that alcohol consumption should decrease during conditions of excess opioid receptor activity e.g. after treatment with morphine, and increase during conditions with deficiencies in opioidergic activity e.g. after imposed abstinence, that alcohol can pharmacologically stimulate directly or indirectly activity of opioid receptors (Koob and Weiss, 1990) and that opioid antagonists should block the reinforcing effects of alcohol and hence decrease alcohol intake e.g. after treatment with naltrexone. Also the observation that β-endorphin in the anterior limbic area is decreased when the desire for cocaine or heroin in rats is high agrees well with the endorphin compensation hypothesis (Sweep et al., 1989). This finding and the enhanced sensitivity to opioid agonists and antagonists after a period of imposed abstinence, suggest that endorphin may be particularly implicated in craving for addictive drugs and relapse of addictive behavior. This collaborates well with the recent findings in human alcoholics showing that both craving and relapse is attenuated after treatment with the opioid antagonist naltrexone (Volpicelli et al., 1992, O'Malley, 1992). This further supports the validity and usefulness of the present animal (rat and monkey) model of addictive behavior.

172

References

Altshuler, H.L., Philips, P.E. and Feinhandler, D.A. (1980) Alteration of ethanol self-administration by naltrexone. *Life Sci.* 26: 679-688.

Bain, G.T. and Kornetsky, C. (1989) Ethanol oral self-administration and rewarding brain stimulation. *Alcohol* 6: 499-503.

Burish, T.G., Maisto, S.A., Cooper, A.M. and Sobell, M.D. (1981) Effects of voluntary short-term abstinence from alcohol on subsequent drinking patterns of college students. *J. Stud. Alc.* 42: 1013-1020.

Barnes, D.M. (1988) Breaking the cycle of addiction. *Science* 241: 1029-1030.

Blum, K. (1983) Alcohol and central nervous system peptides. *Subst. & Alc. Actions/Misuse* 4: 73-87.

Carroll, M.E., Lac, S.T., Walker, S.T., Kragh, M.J. and Newman, T. (1986) Effects of naltrexone on intravenous cocaine self-administration in rats during food satiation and deprivation. *J. Pharmacol. Exp. Ther.* 238: 1-7.

Crabbe, J.C. and Rigter, H. (1980) Learning and the development of alcohol tolerance and dependence. *Trends Neurosci.* 3: 20-22.

De Vry, J., Donselaar, Y. and Van Ree, J.M. (1988) Effects of desglycinamide-(Arg^8)-vasopressin and vasopressin antiserum on the acquisition of intravenous cocaine self-administration. *Life Sci.* 42: 2709-2715.

De Vry, J., Donselaar, I. and Van Ree, J.M. (1989) Food deprivation and acquisition of intravenous cocaine self-administration in rats: Effect of naltrexone and haloperidol. *J. Pharmacol. Exp. Ther* 251: 735-740.

De Wied, D. (1969) Effects of peptide hormones on behavior. In: Ganong, W.F. and Martini, L. (eds.): *Frontiers in Neuroendocrinology*, Oxford University Press, Oxford, pp. 97-140.

De Wied, D. and Jolles, J. (1982) Neuropeptides derived from pro-opiocortin: Behavioral, physiological and neurochemical effects. *Physiol. Rev.* 62: 976-1059.

Dorsa, D.M. and Van Ree, J.M. (1979) Modulation of substantia nigra self-stimulation by neuropeptides related to neurohypophyseal hormones. *Brain Res.* 172: 367-371.

Erickson, C.K. (1990) Reviews and comments on alcohol research. *Alcohol* 7: 557-558.

Ettenberg, A., Pettit, H.O., Bloom, F.E. and Koob, G.F. (1982) Heroin and cocaine intravenous self-administration in rats: mediation by separate neural systems. *Psychopharmacology* 78: 204-209.

Fraenkel, H.M., Van Beek-Verbeek, G., Fabriek, A.J. and Van Ree, J.M. (1983) Desglycinamide-9-arginine-8-vasopressin and ambulant methadone detoxification of heroin addicts. *Alcohol & Alcoholism* 18: 331-335.

Genazzani, A.R., Nappi, G., Facchinetti, F., Mazzella, G.L., Parrini, D., Sinforiani, E., Petraglia, F. and Savoldi, F. (1982) Central deficiency of β-endorphin in alcohol addicts. *J. Clin. Endocrin. Metabol.* 55: 583-586.

Gianoulakis, C., Beliveau, D., Angelogianni, P., Meaney, M., Thavundayil, J., Tawar, V. and Dumas, M. (1989) Different pituitary β-endorphin and adrenal cortisol response to ethanol in individuals with high and low risk for future development of alcoholism. *Life Sci.* 45: 1097-1109.

Goldberg, S.R., Schnindler, C.W. and Lamb, R.J. (1990) Second-order schedules and the analysis of human drug-seeking behavior. *Drug Development Res.* 20: 217-229.

Hoffman, P.L. and Tabakoff, B. (1984) Neurohypophyseal peptides maintain tolerance to the incoordinating effects of ethanol. *Pharmacol. Biochem. Behav.* 12: 539-543.

Hubbell, C.L. and Reid, L.D. (1990) Opioids modulate rats' intake of alcoholic beverages. In: Reid, L.D. (ed.): *Opioids, Bulimia and Alcohol Abuse & Alcoholism*, Springer-Verlag, New York, pp. 145-174.

Koob, G.F. and Weiss, F. (1990) Pharmacology of drug self-administration. *Alcohol* 17: 193-197.

Kornet, M., Goosen, C., Ribbens, L.G. and Van Ree, J.M. (1990a) Analysis of spontaneous alcohol drinking in rhesus monkeys. *Physiol. Behav.* 47: 679-684.

Kornet, M., Goosen, C. and Van Ree, J.M. (1990b) The effect of interrupted alcohol supply on spontaneous alcohol consumption by rhesus monkeys. *Alcohol and Alcoholism* 25: 407-412.

Kornet, M., Goosen, C., Ribbens, L.G. and Van Ree, J.M. (1991a) The effect of desglycinamide-(Arg[8])-vasopressin (DGAVP) on the acquisition of free-choice alcohol drinking in rhesus monkeys. *Alcoholism: Clin. Exp. Res.* 15: 72-79.

Kornet, M., Goosen, C. and Van Ree, J.M. (1991b) The effect of naltrexone on alcohol consumption during chronic alcohol drinking and after a period of imposed abstinence in free-choice drinking rhesus monkeys. *Psychopharmacol.* 104: 367-376.

Kornet, M., Goosen, C., Van Vlaardingen, J.A.M. and Van Ree, J.M. (1992a) Low doses of morphine reduce voluntary alcohol consumption in rhesus monkeys. *Eur. Neuropsychopharmacol.* 2: 73-86.

Kornet, M., Goosen, C., Thijssen, J.H.H. and Van Ree, J.M. (1992b) Endocrine profile during acquisition of free-choice alcohol drinking in rhesus monkeys. Treatment with Desglycinamide-(Arg[8])-vasopressin. *Alcohol Alcoholism* 27: 403-410.

Marks, V. (1979) Biochemical and metabolic basis of alcohol toxicity. In: Mendlewicz, J. and Van Praag, H.M. (eds.): *Advances in Biological Psychiatry, vol. 3: Alcoholism. A Multidisciplinary Approach*, S. Karger, Basel, pp. 88-96.

Mucha, R.F. and Kalant, H. (1979) Effects of desglycinamide[9]-Lysine[8]-vasopressin and prolyl-leucyl-glycinamide on oral ethanol intake in the rat. *Pharmacol. Biochem. Behav.* 10: 229-234.

O'Brien, C.P. (1986) Experimental analysis of conditioning factors in human narcotic addiction. *Pharmacol. Rev.* 27: 533-543.

O'Malley, S.S., Jaffe A.J., Chang, G., Schottenfield, R.S., Meyer, R.E. and Rounsaville, B. (1992) Naltrexone and coping skills therapy for alcohol dependence. A controlled study. *Arch. Gen. Psychiatry* 49: 881-887.

Schaefer, G.J. (1988) Opiate antagonists and rewarding brain stimulation. *Neurosci. Biobehav. Rev.* 12: 1-17.

Schuckit, M.A., Risch, S.C. and Gold, E.O. (1988) Alcohol consumption, ACTH level and family history of alcoholism. *Am. J. Psychiatry* 145: 1391-1395.

Sinclair, J.D., Adkins, J. and Walker, S. (1973a) Morphine-induced suppression of voluntary alcohol drinking in rats. *Nature* 246: 425-427.

Sinclair, J.D., Walker, S. and Jordan, W. (1973b) Behavioral and physiological changes associated with various durations of alcohol deprivation in rats. *Quart. J. Stud. Alcohol.* 34: 744-757.

Sweep, C.G.J., Wiegant, V.M., De Vry, J. and Van Ree, J.M. (1989) β-Endorphin in brain limbic structures as neurochemical correlate of psychic dependence on drugs. *Life Sci.* 44: 1133-1140.

Van Beek-Verbeek, G., Fraenkel, H.M. and Van Ree, J.M. (1983) Des-Gly9-Arg8-vasopressin may facilitate methadone detoxification of heroin addicts. *Subst. Alcohol Actions Misuse* 4: 375-382.

Van Ree, J.M. (1983) Behavioral effects of endorphins: Modulation of opiate reward by neuropeptides related to pro-opiocortin and neurohypophyseal hormones. In: Smith, J.E. and Lane, J.D. (eds.): *The Neurobiology of Opiate Receptors*, Elsevier Biomedical Press, Amsterdam, pp. 109-144.

Van Ree, J.M. (1987) Reward and abuse: Opiates and neuropeptides. In: Engel, J., Oreland, L., Ingvar, D.H., Pernow, B., Rössner, S. and Pellborn, L.A. (eds.): *Brain Reward Systems and Abuse*, Raven Press, New York, pp. 75-88.

Van Ree, J.M. and De Wied, D. (1977a) Modulation of heroin self-administration by neurohypophyseal principles. *Eur. J. Pharmacol.* 43: 199-202.

Van Ree, J.M. and De Wied, D. (1977b) Heroin self-administration is under control of vasopressin. *Life Sci.* 21: 315-320.

Van Ree, J.M., Bohus, B., Versteeg, D.H.G. and De Wied, D. (1978) Neurohypophyseal principles and memory processes. *Biochem. Pharmacol.* 27: 1793-1800.

Van Ree, J.M., Smyth, D.G. and Colpaert, F.C. (1979) Dependence creating properties of lipotropin C-fragment (β-endorphin): evidence for its internal control of behavior. *Life Sci.* 24: 495-502.

Van Ree, J.M., Burbach-Bloemarts, E.M.M. and Wallace, M. (1988) Vasopressin neuropeptides and acquisition of heroin and cocaine self-administration in rats. *Life Sci.* 42: 1091-1099.

Volpicelli, J.R., Davis, M.A. and Olgin, J.E. (1986) Naltrexone blocks the post-shock increase of ethanol consumption. *Life Sci.* 38: 841-847.

Volpicelli, J.R., O'Brien, C.P., Alterman, A.I. and Hayashida, M. (1990) Naltrexone and the treatment of alcohol-dependence: Initial observations. In: Reid, L.D. (ed.): *Opioids, Bulimia and Alcohol Abuse & Alcoholism*, Springer-Verlag, New York, pp. 195-214.

Volpicelli, J.R., Alterman, A.I., Hayashida, M. and O'Brien, C.P. (1992) Naltrexone in the treatment of alcohol dependence. *Arch. Gen. Psychiatry* 49: 876-880.

Wise, R.A. (1978) Catecholamine theories of reward: A critical review. *Brain Res.* 152: 215-247.

Toward a Molecular Basis of Alcohol Use and Abuse
ed. by B. Jansson, H. Jörnvall, U. Rydberg, L. Terenius & B. L. Vallee

The role of adenosine in mediating cellular and molecular responses to ethanol

I. Diamond and A.S. Gordon

Ernest Gallo Clinic and Research Center and the Departments of Neurology and Pharmacology, University of California, San Francisco, Bldg. 1, Rm. 101, San Francisco General Hospital, San Francisco, CA 94110, USA

Summary

We have found that ethanol-induced increases in extracellular adenosine activate adenosine receptors which, in turn, mediate many of the acute and chronic effects of ethanol in the nervous system. Several laboratories have demonstrated the importance of adenosine in mediating the acute and chronic effects of ethanol at multiple levels of investigation in the nervous system. These include genetic selection for ethanol sensitivity in mice, behavioral responses to ethanol in naive and tolerant animals, neurophysiologic responses in hippocampal slices, and at the level of cAMP signal transduction and gene expression in cultured neural cells. In this review we present results from our laboratory which document the role of adenosine in mediating ethanol-induced changes in neural function at a cellular and molecular level. A schematic summary of our findings is:

EtOH ➡ ↓Ado uptake ➡ ↑ Extracellular Ado ➡ Activation of Adenosine A_2 receptor ➡
↑ cAMP ➡ ↑PKA ➡➡➡ Heterologous Desensitization (↓cAMP) ➡➡➡ insensitivity of adenosine uptake to EtOH

Introduction

Adenosine is a major inhibitory neuromodulator in the nervous system. Adenosine is released from virtually all cells and is also produced extracellularly from released precursors (Bruns, 1990; James and Richardson, 1993). The extracellular concentration of adenosine increases with neural activity in the brain (Phillis and Wu, 1981) and in cell culture (Pitchford *et al.*, 1992). Recent evidence suggests that adenosine is a global inhibitor of excitatory synaptic transmission and acts by inhibiting the release of excitatory transmitters from pre-synaptic nerve endings (Yoon and Rothman, 1991; Karunanithi *et al.*, 1992; Lupica *et al.*, 1992; Prince and Stevens, 1992; Scholz and Miller, 1992). In addition, adenosine acting post-synaptically is reported to diminish the response of dopamine (Ferre *et al.*, 1991; 1992) and acetylcholine (Pitchford *et al.*, 1992) receptors. Adenosine is an endogenous anticonvulsant (Dunwiddie, 1985) and protects against ischemic brain damage (Rudolphi *et al.*, 1992). These responses involve activation of adenosine A_1, A_2, and other adenosine receptor subtypes in the cell membrane (Bruns, 1990; Stiles, 1990; 1992; Olah and Stiles, 1992). Blockers of adenosine uptake potentiate cellular responses to adenosine (Dar, 1989; Phillis *et al.*, 1989; Nagy *et al.*, 1990), suggesting that the physiological effects of adenosine are terminated, in part, by re-uptake into the cell. Thus, neural responses inhibited by adenosine are dependent not only on receptor activation, but also on the activity of the adenosine transporter. Adenosine deaminase and other metabolic enzymes also contribute to the regulation of extracellular adenosine levels (Rubio *et al.*, 1989; Bruns, 1990).

An increasing number of studies indicate that adenosine mediates many of the acute and chronic effects of ethanol in the nervous system (Proctor and Dunwiddie, 1984; Proctor *et al.*, 1985; Gordon *et al.*, 1986; Dar *et al.*, 1987; 1989; Diamond *et al.*, 1987; Mochly-Rosen *et al.*, 1988; Nagy *et al.*, 1989; 1990; Dar, 1990a,b; Dar and Clark, 1992; Cullen and Carlen, 1992). Proctor and Dunwiddie (Proctor and Dunwiddie, 1984; Proctor *et al.*, 1985) have shown that sensitivity to adenosine agonists and antagonists correlates with acute ethanol sensitivity in selectively bred mice. Their results suggest that adenosine mediates ethanol-induced sedation in long sleep mice. Studies from Dar's laboratory indicate that adenosine mediates acute ethanol-induced ataxia (Dar *et al.*, 1987; Dar, 1989; 1990a,b; Dar and Clark, 1992). They find that adenosine receptor agonists increase ethanol-induced incoordination and that adenosine receptor antagonists decrease this intoxicating response. In addition, dilazep, which potentiates the effects of adenosine by inhibiting adenosine uptake via nucleoside transporters, exacerbates the intoxicating effects of ethanol (Dar, 1989). Dar's group has also shown cross-tolerance between adenosine agonists and ethanol after chronic exposure to either agent (Dar and Clark, 1992). Based on the dose-response of different ligands, they propose that adenosine A_1 receptors may mediate these effects of ethanol (Dar, 1990a; Dar and Clark, 1992); they have, however, also suggested that A_2 receptors could be involved in this response (Dar, 1990a,b).

The importance of adenosine in mediating effects of ethanol in the nervous system is also apparent at a cellular level in the brain. Recent neurophysiological studies in hippocampal slices by Cullen and Carlen (1992) indicate that adenosine mimics the effects of ethanol in CA1 and dentate granule cells. In addition, they report that an adenosine receptor antagonist, 8-phenyltheophylline, blocks adenosine and ethanol-induced afterhyperpolarization in these cells. In this review we summarize studies from our laboratory which explore the role of adenosine in mediating ethanol-induced changes in neural function at a cellular and molecular level.

Results and Discussion

A schematic representation of our results is as follows:

EtOH ➡ ⬇Ado uptake ➡ ⬆ Extracellular Ado ➡ Activation of Adenosine A$_2$ receptor ➡

⬆ cAMP ➡ ⬆PKA ➡➡➡ Heterologous Desensitization (⬇cAMP) ➡➡➡ insensitivity of

adenosine uptake to EtOH

Adenosine mediates ethanol-induced desensitization

Our studies using cultured cell lines showed that chronic exposure to ethanol results in heterologous desensitization of receptor-dependent cAMP levels (Mochly-Rosen *et al.*, 1988; Nagy *et al.*, 1989). Since heterologous desensitization in other systems is characterized by an initial increase in cAMP levels, we next determined whether acute exposure to ethanol caused an increase in cAMP levels. When NG108-15 cells were incubated with 100 mM ethanol for 10 min, there was a 60% increase in intracellular cAMP levels (Nagy *et al.*, 1989). We explored the possibility that this increase was due to increased extracellular adenosine, a stimulatory agonist that is continually being released and taken up by cells. We found a 2-5-fold increase in extracellular adenosine concentrations after 5-10 min of ethanol exposure in NG108-15 and S49 cells (Nagy *et al.*, 1989; 1990).

This ethanol-induced increase in extracellular adenosine and intracellular cAMP was prevented by adenosine deaminase (ADA) which deaminates adenosine to inosine, a nucleoside with low affinity for the adenosine receptor (Nagy *et al.*, 1989). Moreover, the adenosine receptor antagonists, IBMX and BW A1434U, also blocked ethanol-induced increases in cAMP levels (Nagy *et al.*, 1989). These data suggest that ethanol increased extracellular adenosine and that this extracellular adenosine activated the A$_2$ receptor to stimulate cAMP production. We found no evidence of adenosine A$_1$ receptors in these cells.

In contrast to acute stimulation of cAMP levels by ethanol, chronic exposure to ethanol causes heterologous desensitization of adenosine receptor- and PGE$_1$ receptor-dependent cAMP production (Mochly-Rosen *et al.*, 1988; Nagy *et al.*, 1989). If adenosine were responsible for this effect of ethanol, then ADA should also prevent this response. 1 U/ml ADA, a concentration sufficient to block the acute increase in cAMP, reduced chronic ethanol-induced desensitization of adenosine receptor-stimulated cAMP levels and completely prevented desensitization of the PGE$_1$ receptor (Nagy *et al.*, 1989). These results suggest that an increase in extracellular adenosine is required for ethanol to produce heterologous desensitization. This requirement is not specific for neural cells. S49 lymphoma cells also showed a significant increase in the concentration of extracellular adenosine when exposed to ethanol for 5 min or 24 hr (Nagy *et al.*, 1990). When S49 cells were treated with ethanol for 48 hrs, adenosine receptor and PGE$_1$ receptor-stimulated cAMP levels were reduced to 65 $\pm$ 7 and 36 $\pm$ 6% of control, respectively (Nagy *et al.*, 1989). As in NG108-15 cells, coincubation of S49 cells with ethanol and ADA prevented ethanol-induced heterologous desensitization (Nagy *et al.*, 1989). We also used mutant S49 cells (*kin-*) that lack cAMP-dependent protein kinase (PKA) activity to determine whether activation of PKA by ethanol-induced increases in cAMP was required for heterologous desensitization. We found that chronic ethanol-induced heterologous desensitization does not occur in *kin-* cells, suggesting that activation of PKA is required for heterologous desensitization.

Adenosine transport is required for ethanol-induced heterologous desensitization in S49 cells

If accumulation of extracellular adenosine is required for ethanol-induced heterologous desensitization, then cells which do not release adenosine should not desensitize after chronic exposure to ethanol. In S49 cells, adenosine uptake and release are mediated via a single bidirectional transporter. We used the S49 nucleoside transport-deficient mutants 80-2A6 and 160-D4, which do not transport adenosine, to determine whether adenosine transport is required for ethanol-induced heterologous desensitization. When these mutant cell lines were treated with 200 mM ethanol for 5 min or 100 mM ethanol for 24 hrs, extracellular adenosine was not detectable (Nagy *et al.*, 1989). There was also no desensitization of adenosine receptor- or PGE$_1$ receptor-stimulated cAMP levels after exposure to ethanol for 48 hrs (Nagy *et al.*, 1989). Thus, the adenosine transporter is required for ethanol-induced heterologous desensitization of receptor-dependent cAMP production in S49 cells.

Ethanol increases extracellular adenosine by inhibiting adenosine uptake via the nucleoside transporter

Next, we investigated the mechanism underlying ethanol-induced increases in extracellular adenosine concentration. We found that acute exposure to clinically relevant concentrations of ethanol decreases adenosine uptake by inhibiting influx. We found no effect of ethanol on efflux. The 30-40% decrease in adenosine uptake is sufficient to account for the rate of increase in extracellular adenosine during acute exposure to ethanol (Nagy *et al.*, 1990). The extent of extracellular adenosine accumulation caused by ethanol inhibition of uptake is sufficient to produce heterologous desensitization (Mochly-Rosen *et al.*, 1988; Nagy *et al.*, 1989; 1990). Inhibition of uptake by ethanol appeared to be specific for nucleoside transport. Uptake of another nucleoside, uridine, was decreased to the same extent as adenosine; uptake of isoleucine (Nagy *et al.*, 1990) and glutamine (Krauss *et al.*, 1993) were unaffected by ethanol.

After prolonged exposure to ethanol, S49 cells became tolerant to the acute inhibitory effects of ethanol on adenosine uptake. Ethanol no longer inhibited uptake of [^{3}H]adenosine nor did it increase extracellular adenosine in chronically treated cells (Nagy *et al.*, 1990). These results suggest that the development of cellular tolerance to ethanol might be due to modification of the nucleoside transporter. Since chronic exposure to ethanol results in desensitization of receptor-dependent cAMP production, it is possible that the chronic effects of ethanol on the adenosine transport system are a consequence of decreased receptor-dependent cAMP production. In support of this hypothesis, we have found that ethanol does not inhibit adenosine uptake in mutant S49 cells which are deficient in cAMP signalling.

cAMP-dependent protein kinase regulates inhibition of adenosine transport by ethanol

Loss of ethanol sensitivity of adenosine uptake occurs after chronic ethanol-induced heterologous desensitization of the cAMP signal transduction system. These results suggest a relationship between the effects of ethanol on cAMP signal transduction and inhibition of adenosine uptake (Diamond *et al.*, 1987). Therefore, we examined inhibition of adenosine uptake by ethanol in variants of the S49 cell line that have a functional Ga$_s$ deficiency in receptor-dependent cAMP production (*unc*) or in PKA activity (*kin-*). In S49 wild type cells, acute exposure to ethanol inhibited adenosine uptake and increased extracellular adenosine. In contrast, ethanol did not inhibit adenosine uptake in *kin-* or *unc* cells and did not increase extracellular adenosine (Nagy *et al.*, 1991). In *unc* cells, inhibition of adenosine uptake by ethanol could be restored by bypassing the defect and raising cAMP levels with forskolin. Increasing cAMP levels in *kin-* cells had no effect (Nagy *et al.*, 1991). Because *kin-* cells lack PKA activity, these results suggest that cAMP-dependent phosphorylation is required for ethanol to inhibit adenosine uptake. Taken together, these data suggest that a common ethanol-sensitive component of the transport system, either the nucleoside transporter itself or an associated regulatory component, is regulated by phosphorylation.

After prolonged exposure of S49 wild type cells to ethanol, adenosine uptake is no longer inhibited by rechallenge with ethanol, i.e. nucleoside transport has lost its sensitivity to ethanol. Because this insensitivity is similar to that found in *unc* and *kin-* cells, the insensitivity in cells chronically exposed to ethanol might be due to decreased cAMP levels and, presumably, decreased phosphorylation of the nucleoside transport system. Our data suggest that the adenosine transport system has to be "primed" by phosphorylation before ethanol exposure, i.e., ethanol only binds to the transporter and inhibits uptake if the nucleoside transporter or an associated regulatory component is phosphorylated.

Inhibition of adenosine uptake by ethanol is specific for one type of nucleoside transporter
There are two classes of nucleoside transporters in mammalian cells which are responsible for adenosine uptake: Na^+-independent facilitative carriers, and Na^+-dependent concentrative carriers. The facilitative transporters can be divided into 2 subtypes: those inhibited by the purine analog nitrobenzylmercaptopurine (NBMPR) and those that are resistant to NBMPR inhibition. There are also 2 types of Na^+-dependent transporters that can be distinguished on the basis of substrate affinities. Using human lymphocytes, we have found that uptake via only one type of nucleoside transporter, the NBMPR-sensitive facilitative transporter, is inhibited by ethanol (Krauss *et al.*, 1993) and is thus responsible for extracellular accumulation of adenosine in the presence of ethanol. We studied a wide range of alcohol concentrations and found significant inhibition of nucleoside transport by 25 to 200 mM ethanol. When cultured human lymphocytes were exposed to 200 mM ethanol, only NPMRP-sensitive uptake was inhibited; NBMPR-resistant transport remained unaffected (Krauss *et al.*, 1993). Similar results were obtained for L1210 cells assayed in sodium-free medium to minimize concentrative transport; ethanol inhibited facilitative uptake without affecting NBMPR-resistant transporters. This was also true in N1S1 hepatoma cells where 80% of formycin uptake is resistant to NBMPR. Ethanol did not inhibit NBMPR-resistant uptake. Finally, MA27.1 cells (a mutant of L1210 cells) were used to examine the effect of ethanol on concentrative adenosine transporters, the

only kind of transporter expressed in these cells. Ethanol did not inhibit nucleoside uptake in these cells. Therefore, we conclude that only NBMPR-sensitive transporters are inhibited by ethanol. This is a very specific effect since we find that other nucleoside and amino acid transporters (glutamine, isoleucine) are unaffected by ethanol. Therefore, ethanol inhibition of adenosine uptake is independent of cell type and is only demonstrable in cells with NBMPR-sensitive transporters, including human lymphocytes, S49 cells, and NG108-15 neural cells (Diamond et al., 1987).

In model cell systems, ethanol inhibition of adenosine uptake initiates a cascade of events leading to heterologous desensitization of receptors coupled to adenylyl cyclase via G$_s$ and to reduce cAMP levels (Diamond et al., 1991). We find lymphocytes from actively drinking alcoholics also have reduced receptor-stimulated cAMP levels (Diamond et al., 1987). Thus, an early effect of ethanol on human lymphocytes could be inhibition of adenosine uptake via NBMPR-sensitive facilitative nucleoside transporters. This would lead to heterologous desensitization of cAMP signal transduction in these cells. Identification of NBMPR-sensitive nucleoside transporters as an initial target of ethanol may enable the design of specific agents to prevent adenosine-mediated responses to ethanol.

Acknowledgements

This work was supported by grants from the National Institute of Alcohol Abuse and Alcoholism, the March of Dimes Birth Defects Foundation and the Alcoholic Beverage Medical Research Foundation.

References

Bruns, R.F. (1990) Adenosine receptors: roles and pharmacology. *Ann. NY Acad. Sci.* 603: 211-226.

Cullen, N. and Carlen, P.L. (1992) Electrophysiological actions of acetate, a metabolite of ethanol, on hippocampal dentate granule neurons: interactions with adenosine. *Brain Res.* 588: 49-57.

Dar, M.S. (1989) Central nervous system effects and behavioral interactions with ethanol of centrally administered dilazep and its metabolites in mice. *Eur. J. Pharmacol.* 164: 303-313.

Dar, M.S. (1990a) Central adenosinergic system involvement in ethanol-induced motor in coordination in mice. *J. Pharmacol. Exp. Ther.* 255: 1202-1209.

Dar, M.S. (1990b) Functional correlation between subclasses of brain adenosine receptor affinities and ethanol-induced motor incoordination in mice. *Pharmacol. Biochem. & Behav.* 37: 747-753.

Dar, M.S. and Clark, M. (1992) Tolerance to adenosine's accentuation of ethanol-induced motor incoordination in ethanol-tolerant mice. *Alcohol. Clin. Exp. Res.* 16: 1138-1146.

Dar, M.S., Hardee, M. and Ganey, T. (1989) Brain adenosine modulation of behavioral interactions between ethanol and carbamazepine in mice. *Alcohol* 6: 297-301.

Dar, M.S., Jones, M., Close, G., Mustafa, S.J. and Wooles, W.R. (1987) Behavioral interactions of ethanol and methylxanthines. *Psychopharmacology* 91: 1-4.

Diamond, I., Nagy, L., Mochly-Rosen, D. and Gordon, A. (1991) The role of adenosine and adenosine transport in ethanol-induced cellular tolerance and dependence. *Ann. NY Acad. Sci.* 625: 473-487.

Diamond, I., Wrubel, B., Estrin, E. and Gordon, A.S. (1987) Basal and adenosine receptor-stimulated levels of cAMP are reduced in lymphocytes from alcoholic patients. *Proc. Natl. Acad. Sci. USA* 84: 1413-1416.

Dunwiddie, T.V. (1985) *The physiological role of adenosine in the central nervous system.* Academic Press, Inc., Denver, pp. 63-139.

Ferre, S., Fuxe, K., von Euler, G., Johansson, B. and Fredholm, B.B. (1992) Adenosine-dopamine interactions in the brain. *Neuroscience* 51: 501-512.

Ferre, S., von Euler, G., Johansson, B., Fredholm, B.B. and Fuxe, K. (1991) Stimulation of high-affinity adenosine A_2 receptors decreases the affinity of dopamine D_2 receptors in rat striatal membranes. *Proc. Natl. Acad. Sci. USA* 88: 7238-7241.

Gordon, A.S., Collier, K. and Diamond, I. (1986) Ethanol regulation of adenosine receptor-dependent cAMP levels in a clonal neural cell line: An *in vitro* model of cellular tolerance to ethanol. *Proc. Natl. Acad. Sci. USA* 83: 2105-2108.

James, S. and Richardson, P.J. (1993) Production of adenosine from extracellular ATP at the striatal cholinergic synapse. *J. Neurochem.* 60: 219-227.

Karunanithi, S., Lavidis, N.A. and Bennett, M.R. (1992) The effect of adenosine on spontaneous and evoked quantal secretion from different release sites of amphibian motor-nerve terminals. *Neurosci. Lett.* 147: 49-52.

Krauss, S.W., Ghirnikar, R.B., Diamond, I. and Gordon, A.S. (1993) Inhibition of adenosine uptake by ethanol is specific for one class of nucleoside transporters. *Mol. Pharmacol.*, 44: 1021-1026.

Lupica, C.R., Proctor, W.R. and Dunwiddie, T.V. (1992) Presynaptic inhibition of excitatory synaptic transmission by adenosine in rat hippocampus: Analysis of unitary EPSP variance measured by whole-cell recording. *J. Neurosci.* 12: 3753-3764.

Mochly-Rosen, D., Chang, F.-U., Cheever, L., Kim, M., Diamond, I. and Gordon, A.S. (1988) Chronic ethanol causes heterologous desensitization by reducing α_s mRNA. *Nature* 333: 848-850.

Nagy, L.E., Diamond, I., Casso, D.J., Franklin, C. and Gordon, A.S. (1990) Ethanol increases extracellular adenosine by inhibiting adenosine uptake via the nucleoside transporter. *J. Biol. Chem.* 265: 1946-1951.

Nagy, L.E., Diamond, I., Collier, K., Lopez, L., Ullman, B. and Gordon, A.S. (1989) Adenosine is required for ethanol-induced heterologous desensitization. *Mol. Pharmacol.* 36: 744-748.

Nagy, L.E., Diamond, I. and Gordon, A.S. (1991) cAMP-dependent protein kinase regulates inhibition of adenosine transport by ethanol. *Mol. Pharmacol.* 40: 812-817.

Olah, M.E. and Stiles, G.L. (1992) Adenosine receptors. *Ann. Rev. Physiol.* 54: 211-225.

Phillis, J.W., O'Regan, M.H. and Walter, G.A. (1989) Effects of two nucleoside transport inhibitors, dipyridamole and soluflazine, on purine release from the rat cerebral cortex. *Brain*

Res. 481: 309-316.

Phillis, J.W. and Wu, P.H. (1981) The role of adenosine and its nucleotides in central synaptic transmission. *Prog. Neurobiol.* 16: 187-239.

Pitchford, S., Day, J.W., Gordon, A. and Mochly-Rosen, D. (1992) Nicotinic acetylcholine receptor desensitization is regulated by activation-induced extracellular adenosine accumulation. *J. Neurosci.* 12: 4540-4544.

Prince, D.A. and Stevens, C.F. (1992) Adenosine decreases neurotransmitter release at central synapses. *Proc. Natl. Acad. Sci. USA* 89: 8586-8690.

Proctor, W.R., Baker, R.C. and Dunwiddie, T.V. (1985) Differential CNS sensitivity to PIA and theophylline in long-sleep and short-sleep mice. *Alcohol* 2: 287-291.

Proctor, W.R. and Dunwiddie, T.V. (1984) Behavioral sensitivity to purinergic drugs parallels ethanol sensitivity in selectively bred mice. *Science* 224: 519-521.

Rubio, R., Bencherif, M. and Berne, R.M. (1989) Inositol phospholipid metabolism during the following synaptic activation : role of adenosine. *J. Neurochem.* 52: 797-806.

Rudolphi, K.A., Schubert, P., Parkinson, F.E. and Fredholm, B.B. (1992) Neuroprotective role of adenosine in cerebral ischaemia. *TIPS* 13: 439-445. .

Scholz, K.P. and Miller, R.J. (1992) Inhibition of quantal transmitter release in the absence of calcium influx by a G protein-linked adenosine receptor at hippocampal synapses. *Neuron* 8: 1139-1150.

Stiles, G.L. (1990) Adenosine receptors and beyond : molecular mechanisms of physiological regulation. *Clin. Res.* 88: 10-18.

Stiles, G.L. (1992) Adenosine receptors. *J. Biol. Chem.* 267: 6451-6454.

Yoon, K.-W. and Rothman, S.M. (1991) Adenosine inhibits excitatory but not inhibitory synaptic transmission in the hippocampus. *J. Neurosci.* 11: 1375-1380.

Toward a Molecular Basis of Alcohol Use and Abuse
ed. by B. Jansson, H. Jörnvall, U. Rydberg, L. Terenius & B. L. Vallee

Helicobacter pylori alcohol dehydrogenase

M. Salaspuro

University of Helsinki, Research Unit of Alcohol Diseases, Tukholmankatu 8F, 00290 Helsinki, Finland

Summary

We have recently shown that 34 different *Helicobacter pylori* strains of human and three of animal origin contain alcohol dehydrogenase (ADH). Isoelectric focusing of the enzyme showed activity bands with pI at 7.1 - 7.3, a pattern different from that of gastric mucosal ADHs. The K_m value of *H. pylori* ADH for ethanol oxidation ranges from 64 to 104 mM. Although *H. pylori* ADH was capable of utilizing both NADP and NAD as cofactors in alcohol oxidation, it showed a strong preference for NADP over NAD. At neutral pH *H. pylori* ADH was more effective in aldehyde reduction than in alcohol oxidation. Distinct findings suggest that *H. pylori* ADH could be a metabolic enzyme taking part in ethanol production by fermentation. It is a rather abundant enzyme comprising approx. 0.5% of all bacterial cytosolic proteins. Therefore, the enzyme presumably has a basic role in the functions and maintenance of *H. pylori*. 4-methylpyrazole inhibits *H. pylori* ADH, and suppresses its growth during culture. Bismuth compounds that are commonly used in the treatment of *H. pylori* associated gastric diseases appeared to be potent inhibitors of *H. pylori* ADH. Owing to its high specific activity for ethanol (14 U mg[-1]) under physiological conditions *H. pylori* ADH can also effectively produce acetaldehyde at moderate ethanol levels. This reversed function of the enzyme and the production of the toxic and reactive acetaldehyde could account for at least some of the gastrointestinal morbidity associated with *H. pylori* infection. *H. pylori* lacks aldehyde dehydrogenase activity and can therefore not remove acetaldehyde at least by this pathway.

Introduction

Helicobacter pylori is a gram-negative microaerophilic bacterium which colonizes the upper gastrointestinal tract of more than one in two individuals during their life span (Kosunen *et al.*, 1989). It may be one of the most common bacterial pathogens of humans. *H. pylori* is a major pathogenetic factor behind active chronic gastritis and contributes strongly to peptic ulcer disease (Blaser, 1987; Graham, 1989). Recent evidence has linked it also with the occurrence of gastric cancer (Parsonnet *et al.*, 1991; Nomura *et al.*, 1991; Sipponen *et al.*, 1992; The Eurogast, 1993). A variety of virulence factors have been tentatively identified in *H. pylori* (Blaser, 1990) but the pathogenetic mechanisms whereby bacteria induce gastric injury are so far not well understood.

Many bacteria possess alcohol dehydrogenase (ADH), which enables them to produce energy anaerobically by fermenting different sugars via acetaldehyde to ethanol (Lamed *et al.*, 1981). Bacterial ethanol production is known to occur in the intestine of both humans (Bode *et al.*,

1984) and experimental animals (Krebs and Perkins, 1970; Baraona *et al.*, 1986). We have recently demonstrated that two standard *Helicobacter pylori* strains (NCTC 11637 and 11638) exhibit significant ADH activity both at low and high ethanol concentrations (Roine *et al.*, 1992a; Salmela *et al.*, 1993) (Fig. 1a). The mean ADH activity of these strains was more than ten-fold than that found in *Escherichia coli* and *Campylobacter jejuni* and it was more than 20 times higher than the activity we have found in histologically normal human gastric mucosal biopsy samples (Roine *et al.*, 1992a). In subsequent studies we have been able to show that 34 different *H. pylori* strains of human and three of animal origin contain alcohol dehydrogenase (Höök-Nikanne *et al.*, 1993). In line with the observed for the ADH activity, the cytosols of both helicobacter strains produced in the presence of excess ethanol larger amounts of acetaldehyde than the cytosols of two other ADH-containing bacteria *Escherichia coli* and *Campylobacter jejuni* (Fig. 1b) (Roine *et al.*, 1992a). In sharp contrast to gastric mucosal and liver cells neither *H. pylori* strain studied, however, showed any NAD-linked aldehyde dehydrogenase activity (Salmela *et al.*, 1993). The lack of ALDH in *H. pylori* may favour the accumulation of the very active and toxic acetaldehyde at least in the close vicinity of bacterial

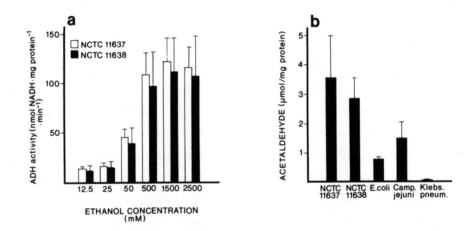

Fig. 1. a. Alcohol dehydrogenase activities of two standard *Helicobacter pylori* strains at different ethanol concentrations (Salmela *et al.*, 1993). b. Acetaldehyde production in the excess of ethanol by two standard *Helicobacter pylori* strains and by three other bacteria (Roine *et al.*, 1992a).

accumulations in gastric mucin layer. Production of acetaldehyde by *H. pylori*, if it also occurs *in vivo*, could account for at least some of the gastroduodenal morbidity associated with the organism.

Characteristics of *Helicobacter pylori* alcohol dehydrogenase

H. pylori alcohol dehydrogenase (HPADH) is active (Salmela *et al.*, 1993) already at an ethanol concentration comparable to the low endogenous ethanol level known to prevail in the stomach for instance during the use of H2-blockers (Bode *et al.*, 1984). The relatively high K_m (65-104mM) of the bacterial enzyme for ethanol oxidation (Roine *et al.*, 1992a; Salmela *et al.*, 1993; Kaihovaara *et al.*, 1994), however, indicates that maximal ethanol oxidizing activity and acetaldehyde production could be observed after drinking of alcoholic beverages. The high optimal pH (9.6) (Salmela *et al.*, 1993) of HPADH is similar to that found for the majority of ADHs from different origins. Significant bacterial ADH activity and acetaldehyde production was, however, detected even at pH 7.4. Accordingly, ADH mediated reactions could be possible also *in vivo* in the neutral mucin layer covering gastric mucosa. Isoelectric focusing of the bacterial enzyme shows activity bands with pI at 7.1-7.3 (Salmela *et al.*, 1993) (Fig. 2), a pattern different from that described previously for gastric mucosal alcohol dehydrogenases (Moreno and Pares, 1991; Yin *et al.*, 1993).

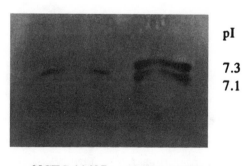

pI

7.3
7.1

NCTC 11637   NCTC 11638

Fig. 2. Isoelectric focusing of ADH of two *H. pylori* strains (NCTC 11637 and NCTC 11638) revealed by activity staining by ethanol (Salmela *et al.*, 1993).

Alcohol dehydrogenase of *Helicobacter pylori* was purified from cytosol of cultured bacteria (strain NCTC 11637) by anion exchange and affinity chromatography (Kaihovaara *et al.*, 1994). On sodium dodecyl sulphate polyacrylamide-gel electrophoresis the 160-fold purified enzyme displayed one protein band with a mobility that corresponded to a M_r of 38.000 (Kaihovaara *et al.*, 1994). Accordingly the subunit size of the enzyme is in the range of the typical values for the medium-chain alcohol dehydrogenases. Purified HPADH represents approximately 0.5% of the bacterial cytosolic protein which supports its metabolic function.

Contrary to human ADH isoenzymes HPADH has a strong preference for NADP ($K_m = 80$ μM) over NAD ($K_m = 4.4$ Mm) as a cofactor in alcohol oxidation (Kaihovaara *et al.*, 1994). Although HPADH has a relatively high K_m for ethanol, it has at the same time several fold higher specific activity for ethanol (32 U mg^{-1}) than that for human gastric isoenzymes (Moreno and Pares, 1991). 4-Methylpyrazole inhibits HPADH in a competitive manner, but in contrast to Class I gastric alcohol dehydrogenases HPADH is inhibited first at millimolar concentrations of 4-methylpyrazole (Salmela *et al.*, 1993). The kinetic differences mentioned above could form the basis for the separation and determination of bacterial and human gastric ADH activities in order to evaluate the possible role of HPADH in gastric "first-pass" metabolism of ethanol. Our preliminary results with a mouse model infected with *Helicobacter felis*, however, indicate that bacterial ADH does not contribute significantly to the total ADH activity of gastric mucosa. This may be due to the rather low bacterial mass in gastric mucin layer as compared to the mass of gastric mucosal cells. On the other hand chronic active gastritis associated with *H. pylori* infection in humans appears to decrease the total ADH activity of gastric mucosa (to be published).

Ethanol and acetaldehyde production by intact *H. pylori*

For the present the properties and amino acid sequences of three NADP-linked alcohol dehydrogenases have been revealed. All these ADHs of *Thermoanaerobium brockii*, *Entamoeba histolytica* and *Clostridium beijerinckii* have been proposed to be metabolic enzymes reducing acetaldehyde to ethanol as the last step in alcoholic fermentation from glucose via pyruvate to ethanol (Lo and Reeves, 1978; Lamed and Zeikus, 1980; Yan *et al.*, 1988). At physiological pH the K_m of HPADH for acetaldehyde (0.9mM, at 25°C) is one order of magnitude lower than the K_m for ethanol (11mM, at 25°C) and the catalytic rate of aldehyde reduction is essentially higher than the rate of alcohol oxidation (Kaihovaara *et al.*, 1994). Accordingly the property of HPADH to reduce acetaldehyde more effectively than to oxidize ethanol under physiological conditions favours its role in alcoholic fermentation. Indeed our most recent results indicate that intact *H. pylori* produce ethanol when grown in micoroaerophilic conditions in broth (Salmela *et al.*, 1994a). However, more information - especially on the redox ratios of NAD/NADH and NADP/NADPH in the cytosol of *H. pylori* - is required before definitive conclusions of the possible metabolic role of HPADH *in vivo* can be drawn.

In the presence of excess ethanol the HPADH catalyzed reaction is reversed and the enzyme produces acetaldehyde (Roine *et al.*, 1992a; Salmela *et al.*, 1993). Also when intact *H. pylori* strains were incubated in sealed vials for two hours at 37°C they formed significant amounts of acetaldehyde already at a low (0.1%) ethanol concentration, and at a higher (2.5%) ethanol concentration acetaldehyde production was even more pronounced (Salmela *et al.*, 1994a). Under physiological conditions (pH 7.4, T 37°C) HPADH has a marked capacity to oxidize ethanol (14 U mg^{-1}) - at least *in vitro* (Kaihovaara *et al.*, 1994). At ethanol concentrations of several hundred millimolar similar to those found in the stomach after ethanol ingestion, HPADH with both a high K_m (65mM) and a high k_{cat} per active site (530 min^{-1}) is fully active (Kaihovaara *et al.*, 1994). Under these conditions the enzyme probably produces acetaldehyde effectively also *in vivo*.

Inhibition of *H. pylori* alcohol dehydrogenase

Although the value of colloidal bismuth subcitrate (CBS) in the treatment of *H. pylori* induced gastric and duodenal diseases is well established less is known about the mechanism by which the drug exerts its effect (Gorbach, 1990). Similarly, omeprazole is known to suppress *H. pylori* (Mainguet *et al.*, 1989), but the mechanism remains speculative. In our hands, CBS already at a drug concentration of 0.01mM inhibited the activity of the cytosolic ADH of the organism by 93% (Roine *et al.*, 1992b). This associated with almost as effective inhibition of acetaldehyde production in the presence of 22mM ethanol (Roine *et al.*, 1992b). In our most recent studies CBS also suppressed acetaldehyde production by intact *H. pylori* (Salmela *et al.*, 1994b). However, rather high drug concentrations were required for this to occur (Salmela *et al.*, 1994b).

Omeprazole inhibited HPADH and suppressed bacterial acetaldehyde formation mediated by the bacterial cytosol to 69% of control at a drug concentration of 0.1mM (Roine *et al.*, 1992b). By contrast, the H_2-receptor antagonists ranitidine and famotidine showed only modest effects on HPADH and acetaldehyde production (Roine *et al.*, 1992b).

As mentioned before 4-methylpyrazole, a well known ADH inhibitor (Salaspuro, 1985), inhibited HPADH and *in vitro* formation of acetaldehyde from ethanol by the organism (Salmela *et al.*, 1993, Kaihovaara *et al.*, 1994). Both phenomena, however, occurred only at relatively high concentrations of the compound (Salmela *et al.*, 1993). Supporting the hypothesis of the role of ADH in the energy metabolism of *H. pylori*, 4-methylpyrazole also inhibited the growth of the bacterium under microaerophilic culture conditions (Salmela *et al.*, 1993). Again, however, as for ADH inhibition, high concentrations of the compound were required for this to occur.

HPADH is an abundant enzyme which prefers NADP(H) as a cofactor contrary to human ADH isoenzymes. The difference in the active site of these enzymes could be used to identify compounds that selectively inhibit bacterial NADP-dependent ADH activities. Furthermore, it would be profitable to study whether some of these compounds, like many other enzyme inhibitors, could be used as antibacterial agents. However, before any major studies on this important subject can be performed more information of the functions of HPADH and especially of its structure is needed. Therefore, one step towards the long-term goal of developing drugs to complement those already available in the treament of *H. pylori* infection, could be the cloning and expression of the HPADH gene.

Potential toxicity of *H. pylori* bacterial acetaldehyde production
Acetaldehyde is both pharmacologically and chemically a very potent and reactive compound, and has been suggested to be a major initiating factor in the pathogenesis of alcoholic liver damage (Salaspuro and Lindroos, 1985; Lieber, 1988; Sorrell and Tuma, 1985; Lauterburg and Bilzer, 1988). Local bacterial acetaldehyde production has been incriminated as a possible pathogenetic factor behind upper respiratory tract cancer in alcoholics (Miyakawa *et al.*, 1986), and in alcohol related rectal carcinogenesis in rats (Seitz *et al.*, 1990). The toxicity of acetaldehyde has been related to several important metabolic and cellular factors.

Covalent binding of acetaldehyde - acetaldehyde adducts
In vitro acetaldehyde has been shown to form adducts with phospholipids (Kenney, 1982) and

proteins (Sorrell and Tuma, 1985) through at least three mechanisms (Lauterburg and Bilzer, 1988). The biochemical background of the possible toxicity associated with the covalent binding of acetaldehyde with tissue proteins is not yet fully understood. Acetaldehyde-protein adducts could inhibit protein secretion (Matsuda *et al.*, 1979), displace pyridoxal phosphate from its binding sites in proteins (Lumeng, 1978) or impair some biological functions of enzyme proteins. According to our preliminary studies similar acetaldehyde-protein adduct formation may occur also in gastric mucosa (Roine *et al.*, 1993). Since acetaldehyde-adduct formation has been shown to occur already at rather low acetaldehyde concentrations, even low endogenous ethanol levels demonstrated in gastric juice (Bode *et al.*, 1984) might initiate the reaction.

Accordingly, conversion of endogenous or exogenous ethanol to acetaldehyde by *H. pylori*, if it occurs also *in vivo*, could lead to the formation of acetaldehyde adducts with gastric mucin or proteins of the gastric epithelial cells in analogy to the adducts formed with hepatic proteins in the liver. This potential acetaldehyde mediated interference with the mucosal defense factors could contribute to *H. pylori* linked gastrointestinal morbidity.

Antibodies against acetaldehyde adducts
It has been shown that mice immunized by acetaldehyde-protein adducts produce specific antibodies against binding epitopes (Israel *et al.*, 1986). Humoral immune response to acetaldehyde-protein adducts is also produced in acute alcoholic liver disease (Hoerner *et al.*, 1988). The immune response has been suggested to contribute to the aggravation or perpetuation of alcoholic liver injury. It remains to be established whether similar immune reactions could occur also in gastric mucosa infected by *H. pylori*.

Acetaldehyde and lipid peroxidation
Under certain *in vitro* conditions the metabolism of acetaldehyde may produce free radicals and mediate lipid peroxidation (Shaw *et al.*, 1981). In addition to the mitochondrial aldehyde dehydrogenase of gastric mucosal cells, acetaldehyde may be oxidized to acetate also via xanthine oxidase (Lewis and Paton, 1982). In this reaction oxygen may be metabolized to the free radical, superoxide, which is a well-known lipid peroxidative agent. Normally free radicals are inactivated by glutathione. Acetaldehyde, however, may bind with glutathione or with cysteine, which is needed in the synthesis of glutathione. As in the liver this might depress

gastric glutathione resulting in a secondary increase of lipid peroxidation (Shaw *et al.*, 1983).

Acetaldehyde and N-nitroso compounds

N-nitroso compounds within gastric juice have been implicated in the etiology of gastric cancer (Pegg, 1984). One of the main precarcinogenic lesions induced by N-nitroso compounds in DNA is the alkylation of guanine at position O^6 (O^6-methylguanine and O^6-ethylguanine)(Kyrtopoulos *et al.*, 1990). On the other hand, gastric mucosal cells have been shown to contain a DNA repair protein, O^6-alkylguanine-DNA-alkyltransferase (AGT), which can eliminate modifications at this premutagenic site (Kyrtopoulos *et al.*, 1990). Human O^6-methylguanine-transferase - as well as other methyltransferases (Garro *et al.*, 1991; Barak and Beckenhauer, 1988) - have, however, been shown to be inhibited by acetaldehyde already at nanomolar concentrations (Espina *et al.*, 1988). Accordingly, acetaldehyde produced by *H. pylori* ADH may severely hamper the AGT mediated repair mechanism by inactivating the gastric enzyme. In rapidly proliferating gastric mucosal cells this effective inactivation of AGT and concomitant inhibition of DNA repair may give rise to a transition mutation - adenine replacing guanine - during cell replication (Loveless, 1969) and ultimately result in the development of gastric cancer. Another possible cocarcinogenic mechanism could be the intragastric nitrate->nitrite->nitrosocarcinogen conversion that has been shown to be enhanced by aldehydes (Hartman, 1982).

Acknowledgements

The author acknowledge permission from W.B. Saunders Co. for Figure 1a and 2, and Pergamon Press for Figure 1a.

References

Barak, A.J. and Beckenhauer, H.C. (1988) The influence of ethanol on hepatic transmethylation. *Alcohol Alcohol* 23: 73-77.

Baraona, E., Julkunen, R., Tannenbaum, L. and Lieber, C.S. (1986) Role of intestinal bacterial overgrowth in ethanol production and metabolism in rats. *Gastroenterology* 90: 103-110.

Blaser, M.J. (1987) Gastric *Campylobacter*-like organisms, gastritis and peptic ulcer disease. *Gastroenterology* 93: 371-383.

Blaser, M.J. (1990) *Helicobacter pylori* and the pathogenesis of gastroduodenal inflammation. *J. Infect. Dis.* 161: 626-633.

Bode, J.C., Rust, S. and Bode, C. (1984) The effect of cimetidine treatment on ethanol formation in the human stomach. *Scand. J. Gastroenterol.* 19: 853-856.

Espina, N., Lima, V., Lieber, C.S. and Garro, A.J. (1988) In vitro and in vivo inhibitory effect of ethanol and acetaldehyde on O6-methylguanine transferase. *Carcinogenesis* 9:761-766.

The Eurogast Study Group: (1993) An international association between *Helicobacter pylori* infection and gastric cancer. *Lancet* 341: 1359-1362.

Garro, A.J., McBeth, D.L., Lima, V. and Lieber, C.S. (1991) Ethanol consumption inhibits fetal DNA methylation in mice: implications for the fetal alcohol syndrome. *Alcohol Clin Exp Res.* 15: 395-398.

Gorbach, S.L. (1990) Bismuth therapy in gastrointestinal diseases. *Gastroenterology* 99:863-875.

Graham, D.Y. (1989) *Campylobacter pylori* and peptic ulcer disease. *Gastroenterology* 96: 615-625.

Hartman, P.E. (1982) Nitrates and nitrites: Ingestion, pharmacodynamics, and toxicology. *Chem. Mutagens* 7: 211-294.

Hoerner, M., Behrens, U.J., Worner, T.M., Blacksberg, I., Braly, L.F., Schaffner, F., et al. (1988) The role of alcoholism and liver disease in the appearance of serum antibodies against acetaldehyde adducts. *Hepatology* 8: 569-574.

Höök-Nikanne, J., Roine, R.P., Salmela, K.S., Kosunen, T.U. and Salaspuro, M. (1993) Alcohol dehydrogenase activities of different Helicobacter strains (abstract). *Gastroenterology* 104: A103.

Israel, Y., Hurwitz, E., Niemela, O. and Arnon, R. (1986) Monoclonal and polyclonal anti bodies against acetaldehyde-containing epitopes in acetaldehyde-protein adducts. *Proc. Natl. Acad. Sci. USA* 83: 7923-7927.

Kaihovaara, P., Salmela, K.S., Roine, R., Kosunen, T.U. and Salaspuro, M. (1994) Purification and characterization of *Helicobacter pylori* alcohol dehydrogenase. *Alcohol Clin. Exp. Res.*, in press.

Kenney, W.C. (1982) Acetaldehyde adducts of phospholipids. *Alcohol Clin. Exp. Res.* 6: 412-416.

Krebs, H.A. and Perkins, J.R. (1970) The physiological role of liver alcohol dehydrogenase. *Biochem. J.* 19: 853-856.

Kosunen, T.U., Höök, J., Rautelin, H.I. and Myllylä, G. (1989) Age-dependent increase of *Campylobacter pylori* antibodies in blood donors. *Scand. J. Gastroenterol.* 24: 110-114.

Kyrtopoulos, S.A., Ampatzi, P., Davaris, P., Haritopoulos, N. and Golematis, B. (1990) Studies in gastric carcinogenesis. IV. O6-methylguanine and its repair in normal and atrophic biopsy specimens of human gastric mucosa. Correlation of O6-alkylguanine-DNA alkyltransferase activities in gastric mucosa and circulating lymphocytes. *Carcinogenesis* 11: 431-436.

Lamed, R. and Zeikus, J.G. (1980) Glucose fermentation pathway of *Thermoanaerobium brockii*. *J. Bacteriol.* 141: 1251-1257.

Lamed, R.J. and Zeikus, J.G. (1981) Novel NADP-linked alcohol-aldehyde/ketone oxido reductase in thermophilic ethanologenic bacteria. *Biochem. J.* 195: 183-190.

Lauterburg, B.H. and Bilzer, M. (1988) Mechanisms of acetaldehyde hepatotoxicity. *J. Hepatol.* 7: 384-390.

Lewis, K.O. and Paton, A. (1982) Could superoxide cause cirrhosis? *Lancet* 2: 188-189.

Lieber, C.S. (1988) Biochemical and molecular basis of alcohol-induced injury to liver and other tissues. *N. Engl. J. Med.* 319: 1639-1650.

Lo, H.-S. and Reeves, R.E. (1978) Pyruvate-to-ethanol pathway in *Entamoeba histolytica*. *Biochem. J.* 171: 225-230.

194

Loveless, A. (1969) Possible relevance of O-6 alkylation of deoxyguanosine to the mutagenicity and carcinogenicity of nitrosamines and nitrosamides. *Nature* 223: 206-207.

Lumeng, L. (1978) The role of acetaldehyde in mediating the deleterious effect of ethanol on pyridoxal 5'-phosphate metabolism. *J. Clin. Invest.* 62: 286-293.

Mainguet, P., Delmee, M. and Debongnie, J.C. (1989) Omeprazole, campylobacter pylori, and duodenal ulcer. *Lancet* 2: 389-390.

Matsuda, Y., Baraona, E., Salaspuro, M. and Lieber, C.S. (1979) Effects of ethanol on liver microtubules and Golgi apparatus. Possible role in altered hepatic secretion of plasma proteins. *Lab. Invest.* 41: 455-463.

Miyakawa, H., Baraona, E., Chang, J.C., Lesser, M.D. and Lieber, C.S. (1986) Oxidation of ethanol to acetaldehyde by bronchopulmonary washings: role of bacteria. *Alcohol Clin. Exp. Res.* 10: 517-520.

Moreno, A. and Pares, X. (1991) Purification and characterization of a new alcohol dehydro genase from human stomach. *J. Biol. Chem.* 266: 1128-1133.

Nomura, A., Stemmermann, G.N., Chyou, P.H., Kato, I., Perez-Perez. G.I. and Blaser, M.J. (1991) Helicobacter pylori infection and the gastric carcinoma among Japanese Americans in Hawaii. *N. Engl. J. Med.* 325: 1132-1136.

Parsonnet, J, Friedman, G.D., Vandersteen, D.P., Chang, Y., Vogelman, J.H., Orentreich, N. and Sibley, R.K. (1991) *Helicobacter pylori* infection and the risk of gastric carcinoma. *N. Engl. J. Med.* 325: 1132-1136.

Pegg, A.E. (1984) Methylation of the O6 position of guanine in DNA is the most likely initiating event in carcinogenesis by methylating agents. *Cancer Invest.* 2: 223-231.

Roine, R.P., Salmela, K.S., Höök-Nikanne, J., Kosunen, T.U. and Salaspuro, M.P. (1992a) Alcohol dehydrogenase mediated acetaldehyde production by Helicobacter pylori - a possible mechanism behind gastric injury. *Life Sci.* 51: 1333-1337.

Roine, R.P., Salmela, K.S., Höök-Nikanne, J., Kosunen, T.U. and Salaspuro, M. (1992b) Colloidal bismuth subcitrate and omeprazole inhibit alcohol dehydrogenase mediated acetaldehyde production by Helicobacter pylori. *Life Sci.* 51: PL195-200.

Roine, R.P., Sillanaukee, P., Itälä, L., Salmela, K.S., Methuen, T., Matysiak-Budnik, T. and Salaspuro, M. (1993) Binding of acetaldehyde to rat gastric mucosa (abstract). *Scand. J. Gastroenterol.* 28: Suppl 197: 72.

Salaspuro, M. (1985) Inhibitors of alcohol metabolism. *Acta. Med. Scand. Suppl.* 703: 219-224.

Salaspuro, M. and Lindros, K. (1985) Metabolism and toxicity of acetaldehyde. In: Seitz, H.K. and Kommerell, B. (eds.): *Alcohol Related Diseases in Gastroenterology.* Berlin: Springer-Verlag, pp. 106-123.

Salmela, K.S., Roine, R.P., Koivisto, T., Höök-Nikanne, J., Kosunen, T.U. and Salaspuro, M. (1993) Characteristics of Helicobacter Pylori alcohol dehydrogenase. *Gastroenterology* 105: 325-330.

Salmela, K.S., Roine, R.P., Höök-Nikanne, J., Kosunen, T.U. and Salaspuro, M. (1994a) Acetaldehyde and ethanol production by *Helicobacter pylori. Scand. J. Gastroenterol.*, 29: in press.

Salmela, K.S., Roine, R.P., Höök-Nikanne, J., Kosunen, T.U. and Salaspuro, M. (1994b) Effect of bismuth and nitecapone on acetaldehyde production by *Helicobacter pylori. Scand. J. Gastroenterol.*, in press.

Seitz, H.K., Simanowski, U.A., Garzon, F.T., Rideout, J.M., Peters, T.J., Koch, A., et al. (1990) Possible role of acetaldehyde in ethanol-related rectal cocarcinogenesis in the rat.

Gastroenterology 98: 406-413.

Shaw, S., Jayatilleke, E., Ross, W.A., Gordon, E.R. and Lieber, C.S. (1981) Ethanol-induced lipid peroxidation: potentiation by long-term alcohol feeding and attenuation by methionine. *J. Lab. Clin. Med.* 98: 417-424.

Shaw, S., Rubin, K.P. and Lieber, C.S. (1983) Depressed hepatic glutathione and increased diene conjugates in alcoholic liver disease. Evidence of lipid peroxidation. *Dig. Dis. Sci.* 28: 585-589.

Sipponen, P., Kosunen, T.U., Valle, J., Riihelä, M. and Seppälä, K. (1992) *Helicobacter pylori* infection and chronic gastritis and gastric cancer. *J. Clin. Pathol.* 45: 319-323.

Sorrell, M.F. and Tuma, D.J. (1985) Hypothesis: alcoholic liver injury and the covalent binding of acetaldehyde. *Alcohol Clin. Exp. Res.* 9: 306-309.

Yan, R.-T., Zhu, C.-X., Golemboski, C. and Chen, J.-S. (1988) Expression of solvent forming enzymes and onset of solvent production in bath cultures of *Clostridium beijerinckii* ("*Clostridium butylicum*"). *Appl. Environ Microbiol.* 54: 642-648.

Yin, S.-J., Chou, F.-J., Chao, S.-F., Tsai, S.-F., Liao, C.-S., Wang, S.-L., Wu, C.-W. and Lee, S.-C. (1993) Alcohol and aldehyde dehydrogenases in human oesophagus: Comparison with stomach enzyme activities. *Alcohol Clin. Exp. Res.* 17: 376-381.

Toward a Molecular Basis of Alcohol Use and Abuse
ed. by B. Jansson, H. Jörnvall, U. Rydberg, L. Terenius & B. L. Vallee
© 1994 Birkhäuser Verlag Basel/Switzerland

Genetic polymorphism of cytochrome P450. Functional consequences and possible relationship to disease and alcohol toxicity

M. Ingelman-Sundberg, I. Johansson, I. Persson, M. Oscarson, Y. Hu, L. Bertilsson[¤],
M.-L. Dahl[¤] and F. Sjöqvist[¤]

Department of Medical Biochemistry and Biophysics, Karolinska institutet, S-171 77 Stockholm, and [¤]Department of Medical Laboratory Sciences and Technology, Karolinska institutet, Huddinge University Hospital, S-141 86 Huddinge, Sweden

Summary

The hepatic cytochrome P450 system participates in the oxidative metabolism of numerous endogenous and exogenous compounds. In total several hundred different P450s have been cloned, but it appears that in humans only about 5-10 isoforms account for the major part of drug metabolism. Some of these are polymorphically distributed in the population. Cytochrome P450 2D6 catalyzes the oxidation of over 25 clinically important drugs, eg neuroleptics, antidepressants and lipophilic β-blockers. Seven % of Caucasians and 1 % of Orientals are defective in this enzyme and clearance of drugs metabolized by the enzyme may be substantially decreased in these individuals, with potentially increased risks for side effects caused by the drug treatment. Some individuals are ultrarapid metabolizers and do not achieve therapeutic drug levels at ordinary doses. The molecular genetic basis of these polymorphisms are presented. Methods for genotyping, which can be of predictive value for a more efficient drug therapy, are discussed. Ethanol-inducible cytochrome P450 2E1 (CYP2E1) oxidizes ethanol and acetaldehyde, in addition to over 80 toxicologically important xenobiotics. Furthermore, this isozyme produces reactive oxy radicals which are implicated in the aetiology of alcoholic liver disease. The gene is polymorphic and a mutation in a putative binding site for HNF1, described to affect gene expression, is more rare among subjects with lung cancer as compared to healthy controls. Further studies might give an answer as to whether any of the polymorphic *CYP2E1* alleles is associated with the sensitivity to obtain alcoholic liver disease.

The cytochrome P450 system

The cytochromes P450 are hemoproteins which participate in the metabolism of numerous endogenous compounds, such as steroids, fatty acids, prostaglandins, neurotransmitters and ketone bodies, as well in the reductive and oxidative metabolism of a huge variety of foreign substances, eg drugs, precarcinogens, solvents, alkaloids and alcohols. In mammals, at least twelve different cytochrome P450 gene families, including 22 subfamilies have been identified. In total, today more than 250 P450 enzymes have been cloned (cf Nelson *et al.*, 1993). It is possible to distinguish two major groups of P-450s: Isozymes in the first group have high specificity for endogenous substrates, are usually noninducible by exogenous compounds and are highly homologous between species. Isozymes belonging to the second group have overlapping substrate specificities, are often inducible by exogenous compounds, participate in the metabolism of xenobiotics and have been poorly conserved during evolution. The latter group consists of members in gene families 1-4, and particularly in gene family 2, numerous structurally similar P450s with different substrate specificities have evolved. The substrate specificities among the xenobiotic metabolizing P450s have been conserved between species only in a few cases, eg CYP1A1, CYP2E1, CYP2D1/6 and CYP3A1/4. This makes extrapolations and interpretations of toxicity studies etc between species difficult.

Genetic polymorphism of human cytochromes P450
Only a limited number of human P450s are important in the metabolism of clinically important drugs. The most important isozymes are CYP1A2, CYP2E1, CYP2D6, CYP3A4 and $CYP2C_{MP}$ and some common drug substrates are listed in Table I. Of these P450 enzymes, CYP2D6 and

Table I. Clinically important cytochromes P450 and their substrates

CYP1A2	Phenacetin	CYP2D6	Metoprolol
	Caffeine		Propranolol
	Paracetamol		Encaimide
	Theophyllin		Amitriptyline
			Desipramine
CYP2E1	Chlorzoxazone		Haloperidol
	Halothane		Perphenazine
	Paracetamol		Zuclopenthixol
	Isoniazid		Codeine
			Dextromethorphan
$CYP2C_{MP}$	Diazepam		
	Hexobarbital	CYP3A4	Cyclosporin
	Mephenytoin		Erythromycin
	Omeprazole		Lidocaine
	Proguanil		Nifedipine

$CYP2C_{MP}$ are polymorphically distributed in the populations. Potentially important polymorphically distributed mutations, affecting enzyme expression have been also described for *CYP1A1* and *CYP2E1* (cf. Ingelman-Sundberg *et al.*, 1992). Among Caucasians, 7 and 3 % are homozygous for defective *CYP2D6* (debrisoquine hydroxylase) and *CYP2C_{MP}* (mephenytoin hydroxylase) genes, respectively, whereas among Orientals the corresponding frequences are 1 and 15-20 % (Bertilsson *et al.*, 1992). The deficiency for drug metabolism is best illustrated by measuring the ratio between the substrate and metabolite in urine (metabolic ratio, MR) collected for a certain period after intake of a probe drug. As illustrated in Fig 1, the MR for debrisoquine, a CYP2D6 substrate, varies more than 10,000-fold between individuals. Poor metabolizers (PM) of debrisoquine have MR > 12.6.

Important pharmacokinetic differences are observed during drug treatment of PM subjects, homozygous for a defective P450 gene, as opposed to extensive metabolizers (EM), heterozygous or homozygous for one functional P450 allele. Thus, drug clearance is substantially reduced among PM for CYP2D6 (Dahl-Puustinen *et al.*, 1989) or $CYP2C_{MP}$ (Andersson *et al.*, 1992), as exemplified using perphenazine and omeprazole as test drugs, respectively. Furthermore, a higher incidence of side effects has been reported among patients PM for CYP2D6 substrates after treatment with antidepressants, as opposed to the situation among subjects belonging to the EM group (cf. Dahl and Bertilsson, 1993).

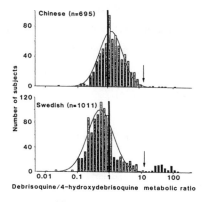

Fig. 1. Distribution of the MR for debrisoquine among 695 Chinese and 1011 Swedish healthy subjects. The arrows indicate MR = 12.6, the antimode between EM and PM. Data from Bertilsson *et al.*, 1992.

Genetic basis for the PM phenotype

The detrimental mutations of $CYP2C_{MP}$ that cause the PM phenotype of mephenytoin hydroxylase have not yet been identified. Isozymes of the *CYP2C* family, in particular CYP2C18 and CYP2C19 catalyze the hydroxylation of the probe drug *S*-mephenytoin. Correlation between phenotype and mRNA is high in the case of expression of *CYP2C18* (Romkes *et al.*, 1991). Further work is needed to clarify this problem and it might be expected that several different types of defective alleles are distributed among Caucasians and Orientals.

By contrast, the genetic basis for the debrisoquine hydroxylase polymorphism has been clarified to a major extent. Cloning of *CYP2D6* and subsequent sequence analysis revealed the occurrence of mainly three defective Caucasian alleles (Gonzalez *et al.*, 1988; Heim and Meyer, 1990; Gaedigk *et al.*, 1991): i) a frame shift mutation (*CYP2D6A*), ii) a mutation causing defect splicing (*CYP2D6B*) and iii) an allele having a deletion of the entire functional gene (*CYP2D6D*). These mutations are the major causes for the lack of functionally active enzyme among the PM individuals (cf. Table II) and genotype analysis can predict with 92-98% accuracy the phenotype of the subjects (Broly *et al.*, 1991, Dahl *et al.*, 1992). The method for genotyping involves allele-specific PCR amplification of the gene region carrying the mutation, followed by mutation specific PCR-analysis (Heim and Meyer, 1990). This two step PCR procedure is required because of the pseudogenes with sequences with very high homology to the active *CYP2D6* gene.

There are major interethnic differences in the distribution of these defective alleles, of which *CYP2D6A* and *CYP2D6B* are virtually absent from Chinese (Johansson *et al.*, 1991, Wang *et al.*, 1993) and an African population (Masimirembwa *et al.*, 1993; cf. Table II). By contrast, the *CYP2D6D* allele, with the active gene deleted, is more evenly spread among the different ethnic groups (Table II). These differences account for the relative absence of poor metabolizers for debrisoquine among Orientals (cf. Kalow and Bertilsson, 1993) and among certain black populations (cf. Masimirembwa *et al.*, 1993).

Table II. Frequency of *wt* and deficient *CYP2D6* alleles in some different ethnic groups

Allele	Functional mutation	Consequence	Allele frequency (%)			
			Caucasians[a]	Chinese[b]	Zimbabw.[c]	Black Americans[d]
wt	-	-	60.3	56	94	86
2D6A	A2637del	Frame shift	2.3	0	0	0.2
2D6B	G1934A	Splicing defect	33.3	0.4	1.8	8.5
2D6D	Gene deletion		4.0	5.7	3.9	6.0

[a]from Dahl *et al.*, 1992; Broly *et al.*, 1991; [b]from Johansson *et al.*, 1991 and Wang *et al.*, 1993; the Chinese also have the slightly deficient *CYP2D6Ch* alleles (38 %); [c]from Masimirembwa *et al.*, 1993; [d]from Evans *et al.*, 1993.

The mean MR for debrisoquine among Chinese EM is higher than observed among Caucasian EM (Fig. 1). Plasma clearance of certain drugs has been described to be lower in Chinese and antidepressants are prescribed at lower doses in China (cf. Kalow and Bertilsson, 1993). The molecular basis for the higher MR is inherent in the high frequency of a haplotype in Chinese, characterized by the presence of two mutated *CYP2D6Ch* genes, yielding a longer fragment upon cleavage with *Xba I* than the normal haplotype (Fig. 2), and less enzyme expressed (Yue *et al.*, 1989, Ingelman-Sundberg *et al.*, 1992, Johansson *et al.*, 1993b). The *CYP2D6Ch* genes have mutations in exons 1 and 9, causing amino acid substitutions, as well as a few mutations in the 5'-flanking region (Johansson *et al.*, 1993b).

Ultrarapid metabolizers of debrisoquine

Certain subjects have very low MR for debrisoquine (cf. Fig 1). Genetic analysis of such individuals, described to require very high doses of antidepressants in order to reach therapeutic plasma levels (Bertilsson *et al.*, 1985), revealed that they carried an allele with two functional *2D6* genes instead of one (Bertilsson *et al.*, 1993). These genes (*CYP2D6L*) had two mutations in exons 6 and 9, causing amino acid substitutions, but the presence of this gene in individuals with one functional *2D6* gene did not influence their MR as compared to subjects carrying the *wt* gene (Johansson *et al.*, 1993a), indicating the lack of importance of these two mutations for enzyme expression and function. Three members of a Swedish family having very low MR for

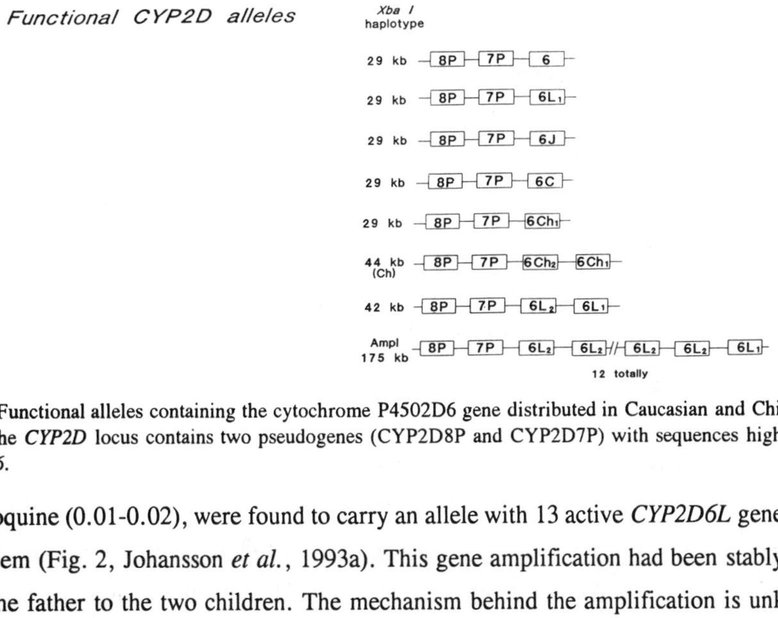

Fig. 2. Functional alleles containing the cytochrome P4502D6 gene distributed in Caucasian and Chinese populations. The *CYP2D* locus contains two pseudogenes (CYP2D8P and CYP2D7P) with sequences highly similar to *CYP2D6*.

debrisoquine (0.01-0.02), were found to carry an allele with 13 active *CYP2D6L* genes arranged in tandem (Fig. 2, Johansson *et al.*, 1993a). This gene amplification had been stably inherited from the father to the two children. The mechanism behind the amplification is unknown, but involves 12-fold duplication of the entire gene. The absence of pressure for its removal could be i) compensation for functions exerted by another defective gene or simply, ii) the absence of any endogenously important function of the *CYP2D6* gene.

In summary, the *CYP2D* locus is highly polymorphic. At least 11 different functional and non-functional alleles are distributed in Caucasians and Chinese (cf. Table II & Fig 2). Using genotype analysis it is possible to predict the metabolic capacity of a subject and this might constitute an efficient tool for a more effective and individualized pharmacotherapy.

Ethanol-inducible cytochrome P4502E1

Ethanol-inducible cytochrome P450 (CYP2E1) has received much attention because of its potentially important role in the toxicity of many chemicals, among them ethanol. Ethanol-inducible cytochrome P450 2E1 (CYP2E1) is mainly found in the liver and here almost entirely in the three to four layers of hepatocytes most proximal to the central vein (Ingelman-Sundberg *et al.*, 1988). The molecular basis for this heterogeneous distribution has been shown to be inherent in a regioselective expression of the *CYP2E1* gene in the perivenous hepatocytes (Johansson *et al.*, 1990). The regioselective hepatic CYP2E1 expression is of interest because ethanol, acetaminophen, *N*-nitrosoamines, solvents and other CYP2E1-specific substrates cause a destruction of mainly the centrilobular liver region (cf. Terelius *et al.*, 1992). Besides ethanol, also acetaldehyde provides an efficient substrate having an affinity about 3 orders of magnitude higher than that of the parent alcohol (Terelius *et al.*, 1991).

CYP2E1 and radical production

CYP2E1 has been found to generate higher amounts of reactive oxygen species than other P450-forms and to effectively oxidize ethanol partially according to a radical-mediated mechanism (See Ingelman-Sundberg *et al.*, 1993). Oxygen radicals generated by CYP2E1 have the capability to initiate membranous lipid peroxidation (Ekström *et al.*, 1989; Castillo *et al.*, 1992; Dai *et al.*, 1993).

CYP2E1 and liver cirrhosis

In the Tsukamoto-French model it is possible to achieve pathophysiological changes of the liver of similar type as registered in the human (Tsukamoto *et al.*, 1985). The role of CYP2E1 in this process can be studied by the use of *eg* different CYP2E1 inhibitors such as isoniazid and diallylsulfide. Upon ethanol-treatment, the rate of microsomal NADPH-dependent lipid peroxidation, inhibitable *in vitro* by anti-CYP2E1 IgG (Ekström and Ingelman-Sundberg, 1989) is increased by 20-200-fold, depending upon the diet (Takahashi *et al.*, 1992). The presence of diallylsulfide in the liquid diet causes destruction of CYP2E1 rather selectively, prevents ethanol induction of this isozyme, inhibits the microsomal NADPH-dependent lipid peroxidation and inhibits to a certain extent the pathology score (Morimoto *et al.*, 1993). Treatment of rats with isoniazid in the liquid ethanol diet, causes inhibition of ethanol-induced elevation of serum transferases (French *et al.*, 1993). This suggests an important role of CYP2E1 generated radicals in the development of alcoholic liver disease. The toxic effects of lipid peroxide products might be exerted through activation of Kupffer cells and Ito cells producing collagen and cytokines (cf. Ingelman-Sundberg *et al.*, 1993).

Regulation of CYP2E1

CYP2E1 is regulated at the transcriptional, pretranslational, translational and posttranslational levels. Transcriptional activation of the gene is seen after extensive starvation (Johansson *et al.*, 1990) and a major basis for the accentuated induction of CYP2E1 during chronic ethanol treatment appears to be connected with increased rate of gene transcription seen at high ethanol concentrations (Badger *et al.*, 1993; Ronis *et al.*, 1993). Continuous infusion of ethanol causes the blood (BAC) and urinary alcohol (UAC) concentrations to fluctuate in a cyclic manner (Tsukamoto *et al.*, 1985; Badger *et al.*, 1993) and CYP2E1 levels follow in a proportional manner up to at least BAC = 400 mg/dl. At BAC < 200 mg/dl this increase is entirely caused by posttranslational stabilization, whereas at BAC > 200 mg/dl transcriptional gene activation is seen (Badger *et al.*, 1993; Ronis *et al.*, 1993). The posttranslational mechanism is mediated mainly by stabilization of the enzyme by isozyme specific substrates (Eliasson *et al.*, 1990). The mechanism involves substrate-regulation of a cAMP-dependent phosphorylation of CYP2E1 on Ser^{129} (Eliasson *et al.*, 1990) and subsequent degradation in the endoplasmic reticulum (Eliasson *et al.*, 1992, Zhukov *et al.*, 1993).

Genetic polymorphism of CYP2E1

The *CYP2E1* gene is polymorphically distributed as evident from RFLP and the restriction enzymes *Taq I, Rsa I* and *Dra I* (McBride *et al.*, 1987; Uematsu *et al.*, 1991). The RFLPs using these three restriction enzymes are only partially linked and the frequencies of the rare alleles in Caucasians are in the range of 0.10 to 0.18 (Table III). In addition, a polymorphic site at -1019bp, within a nuclear protein binding region, partially linked to the *Dra I* polymorphism (Persson *et al.*, 1993) and being suggested to affect the extent of *CYP2E1* gene expression, has been described (Hayashi *et al.*, 1991).

Table III. Genetic polymorphism of ethanol-inducible cytochrome P450 2E1 (CYP2E1)

Polymorphism Alleles	Location	Polymorphic site	Frequency of rare allele[a]	Comments
Rsa I/Pst I c1→c2	5'-flanking	-1019 CT[b] -1259 GC	0.05	The c2 allele causes higher expression in *in vitro* transcr. systems[b]; less frequent among subjects with lung cancer[a]
Rsa I H→I	intron 5[c]		0.18	
Dra I D→C	intron 6[c]	7668 TA[a]	0.10	The C-allele has been suggested to be more rare among Japanese lung cancer patients as compared to controls. This is not the case among Caucasians.[a]
Taq I A_1→A_2	intron 7[d]	9931-4	0.12	

[a]data from Persson *et al.*, 1993; the allele frequencies concern a Swedish population, whereas among Orientals the rare alleles are 2-5-fold more common; [b]from Hayashi *et al.*,1991; [c]from Uematsu *et al.*, 1991; [d]from McBride *et al.*, 1987.

The rare allele (c2) has a mutation in a putative binding site for HNF1 (Hayashi *et al.*, 1991) and has been found to be less common among subjects with lung cancer, as compared to controls (Persson *et al.*, 1993).

Conclusions

The distribution of some clinically important cytochrome P450 genes is highly polymorphic. Genotyping of patients for alleles indicative for very low or very high rate of drug metabolism might be of importance for a more effective drug therapy, with less side effects and better treatment efficiency as benefits. With respect to CYP2E1, phenotype studies for inducibility and further genetic analysis are required before it might be possible to relate specific allelic forms with increased risk for cancer and alcoholic liver disease.

Acknowledgements

The work in the authors laboratory is supported by grants from the Swedish Alcohol Research Fund, the Swedish Natural Science Research Council and from the Swedish Medical Research Council (grants to MIS and FS).

References

Andersson, T., Regårdh, C.G., Lou, Y.-C., Zhang, Y., Dahl, M.-L. and Bertilsson, L. (1992) Polymorphic hydroxylation of *S*-mephenytoin and omeprazole metabolism in Caucasian and

Chinese subjects. *Pharmacogenetics* 2: 25-31.

Badger, T.M., Huang, J., Ronis, M. and Lumpkin, C.K. (1993) Induction of cytochrome P450 2E1 during chronic ethanol exposure occurs via transcription of the CYP2E1 gene when blood alcohol concentrations are high. *Biochem. Biophys. Res. Commun.* 190:780-785.

Bertilsson, L., Åberg-Wistedt, A., Gustafsson, L.L. and Nordin, C. (1985) Extremely rapid hydroxylation of debrisoquine: a case report with implications for treatment with nortriptyline and other tricyclic antidepressants. *Ther. Drug Monit.* 7: 478-480.

Bertilsson, L., Lou, Y.Q., Du, Y.L., Liu, Y., Kuang, T.Y., Liao, X.M., Wang, K.Y., Reviriego, J., Iselius, L. and Sjöqvist F. (1992) Pronounced differences between native Chinese and Swedish populations in the polymorphic hydroxylations of debrisoquine and S-mephenytoin. *Clin. Pharmacol. Ther.* 51: 388-397.

Bertilsson, L., Dahl, M-L., Sjöqvist, F., Åberg-Wistedt, A., Humble, M., Johansson, I., Lundqvist, E. and Ingelman-Sundberg, M. (1993) Molecular basis for rational megaprescribing in ultrarapid hydroxylators of debrisoquine. *Lancet* 341: 63.

Broly, F., Gaedigk, A., Heim, M., Eichelbaum, M., Morike, K. and Meyer, U.A. (1991) Debrisoquine/sparteine hydroxylation genotype and phenotype: Analysis of common mutations and alleles of *CYP2D6* in European population. *DNA Cell Biol.* 10: 545-558.

Castillo, T., Koop, D.R., Kamimura, S., Triadafilopoulos, G. and Tsukamoto, H. (1992) Role of cytochrome P-450 2E1 in ethanol-, carbon tetrachloride- and iron-dependent microsomal lipid peroxidation. *Hepatology* 16: 992-996.

Dahl-Puustinen, M.L., Lidén, A., Alm, C., Nordin, C. and Bertilsson, L. (1989) Disposition of perphenazine is related to polymorphic debrisoquine hydroxylation in human beings. *Clin. Pharmacol. Ther.* 46: 78-81.

Dahl, M.L., Johansson, I., Porsmyr-Palmertz, M., Ingelman-Sundberg, M. and Sjöqvist, F. (1992) Analysis of the *CYP2D6* gene in relation to debrisoquine and desipramine hydroxylation in a Swedish population. *Clin. Pharmacol. Ther.* 52: 12-17.

Dahl, M.L. and Bertilsson (1993) Genetically variable metabolism of antidepressants and neuroleptic drugs in man. *Pharmacogenetics* 3: 61-70.

Dai, Y., Rashba-Step, J. and Cederbaum, A. (1993) Stable expression of human cytochrome P4502E1 in HepG2 cells: Characterization of catalytic activities and production of reactive oxygen intermediates. *Biochemistry* 32: 6928-6937.

Ekström, G. and Ingelman-Sundberg, M. (1989) Rat liver microsomal NADPH-supported oxidase activity and lipid peroxidation dependent on ethanol-inducible cytochrome P-450. *Biochem. Pharmacol.* 38: 1313-1319.

Eliasson, E., Johansson, I. and Ingelman-Sundberg, M. (1990) Substrate, homone and cAMP-dependent regulation of cytochrome P450 degradation. *Proc. Natl. Acad. Sci. USA* 87: 3225-3229.

Eliasson, E., Mkrtchian, S. and Ingelman-Sundberg, M. (1992) Hormone- and substrate regulated intracellular degradation of cytochrome P450 (2E1) involving MgATP-activated rapid proteolysis in the endoplasmic reticulum membranes. *J. Biol. Chem.* 267: 15765-15769.

Evans, W.E., Relling, M.V., Rahman, A., McLeod, H.L., Scott, E.P. and Lin, J.S. (1993) Genetic basis for low prevalence of deficient CYP2D6 oxidative drug metabolism in black Americans. *J. Clin. Invest.* 91: 2150-2154.

French, S.W., Wong, K., Jui, L., Albano, E., Hagbjörk, A.-L. and Ingelman-Sundberg, M. (1993) Effect of ethanol on cytochrome P450 (CYP2E1), lipid peroxidation and serum protein adduct formation in relation to liver pathology pathogenesis. *Exp. Mol. Pathol.* 58: 61-75.

206

Gaedigk, A., Blum, M., Gaedigk, R., Eichelbaum, M. and Meyer, U.A. (1991) Deletion of the entire cytochrome P450 CYP2D6 gene as a cause of impaired drug metabolism in poor metabolizers of the debrisoquine/sparteine polymorphism. *Am. J. Hum. Genet.* 48: 943-950.

Gonzalez, F.J., Skoda, R.C., Kimura, S., Umeno, M., McBride, O.W., Gelboin, H.V., Hardwick, J.P. and Meyer, U.A. (1988) Characterization of the common genetic defect in humans deficient in debrisoquine metabolism. *Nature* 331: 442-446.

Hayashi, S.-I., Watanabe, J. and Kawajiri, K. (1991) Genetic polymorphism in the 5'-flanking region change the transcriptional regulation of the human cytochrome P450IIE1 gene. *J. Biochem.* 110: 559-565.

Heim, M.H. and Meyer, U.A. (1990) Genotyping of poor metabolisers of debrisoquine by allele-specific PCR amplification. *Lancet* 336: 529-532.

Ingelman-Sundberg, M., Johansson, I., Penttilä, K.E., Glaumann, H. and Lindros, K.O. (1988) Centrilobular expression of ethanol-inducible cytochrome P-450 (IIE1) in rat liver. *Biochem. Biophys. Res. Commun.* 157: 55-60.

Ingelman-Sundberg, M., Johansson, I., Persson, I., Lundqvist, E., Dahl. M.-L., Bertilsson, L. and Sjöqvist, F. (1992) Genetic polymorphisms of cytochromes P450: Interethnic differences and relationship to incidence of lung cancer. *Pharmacogenetics* 2: 264-271.

Ingelman-Sundberg, M., Johansson, I., Hu, Y., Terelius, Y., Eliasson, E., Clot, P. and Albano, E. (1993) Ethanol-inducible cytochrome P4502E1: Genetic polymorphism, regulation and possible role in the etiology of alcohol-induced liver disease. *Alcohol,* in press.

Johansson, I., Lindros, K.O., Eriksson, H. and Ingelman-Sundberg, M. (1990) Transcriptional control of CYP2E1 in the perivenous liver region and during starvation. *Biochem. Biophys. Res. Commun.* 173: 331-338.

Johansson, I., Yue, Q.Y., Dahl, M.L., Heim, M., Säwe, J., Bertilsson, L., Meyer, U.A., Sjöqvist, F. and Ingelman-Sundberg, M. (1991) Genetic analysis of the interethinic difference between Chinese and Caucasians in the polymorphic metabolism of debrisoquine and codeine. *Eur. J. Clin. Pharmacol.* 40: 553-556.

Johansson, I., Lundqvist, E., Bertilsson, L., Dahl, M.L., Sjöqvist, F. and Ingelman-Sundberg, M. (1993a) Inherited amplification of an active gene in the cytochrome P4502D-locus as a cause of ultrarapid metabolism of debrisoquine. *Proc. Natl. Acad. Sci. USA,* in press.

Johansson, I., Oscarsson, M., Bertilsson, L., Yue, Q.Y., Dahl, M.-L., Sjöqvist, F. and Ingelman-Sundberg, M. (1993b) Genetic analysis of the Chinese *CYP2D* locus. Characterization of variant *CYP2D6* genes present in subjects with diminished capability for debrisoquine hydroxylation. Manuscript.

Kalow, W. and Bertilsson, L. (1993) Interethnic factors affecting drug responses. In: Testa, B. and Meyer, U.A. (eds.): *Advances in Drug Research,* Academic Press Inc, London 1993, in press.

Masimirembwa, C.M., Johansson, I., Hasler, J.A. and Ingelman-Sundberg, M. (1993) Genetic polymorphism of cytochrome P450 2D6 in Zimbabwean population. *Pharmacogenetics,* in press.

McBride, O.W., Umeno, M., Gelboin, H.V. and Gonzalez, F.J. (1987) A *Taq I* polymorphism in the human P450IIE1 gene on chromosome 10 (CYP2E) *Nucleic Acids Res.* 15: 10071.

Morimoto, M., Hagbjörk, A.-L., Nanji, A.A., Ingelman-Sundberg, M., Lindros, K. O., Albano, E. and French, S.W. (1993) Role of CYP2E1 in alcoholic liver disease pathogenesis. *Alcohol,* in press.

Nelson, D.R., Kamataki, T., Waxman, D.J., Guengerich, F.P., Estabrook, R.W., Feyereisen,

R., Gonzalez, F.J., Coon, M.J., Gunsalus, I.C., Gotoh, O., Okuda, K. and Nebert, D.W. (1993) The P450 superfamily: Update on new sequences, gene mapping, accession numbers, early trivial names of enzymes and nomenclature. *DNA and Cell Biol.* 12: 1-51.

Persson, I., Johansson, I., Bergling, H., Dahl, M.-L., Högberg, J., Rannug, A. and Ingelman-Sundberg, M. (1993) Genetic polymorphism of *CYP2E1* in a Swedish population. Relationship to the occurrence of lung cancer. *FEBS Lett.* 319: 207-211.

Romkes, M., Faletto, M.B., Blaisdell, J.A., Rauchy, J.L. and Goldstein, J.A. (1991) Cloning and expression of complementary DNAs for multiple members of the human cytochrome P450IIC subfamily. *Biochemistry* 30: 3247-3255.

Ronis, M.J.J., Crough, J., Mercado, C., Irby, D., Valentine, C., Lumpkin, C.K., Ingelman-Sundberg, M. and Badger, T.M. (1993) Cytochrome P450 CYP2E1 induction during alcohol exposure occurs by a two step mechanism associated with blood alcohol concentrations in rats, *J. Pharm. Exp. Ther.* 264: 944-950.

Takahashi, H., Johansson, I., French, S.W. and Ingelman-Sundberg, M. (1992) Effects of dietary fat consumption on activities of the microsomal ethanol oxidizing system and ethanol-inducible cytochrome P450 in the liver of rats chronically fed ethanol. *Pharmacol. Toxicol.* 70: 347-352.

Terelius, Y., Norsten-Höög, C., Cronholm, T. and Ingelman-Sundberg, M. (1991) Acetaldehyde as an efficient substrate for ethanol-inducible cytochrome P450 (CYP2E1). *Biochem. Biophys. Res. Commun.* 179: 689-694.

Terelius, Y., Lindros, K. O., Albano, E. and Ingelman-Sundberg, M. (1992) Isozyme-specificity of cytochrome P450-mediated hepatotoxicity. In: Rein, H. and Ruckpaul, K. (eds.): *Frontiers of Biotransformation*, vol 8, Akademie Verlag, Berlin, pp.187-232.

Tsukamoto, H., French, S.W., Benson, N., Delgado, G., Rao, G.A., Larkin, E.C. and Largman, C. (1985) Severe and progressive steatosis and focal necrosis in rat liver induced by continuous intragastric infusion of ethanol and low fat diet. *Hepatology* 5: 224-232.

Uematsu, F., Kikuchi, H., Ohmachi, T., Sagami, I., Motomiya, M., Kamataki, T., Komori, M. and Watanabe, M. (1991) Restriction fragment length polymorphism of the human cytochrome P450IIE1 gene. *Nucleic Acids Res.* 19: 2803.

Wang, S.L., Huang, J., Lai, M.D., Liu, B.H. and Lai, M.L. (1993) Molecular basis genetic variation in debrisoquine hydroxylation in Chinese subjects: Polymorphism in RFLP and DNA sequence of CYP2D6. *Clin. Pharmacol. Ther.* 53: 410-418.

Yue, Q.Y., Bertilsson, L., Dahl-Paustinen, M.L., Säwe, J., Sjöqvist, F., Johansson, I. and Ingelman-Sundberg, M. (1989) Dissociation between debrisoquine hydroxylation phenotype and genotype among Chinese. *Lancet* ii: 870.

Zhukov, A., Werlinder, V. and Ingelman-Sundberg, M. (1993) Purification and characterization of two membrane bound serine proteinases from rat liver microsomes active in degradation of cytochrome P450. *Biochem. Biophys. Res. Commun.*, in press.

Toward a Molecular Basis of Alcohol Use and Abuse
ed. by B. Jansson, H. Jörnvall, U. Rydberg, L. Terenius & B. L. Vallee
© 1994 Birkhäuser Verlag Basel/Switzerland

Serotonin-altering medications and desire, consumption and effects of alcohol - treatment implications

C.A Naranjo and K.E. Bremner

Psychopharmacology Research Program, Sunnybrook Health Science Centre, 2075 Bayview Ave, Room E246, Toronto, Ontario M4N 3M5, Canada, and Departments of Pharmacology, Psychiatry and Medicine, University of Toronto and Addiction Research Foundation, Toronto

Summary

The relationship between serotonin neurotransmission and alcohol consumption (AC) was first determined in preclinical studies. AC generally increases following treatments which decrease serotonin activity, and levels of 5-HT and metabolites are low in some brain regions of alcohol-preferring rats. Pharmacological treatments which enhance serotonergic neurotransmission (uptake inhibitors, releasers, agonists) consistently reduce AC in rats. Serotonin uptake inhibitors (SUI; e.g., citalopram, fluoxetine) have been studied extensively in humans. In several double-blind randomized, placebo-controlled trials, SUI consistently decreased short-term (2-4 weeks) AC by averages of 15% to 20% in nondepressed mildly/moderately dependent alcoholics who received no other treatment. Some subjects decreased AC by up to 60%. The effects of SUI on AC were dose-dependent and not related to side effects (few and mild) or changes in anxiety or depression (not observed). SUI decreased desire to drink and liking for alcohol, suggesting a mechanism of action, to be considered in the development of treatments to reduce AC and prevent relapse. However, while an adjunctive brief psychosocial intervention enhanced the short-term effect of a SUI, the long-term (12-week) effects of SUI and placebo were similar. Other drugs acting on the 5-HT system have been tested in humans, but results are inconclusive. For example, buspirone, a 5-HT_{1A} receptor partial agonist, reduced anxiety and alcohol craving, but not AC; a 5-HT partial agonist, m-CPP, increased craving in abstinent alcoholics; modest reductions in AC were observed with a 5-HT_3 antagonist, ondansetron (0.5 mg/day, but not 4 mg/day). Ritanserin, a 5-HT_2 antagonist, reduced desire to drink and prevented relapse in a small (n=5) study, and there was some indication that it reduced desire to drink and enhanced alcohol effects without reducing AC, in another study. The therapeutic potential of these medications is being studied. SUI and other serotonin-altering medications are promising new neuropharmacological treatments for AC.

Parts of this manuscript have been published in abstracts and papers by the authors.

The views expressed are those of the authors and do not necessarily reflect those of the Addiction Research Foundation.

Introduction

The relationship between central serotonergic neurotransmission and alcohol intake was first determined in preclinical studies. Many agents have been tested and the results in general suggest that treatments which increase serotonin activity decrease ethanol intake. For example, the serotonin precursors, tryptophan and 5-hydroxytryptophan, and serotonin agonists, such as quipazine and MK212, reduced alcohol intake in rats in free-choice paradigms (Zabik *et al.*, 1985; Lawrin *et al.*, 1986; Amit and Smith, 1992). In particular, all selective serotonin uptake inhibitors tested, including zimeldine, citalopram, fluvoxamine and fluoxetine, consistently reduced alcohol intake in rats (Lawrin *et al.*, 1986; Engel *et al.*, 1992; Higgins *et al.*, 1992). Conversely, manipulations which reduce central serotonin activity usually resulted in increased alcohol intake (Engel *et al.*, 1992; Higgins *et al.*, 1992). Furthermore, low levels of serotonin and/or 5-hydroxyindoleacetic acid have been found in some brain regions of alcohol-preferring rats (McBride *et al.*, 1992) and in cerebrospinal fluid of alcoholics (Ballenger *et al.*, 1979).

These results led to the clinical testing of serotonergic medications for decreasing alcohol consumption. The first clinical trials were conducted with the serotonin uptake inhibitor zimeldine (Naranjo *et al.*, 1984a; Amit *et al.*, 1985). The results indicated the therapeutic potential of these medications, several of which are now available for human use. Therefore we systematically tested the effects of other serotonin uptake inhibitors, citalopram, viqualine and fluoxetine, on short-term ethanol intake and other addictive behaviors in humans with mild to moderate alcohol dependence. Also, citalopram was combined with a brief psychosocial intervention in order to potentiate its effects and evaluate the long-term outcome.

The results of these and other studies confirmed the short-term efficacy of serotonin uptake inhibitors. Decreases in desire to drink were also reported in some clinical trials (Amit *et al.*, 1985; Gorelick and Paredes, 1992; Naranjo *et al.*, 1989), leading to the testing of citalopram and fluoxetine in an experimental paradigm designed to assess interest, desire and liking of alcohol. The results indicated that decreases in interest and desire (urge to drink) and liking (reinforcing effects) of alcohol may mediate the effect of serotonin uptake inhibitors (Naranjo *et al.*, 1992a; 1992b).

Recent interest in serotonin and the discovery of receptor subtypes have led to studies of the effects of various serotonin agonists and antagonists on alcohol intake, craving, desire, and reinforcing effects. The results have not been consistent and the receptor subtypes involved in the regulation of alcohol intake and effects have not yet been identified. However, research with serotonin-altering medications has contributed to progress in the pharmacotherapy of alcoholism, and current and future studies may lead to improvements in treatment.

Effects of serotonin uptake inhibitors on alcohol intake in humans

The methodology and main results of our initial studies with serotonin uptake inhibitors are described in detail elsewhere (Naranjo et al., 1984a; 1987; 1989; 1990a). Subjects had mild to moderate alcohol dependence (American Psychiatric Association 1980; 1987), and were non-depressed, socially stable, and did not abuse or only occasionally abused other drugs (other than cigarette smoking).

After the two-week baseline periods, subjects were randomized, double-blind, to receive a serotonin uptake inhibitor (zimeldine 200 mg/day, citalopram 20 or 40 mg/day, viqualine 100 or 200 mg/day, or fluoxetine 40 or 60 mg/day) or placebo for two or four weeks according to the study protocols. They attended weekly/bi-weekly assessments but no other treatment or advice was offered. Several objective measures confirmed compliance with medication and accurate reporting of alcohol consumption. None of the lower doses of serotonin uptake inhibitor significantly changed alcohol intake compared with placebo or baseline. Zimeldine 200 mg/day and citalopram 40 mg/day, viqualine 200 mg/day, and fluoxetine 60 mg/day all decreased short-term (2 to 4 weeks) alcohol intake by averages of 14% to 20% from baseline, with some subjects decreasing their alcohol intake by up to 60%. The placebos did not change alcohol intake significantly from baseline (less than 2% average).

Consistent results in several human studies are required to establish the potential clinical usefulness of a new medication. Our results with zimeldine were confirmed in another study (Amit et al., 1985), in which 12 social drinkers reduced their alcohol intake during inpatient experimental drinking sessions. Although zimeldine was later withdrawn from the market because of hepatic and neurologic toxicity (Fagius et al., 1985; Naranjo et al., 1990b), other serotonin uptake inhibitors, such as fluoxetine, citalopram, sertraline and paroxetine, have been approved for clinical use as antidepressants in many countries. Recently some of these drugs have been tested for their effects on alcohol consumption. For example, a double-blind, placebo-controlled inpatient study was conducted with fluoxetine (Gorelick and Paredes, 1992). Twenty alcohol-dependent males, staying on a locked hospital ward with measured alcoholic drinks available in a fixed interval drinking decision paradigm 13 times each day, received fluoxetine 20 to 60 mg/day (n=10) or placebo daily (n=10) for 28 days after a 3-day baseline period. The subjects in the fluoxetine group decreased their alcohol intake during only the first week of treatment, by 14% from baseline. During the last 3 weeks, their alcohol intake was not significantly different from baseline. The placebo group showed no significant changes at any time period (less than 4% from baseline). Therefore, the short-term effects of fluoxetine were confirmed.

Another study combined a serotonin uptake inhibitor with a standardized nonpharmacological treatment for alcohol abuse in order to maximize the pharmacological effect and assess long-term effects. Mildly/moderately dependent alcoholics received, double-blind, citalopram 40 mg/day (n=31) or placebo (n=31) in conjunction with a brief cognitive-behavioral treatment program (Sanchez-Craig and Wilkinson, 1987). The short-term reduction in alcohol intake by citalopram (21.8%) (p<0.05 compared with placebo) was potentiated (to 39.2%) by the nonpharmacological intervention (Naranjo *et al.*, 1992c). However, during the entire 12-week treatment period the effects of citalopram and placebo were not significantly different (decreases of 35% and 39%, respectively, from baseline alcohol intake). The reductions in alcohol intake were maintained into the fifth to eighth week post-treatment for subjects in both treatment groups; average decreases from baseline were 46% (post-citalopram) and 53% (post-placebo). Concomitant decreases in alcohol dependence (Alcohol Dependence Scale scores) and alcohol-related problems (MAST scores) were observed (Naranjo and Bremner, 1993a).

The mechanism of the effect of serotonin uptake inhibitors on alcohol intake is not fully understood. Subjects in our studies were not clinically depressed or anxious and changes in depression or anxiety were not observed. No consistent temporal relationship between the few side effects and decreases in drinking could be discerned. In addition, serotonin uptake inhibitors do not produce an alcohol-sensitizing reaction and have not been found to have any adverse pharmacokinetic interactions with alcohol (Naranjo *et al.*, 1984b; Lemberger *et al.*, 1985; Lader *et al.*, 1986; Sullivan *et al.*, 1989; Shaw *et al.*, 1989).

While serotonin has been shown to be related to several psychiatric disorders and human behaviors (Stahl, 1992), the effect of serotonin uptake inhibitors appears to be fairly specific to alcohol intake. For example, there is no evidence from the studies described above that these drugs decrease all fluid intake or other consummatory or addictive behaviors because non-alcoholic drinks and cigarette smoking did not change (Naranjo and Bremner, 1992; Sellers *et al.*, 1987). Weight loss was observed in most of the male heavy drinkers who received citalopram, viqualine and fluoxetine. The average weight loss per 2 weeks was 0.3 to 1.2 kg and could not be accounted for by decreases in calories from alcohol. Moreover, the lower doses of serotonin uptake inhibitors decreased body weight, but not alcohol intake. Changes in body weight and alcohol intake did not correlate, suggesting that these effects are independent (Naranjo and Bremner, 1992). As loss of appetite was frequently reported during citalopram 40 mg/day and fluoxetine 60 mg/day (p's < 0.05 compared with placebo), food intake probably decreased. Therefore, serotonin uptake inhibitors may decrease appetite, or desire to eat, and also decrease desire to drink alcohol. In the outpatient studies, increases in abstinent days were frequently observed (Naranjo *et al.*, 1984b; 1987), indicating that these drugs may exert their main effect before the initiation of drinking. Consistent with these findings, decreases in desire to drink were also reported with zimeldine (Amit *et al.*, 1985), fluoxetine (Gorelick and Paredes, 1992) and viqualine (Naranjo *et al.*, 1989).

In order to determine the importance of this mechanism, the effects of citalopram (40 mg/day) and fluoxetine (60 mg/day) on desire to drink were directly assessed in two recent studies. Immediately following one or two outpatient weeks of serotonin uptake inhibitor and placebo treatments, subjects (n = 16 per study) participated in experimental drinking sessions, in which they were required to consume as many of 18 minidrinks as possible (equivalent to six standard drinks) at 5-minute intervals and rate their desire and liking for alcohol, intoxication and mood. There was some indication that citalopram decreased the desirability of alcohol early in the experimental drinking session (Naranjo *et al.*, 1992a). Fluoxetine almost completely suppressed desire for alcohol and also decreased liking for alcohol, as rated in the experimental drinking session, compared with placebo (Naranjo *et al.*, 1992b). These results indicate that decreases in desire (urge to drink) and liking (reinforcing effects) of alcohol may mediate the effects of serotonin uptake inhibitors (Naranjo and Bremner, 1993b).

This mechanism of action should be considered in the development of treatments to reduce alcohol intake and prevent relapse. Desire or craving for alcohol is an internal (neurobiological) cue which increases the disposition to drink, especially in situations which are "high-risk" for an individual (Annis, 1990; Naranjo and Kadlec, 1991). Serotonin uptake inhibitors may be effective pharmacological probes to help to determine the degrees to which internal and situational factors lead to relapse, and thus, to direct individual patients into appropriate treatment. Furthermore, they may, by decreasing desire to drink, increase feelings of self-efficacy or ability to cope in high-risk situations and thereby provide both pharmacological and psychological benefits. Future research is required to assess the role of serotonin uptake inhibitors in relapse prevention therapy.

Effects of other serotonin-altering medications

Recent advances in molecular biology have led to frequent and significant changes in the classification and nomenclature of serotonin receptors (Humphrey et al., 1993). At least four groups of receptors (5-HT$_1$, 5-HT$_2$, 5-HT$_3$, 5-HT$_4$) and several subtypes (e.g., 5-HT$_{1A}$) have been identified, each recognizing different ligands that act as agonists or antagonists (Beer et al., 1993; Humphrey et al., 1993; Watson and Girdlestone, 1993). Many serotonergic drugs have been classified as selective agonists or antagonists for these receptor subtypes and studies have attempted to determine which receptors are most important in the regulation of alcohol intake in animal models (Sellers et al., 1992). We will briefly review some of the recent clinical studies. However, the results have not been conclusive and no coherent picture has emerged.

Serotonin agonists

While several serotonin agonists have decreased alcohol consumption in animals, few have been tested in humans and the results have been inconsistent. For example, the serotonin$_{1B}$ agonist, 1-(3-chlorophenyl)piperazine (m-CPP), and the serotonin$_{1C}$ agonist, 1-(6-chloro-2-pyrazino) piperazine (MK212), decreased oral ethanol self-administration in rats (Lawrin et al., 1986; Higgins et al., 1992), but when m-CPP was administered to 21 abstinent male alcoholics, 11 patients reported an ethanol-like "high" feeling and 7 reported an urge to drink (Benkelfat et al., 1991). These results were replicated in a randomized, double-blind, placebo-controlled study, in which m-CPP produced alcohol-like effects and alcohol craving in recently detoxified alcoholics (Krystal et al., 1993). Alcohol consumption was not assessed in either study.

Buspirone, a serotonin$_{1A}$ agonist, reduced alcohol intake in rats (Engel et al., 1992). In clinical studies, this nonbenzodiazepine anxiolytic reduced alcohol craving, anxiety and depression scale scores, but not the amount of alcohol consumed, compared with placebo (Bruno, 1989). The buspirone-treated subjects reduced their alcohol intake by an average of 57%

from baseline, and had a significantly better response rate (where a "responder" reduced alcohol intake by > 50% and completed the 8-week treatment period). Because of the small sample size and high discontinuation rate among the patients who received placebo, these results should be interpreted with caution. In another study, buspirone was an effective anxiolytic which reduced alcohol craving and improved clinical global impression scores in recently detoxified anxious alcoholics (Tollefson et al., 1992). Thus, buspirone may be effective only in anxious alcoholics (Kranzler and Meyer, 1989) and its effects on alcohol intake may be secondary to alleviation of anxiety. As alcoholism and anxiety are frequently comorbid, buspirone may be an effective pharmacotherapy for this subgroup of patients.

Serotonin antagonists

Many serotonin antagonists have had no effects or inconsistent effects on alcohol intake in rats (McBride et al., 1992; Engel et al., 1992). The serotonin$_2$ antagonist ritanserin reduced alcohol intake and preference in rats in a dose dependent manner in one study (Meert and Janssen, 1991), but failed to do so in others (Engel et al., 1992; Sellers et al., 1991a). In five male abstinent alcoholics, ritanserin 10 mg/day reportedly reduced desire to drink and helped to maintain abstinence during the 28-day treatment period and a 15-day follow-up placebo period (medication administered single-blind) (Monti and Alterwain, 1991). However, in a double-blind placebo-controlled study with 39 mildly/moderately dependent alcoholics, ritanserin 5 mg/day decreased desire and craving for alcohol during the 2-week outpatient treatment period compared with a one-week baseline, but not compared with during the 2-week ritanserin 10 mg/day or placebo treatment periods. Neither dose of ritanserin significantly reduced alcohol intake. Experimental drinking sessions were conducted immediately following baseline and treatment periods. Ritanserin 5 mg/day reduced desire to drink early in the session compared with ritanserin 10 mg/day. Also ritanserin 10 mg/day increased some of the subjective effects of alcohol, such as self-ratings of intoxication and friendliness, during the experimental drinking sessions (Naranjo et al., 1993a,b). The serotonin$_3$ antagonist ondansetron (0.1 mg/kg) reduced ethanol intake in rats and ondansetron 0.25 mg b.i.d. reduced alcohol consumption in male alcohol abusers during the latter half of a 6-week treatment period and during post-treatment compared with ondansetron 2 mg b.i.d. or placebo (Sellers et al., 1991b). Further research is required to confirm the effects and mechanism of action of serotonin antagonists. In animals, a serotonin₃ receptor antagonist reduced some of the symptoms of anxiety usually observed with alcohol withdrawal (Costall *et al.*, 1993). Other serotonin₃ antagonists blocked the discriminitive stimulus properties of ethanol in pigeons (Grant and Barrett, 1991). Therefore, serotonin antagonists may treat withdrawal symptoms, and thus reduce the risk of relapse to drinking. Alternatively, or also, they may block the rewarding effects of alcohol by interacting with the dopamine system (Engel *et al.*, 1992) so that drinking behavior is no longer reinforced. Further research is required to determine the clinical effects and potential therapeutic application of serotonin antagonists.

In conclusion, many serotonin-altering drugs modify alcohol consumption and effects. Full exploration of their therapeutic potential is required.

Acknowledgements

We wish to thank Drs. E.M. Sellers, J.T. Sullivan, V. Khouw, P. Sanhueza, H. Valencia, Z. Ul Hassan, M. Umana and S.L. Lee, Mrs. T. Fan, Mr. W. Juzytsch, Mrs. D. Woodley-Remus, Mrs. G. Kennedy, Miss M. McPhee and Mr. M. Paunil for their help in conducting our studies, which were supported by the Intramural Grant Program of the Addiction Research Foundation of Ontario. We also thank Mrs. Linda Neuman for typing the manuscript.

References

American Psychiatric Association (1980) DSM-III-R Diagnostic and Statistical Manual of Mental Disorders (Third Edition), Washington, D.C.

American Psychiatric Association (1987) DSM-III-R Diagnostic and Statistical Manual of Mental Disorders (Third Edition-Revised), Washington, D.C.

Amit, Z., Brown, Z., Sutherland, A., Rockman, G., Gill, K. and Selvaggi, N. (1985) Reduction in alcohol intake in humans as a function of treatment with zimelidine: Implications for treatment. In: Naranjo, C.A. and Sellers, E.M. (eds.): *Research Advances in New Psychopharmacological Treatments for Alcoholism*, Elsevier Science Publishers B.V., Amsterdam, pp. 189-198.

Amit, Z. and Smith, B.R. (1992) Neurotransmitter systems regulating alcohol intake. In: Naranjo, C.A. and Sellers, E.M. (eds.): *Novel Pharmacological Interventions for Alcoholism*, Springer-Verlag, New York, pp. 161-183.

Annis, H.M. (1990) Relapse to substance abuse: Empirical findings within a cognitive-social learning approach. *J. Psychoactive Drugs* 22: 117-124.

Ballenger, J.F., Goodwin, L., Major, W. and Brown G. (1979) Alcohol and central serotonin metabolism in man. *Arch. Gen. Psychiatry* 36: 222-227.

Beer, M.S., Middlemiss, D.N. and McAllister, G. (1993) 5-HT₁-like receptors: six down and still counting. *Trends in Pharmacological Sciences* 14: 228-231.

Benkelfat, C., Murphy, D.L., Hill, J.L., George, D.T., Nutt, D. and Linnoila, M. (1991)

Ethanollike properties of the serotonergic partial agonist m-Chlorophenylpiperazine in chronic alcoholic patients. *Arch. Gen. Psychiatry* 48: 383.

Bruno, F. (1989) Buspirone in the treatment of alcoholic patients. *Psychopathology* 22 (suppl.1): 49-59.

Costall, B., Domeney, A.M., Kelly, M.E., Tomkins, D.M., Naylor, R.J., Wong, E.H.F., Smith, W.L., Whiting, R.L. and Eglen, R.M. (1993) The effect of the 5-HT$_3$ receptor antagonist, RS-42358-197, in animal models of anxiety. *Eur. J. Pharmacol.* 234: 91-99.

Engel, J.A., Enerback, C., Fahlke, C., Hulthe, P., Hard, E., Johannessen, K., Svensson, L. and Söderpalm, B. (1992) Serotonergic and dopaminergic involvement in ethanol intake. In: Naranjo, C.A. and Sellers, E.M. (eds.): *Novel Pharmacological Interventions for Alcoholism*, Springer-Verlag, New York, pp. 68-82.

Fagius, J., Osterman, P.O., Siden, A. and Wiholm, B.-E. (1985) Guillain-Barre syndrome following zimeldine treatment. *Neurol. Neurosurg. Psychiatry* 48: 65-69.

Gorelick, D.A. and Paredes, A. (1992) Effect of fluoxetine on alcohol consumption in male alcoholics. *Alcohol Clin. Exp. Res.* 16: 261-265.

Grant, K.A. and Barrett, J.E. (1991) Blockade of the discriminative stimulus effects of ethanol with 5-HT$_3$ receptor antagonists. *Psychopharmacology* 104: 451-456.

Higgins, G.A., Lawrin, M.O. and Sellers, E.M. (1992) Serotonin and alcohol consumption. In: Naranjo, C.A. and Sellers, E.M. (eds.): *Novel Pharmacological Interventions for Alcoholism*, Springer-Verlag, New York, pp. 83-91.

Humphrey, P.P.A., Hartig, P. and Hoyer, D. (1993) A proposed new nomenclature for 5-HT receptors. *Trends in Pharmacological Sciences* 14: 233-236.

Kranzler, H.R. and Meyer, R.E. (1989) An open trial of buspirone in alcoholics. *J. Clin. Psychopharmacol.* 9: 379-380.

Krystal, J.H., Webb, E., Cooney, N., Kranzler, H. and Charney, D.S. (1993) Specificity of ethanol-like effects elicited by serotonergic and noradrenergic mechanims: m-CPP and yohimbine effects in recently detoxified alcoholics [abstract], in press.

Lader, M., Melhuish, A., Freka, G., Overo, K.F. and Christensen, V. (1986) The effect of citalopram in single and repeated doses and with alcohol on physiological and psychological measures in healthy subjects. *Eur. J. Clin. Pharmacol.* 31: 183-190.

Lawrin, M.O., Naranjo, C.A. and Sellers, E.M. (1986) Identification and testing of new drugs for modulating alcohol consumption. *Psychopharmacol. Bull.* 22: 1020-1025.

Lemberger, L., Rowe, H., Bergstrom, R.F., Farid, K.Z. and Enas, G.G. (1985) Effect of fluoxetine on psychomotor performance, physiologic response and kinetics of ethanol. *Clin. Pharmacol. Ther.* 37: 658-664.

McBride, W.J., Murphy, W.J., Lumeng, L. and Li, T.-K. (1992) Serotonin and alcohol consumption. In: Naranjo, C.A. and Sellers E.M. (eds.): *Novel Pharmacological Interventions for Alcoholism*, Springer-Verlag, New York, pp. 59-67.

Meert, T.F. and Janssen, P.A.J. (1991) Ritanserin, a new therapeutic approach for drug abuse. Part I: Effects on alcohol. *Drug Dev. Res.* 24: 235-249.

Monti, J.M. and Alterwain, P. (1991) Ritanserin decreases alcohol intake in chronic alcoholics. *Lancet* 337: 60.

Naranjo, C.A. and Kadlec, K.E. (1991) Possible pharmacological probes for predicting and preventing relapse in treated alcoholics. *Alcohol & Alcoholism* Suppl. 1: 523-526.

Naranjo, C.A. and Bremner, K.E. (1992) Evaluation of serotonin uptake inhibitors in alcoholics: a review. In: Naranjo, C.A. and Sellers, E.M. (eds.): *Novel Pharmacological*

Interventions for Alcoholism, Springer-Verlag, New York, pp. 105-117.

Naranjo, C.A. and Bremner, K.E. (1993a) Treatment-related attenuations of alcohol intake, alcohol dependence (AD) and problems [abstract]. *Clin. Pharmacol. Ther.* 53: 176.

Naranjo, C.A. and Bremner, K.E. (1993b) Clinical pharmacology of serotonin-altering medications for decreasing alcohol consumption. *Alcohol and Alcoholism*, Suppl. 2: 221-229.

Naranjo, C.A., Sellers, E.M., Roach, C.A., Woodley, D.V., Sanchez-Craig, M. and Sykora, K. (1984a) Zimelidine-induced variations in alcohol intake by non-depressed heavy drinkers. *Clin. Pharmacol. Ther.* 35: 374-381.

Naranjo, C.A., Sellers, E.M., Kaplan, H.L., Hamilton, C. and Khouw, V. (1984b) Acute kinetic and dynamic interactions of zimelidine and ethanol. *Clin. Pharmacol. Ther.* 36: 654-660.

Naranjo, C.A., Sellers, E.M., Sullivan, J.T., Woodley, D.V., Kadlec, K. and Sykora, K. (1987) The serotonin uptake inhibitor citalopram attenuates ethanol intake. *Clin. Pharmacol. Ther.* 41: 266-274.

Naranjo, C.A., Sullivan, J.T., Kadlec, K.E., Woodley-Remus, D.V., Kennedy, G. and Sellers, E.M. (1989) Differential effects of viqualine on alcohol intake and other consummatory behaviors. *Clin. Pharmacol. Ther.* 46: 301-309.

Naranjo, C.A., Kadlec, K.E., Sanhueza, P., Woodley-Remus, D. and Sellers, E.M. (1990a) Fluoxetine differentially alters alcohol intake and other consummatory behaviors in problem drinkers. *Clin. Pharmacol. Ther.* 47: 490-498.

Naranjo, C.A., Lane, D., Ho-Asjoe, M. and Lanctot, K.L. (1990b) A Bayesian assessment of idiosyncratic adverse reactions to new drugs: Guillain-Barre syndrome and zimeldine. *J. Clin. Pharmacol.* 30: 174-180.

Naranjo, C.A., Poulos, C.X., Kadlec, K.E. and Lanctot, K.L. (1992a) Citalopram decreases desirability, liking and consumption of alcohol in alcohol-dependent drinkers. *Clin. Pharmacol. Ther.* 51: 729-739.

Naranjo, C.A., Bremner, K.E., Poulos, C.X. and Lanctot, K.L. (1992b) Fluoxetine decreases desire for alcohol [abstract]. *Clin. Pharmacol. Ther.* 51: 168.

Naranjo, C.A., Bremner, K.E. and Lanctot, K.L. (1992c) Short- and long-term effects of citalopram (C) combined with a brief psychosocial intervention (BPI) for alcoholism [abstract]. *Clin Pharmacol Ther.*, 51: 168.

Naranjo, C.A., Poulos, C.X., Umana, M., Lanctot, K.L. and Bremner, K.E. (1993a) Effects of ritanserin (R) on desire to drink and consummatory behaviours (CB) in heavy alcohol drinkers [abstract]. *Clin. Pharmacol. Ther.* 53: 176.

Naranjo, C.A., Poulos, C.X., Bremner, K.E., Lanctot, K.L., Kwok, M. and Umana, M. (1993b) Ritanserin-induced variations in desire to drink, alcohol intake and alcohol effects. In: NIDA Monograph of the Proceedings of the 55th Annual Scientific Meeting, College on Problems of Drug Dependence, Toronto, Canada, June 12-17.

Sanchez-Craig, M. and Wilkinson, D.A. (1987) Treating problem drinkers who are not severely dependent on alcohol. *Drugs and Society* 1: 39-67.

Sellers, E.M., Naranjo, C.A. and Kadlec, K. (1987) Do serotonin uptake inhibitors decrease smoking? Observations in a group of heavy drinkers. *J. Clin. Psychopharmacol.* 7: 417-420.

Sellers, E.M., Romach, M.K., Frecker, R.C. and Higgins, G.A. (1991) Efficacy of the 5-HT$_3$ antagonist ondansetron in addictive disorders, In: Racagni, G., Brunello, N. and Fukudo, T. (eds.): *Proceedings of the 5th World Congress of Biological Psychiatry, Vol. 2*, Elsevier Science Publishers B.V., Amsterdam, pp. 894-897.

Sellers, E.M., Higgins, G.A. and Sobell, M.B. (1992). 5-HT and alcohol abuse. *Trends in Pharmacological Sciences* 13: 69-75.

Shaw, C.A., Sullivan, J.T., Kadlec, K.E., Kaplan, H.L., Naranjo, C.A. and Sellers, E.M. (1989) Ethanol interactions with serotonin uptake selective and non-selective antidepressants: fluoxetine and amitriptyline. *Human Psychopharmacology* 4: 113-120.

Stahl, S.M. (1992) Serotonin neuroscience discoveries usher in a new era of novel drug therapies for psychiatry. *Psychopharmacol. Bull.* 28: 3-9.

Sullivan, J.T., Naranjo, C.A., Shaw, C.A., Kaplan, H.L., Kadlec, K.E. and Sellers, E.M. (1989) Kinetic and dynamic interactions of oral viqualine and ethanol in man. *Eur. J. Clin. Pharmacol.* 36: 93-96.

Tollefson, G.D., Montague-Clouse J. and Tollefson, S.L. (1992) Treatment of comorbid generalized anxiety in a recently detoxified alcoholic population with a selective serotonergic drug (Buspirone). *J. Clin. Psychopharmacol.* 12: 19-26.

Watson, S. and Girdlestone, D. (1993) Receptor nomenclature supplement. *Trends in Pharmacological Sciences* 14: 21-22.

Weingartner, H., Rudorfer, M.V., Buchsbaum, M.S. and Linnoila, M. (1983) Effects of serotonin on memory impairment produced by ethanol. *Science* 221: 472-474.

Zabik, J.E., Binkerd, K. and Roache, J.D. (1985) Serotonin and ethanol aversion in the rat. In: Naranjo, C.A. and Sellers, E.M. (eds): *Research Advances in New Pharmacological Treatments for Alcoholism.* Elsevier Science Publishers, Amsterdam, pp. 87-101.

Toward a Molecular Basis of Alcohol Use and Abuse
ed. by B. Jansson, H. Jörnvall, U. Rydberg, L. Terenius & B. L. Vallee
© 1994 Birkhäuser Verlag Basel/Switzerland

The alcohol dehydrogenase system

Hans Jörnvall

Department of Medical Biochemistry and Biophysics, Karolinska Institutet, S-171 77 Stockholm, Sweden

Summary

Alcohol dehydrogenases constitute a complex system of enzymes, classes, isozymes, and allelic variants. The zinc containing, well-known liver enzyme is a class I medium-chain alcohol dehydrogenase. Other classes of this family include the class II protein, the glutathione-dependent formaldehyde dehydrogenase (the class III enzyme), the stomach-expressed class IV form, and the recently defined class V protein. Characterized forms suggest that the glutathione-dependent formaldehyde dehydrogenase is the original ancestor, defining a role for the whole protein family in cellular defense mechanisms. The isozyme-multiple class I protein is derived from an early gene duplication, allowing sub-specialization in vertebrates. Class IV is the one most ethanol-active and appears to be derived from the class I line. Allelic variants within class I, in association with aldehyde dehydrogenase variants, correlate with population differences in ethanol metabolism and hence with susceptibility to develop alcohol-related diseases. The structures also correlate with functional properties and define molecular building units for the whole family.

Introduction

Our view on the multiplicity of mammalian alcohol dehydrogenase has constantly changed. The activity was initially known as one enzyme (purified from horse liver), then as different isozymes, and soon also as classes, three of which were established structurally in the 80-ies (Kaiser *et al.*, 1988). Further classes have been added and characterized recently (Yasunami *et al.*, 1991; Parés *et al.*, 1994), and additional mammalian proteins, sorbitol dehydrogenase (Jörnvall *et al.*, 1981) and ζ-crystallin (Borrás *et al.*, 1989), have been found also to belong to this family. Outside the mammalian system, still more alcohol dehydrogenases exist, including those of short-chain dehydrogenases and other families.

In humans, the short-chain dehydrogenase family includes several steroid, prostaglandin and other dehydrogenases (Persson *et al.*, 1991), but a short-chain ethanol dehydrogenase has not yet been identified in human tissues. Instead, the liver enzyme, containing catalytic zinc and belonging to the medium-chain alcohol dehydrogenase family, constitutes the major ethanol metabolizing enzyme in humans. This family, with its enzymes, classes, and isozymes (Table I), originating from ancestral gene duplications at separate levels, gives new insight into the rules of protein chemistry at large and enzyme evolution. A major impact, though, of the new

Table I. Multiplicities of the human alcohol dehydrogenase (ADH) system

Level of multiplicity	Characteristics of major form in human liver	Characteristics in relation to other forms
Families	Medium-chain ADH family	One of several ADH families in nature
Enzymes	Alcohol dehydrogenases	One of several related enzymes in humans
Classes	I	One of five classes (I-V) in humans
Isozymes	β and γ subunits (adults)	Two of three subunit types (α, β, γ)

information, is the insight it gives into the basic metabolic function of the alcohol dehydrogenase system, as summarized below after a presentation of the system.

Family multiplicity

Alcohol dehydrogenase activities, although ascribed one EC number in common (EC 1.1.1.1), represent protein families with different molecular architecture, catalytic mechanisms and subunit sizes. The constituents of the medium-chain alcohol dehydrogenase family are the ones analyzed most thoroughly and include the zinc-containing liver enzymes. Tertiary structures are known for one horse and one human isozyme of this type (Eklund et al., 1976, Hurley et al., 1991) as outlined in another chapter (Eklund et al., this volume).

For the short-chain dehydrogenase family, two tertiary structures are also known (Ghosh et al., 1991; Varughese et al., 1992). Outside the mammalian system, an alcohol dehydrogenase of the short-chain type is the Drosophila alcohol dehydrogenase, for which conformational data are emerging (Ladenstein, Gonzàlez-Duarte, and Atrian, personal communication). Reaction mechanisms, active sites, domain arrangements, and metal requirements are fundamentally different between the short- and medium-chain families, although common aspects of the coenzyme-binding fold exist, as between all dehydrogenases. Additional families, like long-chain alcohol dehydrogenases (Inoue et al., 1989) and iron-activated alcohol dehydrogenases (Scopes, 1983; Williamson and Paquin, 1987), known from prokaryotic and fungal sources, are thus far considerably less well defined but appear distinct. Many other special alcohol dehydrogenases have also been reported, but may turn out to derive from one of the other families. For example, the "factor dependent" prokaryotic alcohol dehydrogenase is structurally related to the medium-chain alcohol dehydrogenases (van Opheim *et al.*, 1992).

In summary, separate protein families have given rise to alcohol dehydrogenase enzymes, operating by means of different catalytic mechanisms and distinct structures. Human ethanol-metabolizing dehydrogenases belong to the medium-chain, zinc-containing family, while short-chain enzymes, although common, exhibit other specificities in human tissues.

Enzyme multiplicity

For the short-chain family, more than 50 different enzymes have now been characterized, ranging from insect alcohol dehydrogenase and a number of prokaryotic enzymes to mammalian steroid and prostaglandin dehydrogenases. For the medium-chain family, the discovery of ζ-crystallin, once considered a taxon-specific lens protein, as well as its reductase activity (Rao and Zigler, 1991), liver expression in vertebrates in general (Huang *et al.*, 1990; Rao and Zigler, 1992), and a molecular architecture like that for medium-chain enzymes (Jörnvall *et al.*, 1993) open a wide interest in this protein type.

Class multiplicity

The distinction of separate classes (Vallee and Bazzone, 1983) of the liver alcohol dehydro-genases has been extended (Yasunami *et al.*, 1991; Parés *et al.*, 1994) and now recognizes five classes of alcohol dehydrogenase expressed in tissues of humans and many vertebrates. These five classes all derive from gene duplications (Fig. 1), which range down to early vertebrate times. Presently, class IV appears to be the one of most recent origin (cf. Parés *et al.*, 1994), but is present at least from the level of amphibians, suggesting similar arrangements of these five classes in all mammals.

Class I represents the classical liver enzyme, exhibiting considerable activity toward ethanol, sensitivity to pyrazole inhibition, and frequent existence of isozymes. Class III is equivalent to glutathione-dependent formaldehyde dehydrogenase (Koivusalo *et al.*, 1989), of widespread occurrence, and appears to be the form of most distant origin and defined function (Kaiser *et al.*, 1993). Class IV is specifically expressed in stomach, and has the highest ethanol activity of all classes thus far characterized (Parés *et al.*, 1994). Classes II and V have been studied less thoroughly and each has not been characterized in more than single species.

The classes behave like separate enzymes, with expressions, regulations, and specificities which are not identical. They do not cross-hybridize and, hence, form only class-specific

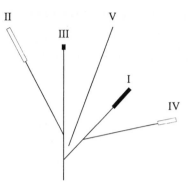

Fig. 1. Structural relationships of the five mammalian classes of alcohol dehydrogenase presently known. Line lengths and branch points drawn to scale as obtained with programs for construction of phylogenetic trees from all known human and rat forms of the class I-V enzymes. Data from Parés *et al.* (1993). Bars at the ends of the lines give the extent of species variation (human/rat), when known (filled; established structures in both species) or estimated (unfilled; incomplete structures in either species), relative to the class variations. The constant nature of the class III protein versus the class I form is obvious from the relative length differences of the bars, showing the species variation to be much smaller in III than I, while still tentative in II and IV, and unknown in V.

dimeric molecules. This is attributable to variable segments in the molecules, defining molecular building units (Persson *et al.*, 1993). However, all classes have a basic fold in common, and at least some activity toward a substrate in common, ethanol. Thus, classes are truely intermediate between completely different enzymes and typical isozymes.

Isozyme multiplicity

Isozymes appear to be most frequent in the liver alcohol dehydrogenases of class I. In humans, three isozyme subunits, α, β, and γ, associate freely into all possible dimeric combinations (Smith *et al.*, 1971). The three subunits correspond to closely associated genes, *ADH 1-3*, derived from gene duplications during mammalian radiation. Hence, isozyme patterns are not identical among different mammals. Furthermore, the expression patterns of human isozyme subunits differ, with the α-subunit being largely a fetal form (Smith *et al.*, 1971). Both the β and γ types occur in allelic variants, three of which have been characterized structurally for the human β-chain (Jörnvall *et al.*, 1984; Burnell *et al.*, 1987) and two for the human γ-chain (Höög *et al.*, 1986). These allelic variants are associated with functional differences in coenzyme binding and, hence, with catalytic differences, reflecting population differences in sensitivity to alcohol intake (below).

Functional conclusions

The major functional impact of the molecular scheme now discernible is two-fold: On the one hand, it shows that the most constant and ancestral form characterized hitherto among all alcohol dehydrogenases is the class III enzyme with its glutathione-dependent formaldehyde dehydrogenase activity. It is the only form found in invertebrates (Kaiser *et al.*, 1993), of ancient origin, and with a highly constant protein. In contrast, the ethanol activity has been generated repeatedly in nature (Fig. 2) and is variable in structure, including at the isozyme level in humans. Hence, the original metabolic role of the alcohol dehydrogenase system seems to derive from the basic property of glutathione-coupled formaldehyde elimination. This positions the enzyme system as part of the cellular defense mechanisms. It appears significant that other enzymes with extensive multiplicity derived from successive duplications also have such functions, e.g. cytochromes P450 and glutathione transferases. The formaldehyde elimination gives a defined function for the original form of the alcohol dehydrogenase system.

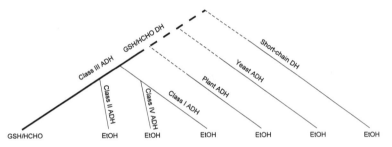

Fig. 2. Repeated formation of ethanol dehydrogenase (EtOH) activity in nature through several evolutionary lines (thin), emphasizing the continuous presence of the formaldehyde dehydrogenase (GSH/HCHO) class III line (thick). Relative positions of branch points appears to be as shown, but branches are given in arbitrary spacing and angles, with dashed lines where details of the relationships are unknown.

The second impact of this scheme, is the sub-specialization noticed in vertebrates/mammals. The various forms have separate genetic regulations; the basic class III form is essentially constitutive and largely ubiquitous, whereas the other classes which originated later appear to reflect further regulation and more distinct tissue distributions. The emerging picture is that the property evolving with this enzyme system is an increased sub-specialization in vertebrates, allowing for additional regulation. Perhaps, one of these additions may be associated with vertebrate differentiation, since retinol is one of the substrates for the emerging class I forms and retinoic acid is a general differentiation factor in vertebrates (Roberts and Sporn, 1984).

This aspect is outlined further in another chapter (Duester, this volume).

Direct correlation with altered alcohol intake and with alcoholism

In addition to the overriding functional principles discernible from the organisation outlined above, explanations to different alcohol intakes in populations have been possible to explain by the molecular insight. The β_1, β_2, and β_3 allelo-types of human alcohol dehydrogenase class I, as well as the γ_1 and γ_2 allelo-types, have amino acid substitutions corresponding to residues affecting coenzyme binding. Since coenzyme dissociation is a rate-limiting factor in the enzymatic reaction, ethanol turnover is influenced by the allelic variants. β_1, the common β-subtype in Caucasians, has Arg-47, resulting in tight coenzyme binding and low turnover rate. The Oriental subtype, β_2, has His-47, with consequently less tight coenzyme binding and increased turnover rate. A similar explanation is discernible also for the β_3 type (Cys/Arg difference at position 369) and γ_1/γ_2 types (exchanges at positions 271 and 349). Especially when together with an isozyme difference in the next enzyme (aldehyde dehydrogenase), where an Oriental allele produces an inactive form of the mitochondrial enzyme, these isozyme differences correlate with tendencies to develop alcoholism and alcohol-related diseases (cf. Bosron and Li, 1986; Shibuya et al., 1989; Thomasson et al., 1991). This constitutes a direct correlation between alcohol related diseases and specific molecular forms of genes and enzymes.

Conclusions and future perspectives

The knowledge about alcohol-metabolizing enzymes has increased rapidly. Basic functional mechanisms and sub-specialization has been discerned, as well as links with sensitivity to develop alcoholism and alcohol-related diseases. Much of this progress derives from the ever increasing structural characterization of naturally occurring subforms and their relationships. Further insight is to be expected at all levels of these studies.

One is the continued characterization of novel classes and their roles. Essentially, thus far class V is largely unstudied, class IV just characterized, and class II little known. The immediate future will probably give increasing insight into these classes and their links with alcohol metabolism and disease states at the molecular level.

For the more different enzymes, prosperous insights are also anticipated: in particular, the fact that ζ-crystallin, of the same protein family as the liver alcohol dehydrogenases, is a reductase in liver (Rao and Zigler, 1991) opens new avenues. Several reductases are fundamental in basic metabolism. Ethanol intake increases the hepatic NADH load, resulting in additional metabolic consequences. Long-term alcohol intake is associated with hepatic lipid deposits, and lipid biosynthesis is associated with reductase activities. If these activities could be linked through related reductases, a field of further insight into the interactions at the metabolic level can be anticipated. Future characterization of all the reductase activities appears promising.

Finally, regarding still more distant protein relationships, the finding of separate enzyme families is of great interest. Characterization of each family, even those thus far not yet known to be linked directly to alcohol turnover in humans, like short-chain dehydrogenases, may be productive for establishment of additional links between molecular architecture, metabolism, and sensitivity to alcohol related diseases.

Acknowledgements

Support by the Swedish Medical Research Council (13X-3532) is gratefully acknowledged.

References

Borrás, T., Persson, B. and Jörnvall, H. (1989) Eye lens ζ-crystallin relationships to the family of "long-chain" alcohol/polyol dehydrogenases. Protein trimming and conservation of stable parts. *Biochemistry* 28: 6133-6139.

Bosron, W.F. and Li, T.-K. (1986) Genetic polymorphism of human liver alcohol and aldehyde dehydrogenases, and their relationship to alcohol metabolism and alcoholism. *Hepatology* 6: 502-510.

Burnell, J.C., Carr, L.G., Dwulet, F.E., Edenberg, H.J., Li, T.-K. and Bosron, W.F. (1987) The human β_3 alcohol dehydrogenase subunit differs from β_1 by a Cys for Arg-369 substitution which decreases NAD(H) binding. *Biochem. Biophys. Res. Commun.* 146: 1227-1233.

Eklund, H., Nordström, B., Zeppezauer, E., Söderlund, G., Ohlsson, I., Boiwe, T., Söderberg, B.-O., Tapia, O., Brändén, C.-I. and Åkeson, Å. (1976) Three-dimensional structure of horse liver alcohol dehydrogenase at 2.4 Å resolution. *J. Mol. Biol.* 102: 27-59.

Ghosh, D., Weeks, C.M., Grochulski, P., Duax, W.L., Erman, M., Rimsay, R.L. and Orr, J.C. (1991) Three-dimensional structure of holo $3\alpha,20\beta$-hydroxysteroid dehydrogenase: A member of a short-chain dehydrogenase family. *Proc. Natl. Acad. Sci. USA* 88: 10064-10068.

Höög, J.-O., Hedén, L.-O., Larsson, K., Jörnvall, H. and von Bahr-Lindström, H. (1986) γ_1 and γ_2 subunits of human liver alcohol dehydrogenase. cDNA structures, two amino acid replacements, and compatibility with changes in the enzymatic properties. *Eur. J. Biochem.* 159, 215-218.

Huang, Q.-L., Du, X.-Y., Stone, S.H., Amsbaugh, D.F., Datiles, M., Hu, T.-S. and Zigler, Jr., J.S. (1990) Association of hereditary cataracts in strain 13/N guinea-pigs with mutation of the gene for ζ-crystallin. *Exp. Eye Res.* 50: 317-325.

Hurley, T.D., Bosron, W.F., Hamilton, J.A. and Amzel, L.M. (1991) Structure of human $\beta\beta$ alcohol dehydrogenase: Catalytic effects of non-active site substitutions. *Proc. Natl. Acad. Sci. USA* 88: 8149-8153.

Inoue, T., Sunagawa, M., Mori, A., Imai, C., Fukuda, M., Takagi, M. and Yano, K. (1989) Cloning and sequencing of the gene encoding the 72-kilodalton dehydrogenase subunit of alcohol dehydrogenase form *Acetobacter aceti. J. Bacteriol.* 171: 3115-3122.

Jörnvall, H., Persson, M. & Jeffery, J. (1981) Alcohol and polyol dehydrogenases are both divided into two protein types, and structural properties cross-relate the different enzyme activities within each type. *Proc. Natl. Acad. Sci. USA* 78: 4226-4230.

Jörnvall, H., Hempel, J., Vallee, B.L., Bosron, W.F. and Li, T.-K. (1984) Human liver alcohol dehydrogenase: amino acid substitution in the $\beta_2\beta_2$ Oriental isozyme explains functional properties, establishes an active site structure, and parallels mutational exchanges in the yeast enzyme. *Proc. Natl. Acad. Sci. USA* 81: 3024-3028.

Jörnvall, H., Persson, B., Du Bois, G.C., Lavers, G.C., Chen, J.H., Gonzalez, P., Rao, P.V. and Zigler, Jr., J.S. (1993) ζ-Crystallin versus other members of the alcohol dehydrogenase super-family. Variability as a functional characteristic. *FEBS Lett.* 322: 240-244.

Kaiser, R., Holmquist, B., Hempel, J., Vallee, B.L. & Jörnvall, H. (1988) Class III human liver alcohol dehydrogenase: a novel structural type equidistantly related to the class I and class II enzymes. *Biochemistry* 27: 1132-1140.

Kaiser, R., Fernández, M.R., Parés, X. and Jörnvall, H. (1993) Origin of the human alcohol dehydrogenase system: implications from the structure and properties of the octopus protein. *Proc. Natl. Acad. Sci. USA* 90: in press.

Koivusalo, M., Baumann, M. and Uotila, L. (1989) Evidence for the identity of glutathione-dependent formaldehyde dehydrogenase and class III alcohol dehydrogenase. *FEBS Lett.* 257: 105-109.

Parés, X., Cederlund, E., Moreno, A., Hjelmqvist, L., Farrés, J. and Jörnvall, H. (1994) Mammalian class IV alcohol dehydrogenase (stomach ADH): structure, origin and correlation with enzymology. *Proc. Natl. Acad. Sci. USA*, in press.

Persson, B., Krook, M. and Jörnvall, H. (1991) Characteristics of short-chain alcohol dehydrogenases and related enzymes. *Eur. J. Biochem.* 200: 537-543.

Persson, B., Bergman, T., Keung, W.M., Waldenström, U., Holmquist, B., Vallee, B.L. and Jörnvall, H. (1993) Basic features of class-I alcohol dehydrogenase: variable and constant segments coordinated by inter-class and intra-class variability. *Eur. J. Biochem.* 216: 49-56.

Rao, P.V. and Zigler, Jr., J.S. (1991) ζ-Crystallin from guinea pig lens is capable of functioning catalytically as an oxidoreductase. *Arch. Biochem. Biophys.* 284: 181-185.

Rao, P.V. and Zigler, Jr., J.S. (1992) Purification and characterization of ζ-crystallin/quinone reductase from guinea pig liver. *Biochim. Biophys. Acta* 1117: 315-320.

Roberts, A.B. and Sporn, M.B. (1984) Cellular biology and biochemistry of the retinoids. In: Sporn, M.B., Roberts, A.B. and Goodman, D.S. (eds.): *The Retinoids*, Vol 2, Academic, Orlando, pp. 209-286.

Scopes, R.K. (1983) An iron-activated alcohol dehydrogenase. *FEBS Lett.* 156: 303-306.

Shibuya, A., Yasunami, M. and Yoshida, A. (1989) Genotypes of alcohol dehydrogenase and aldehyde dehydrogenase loci in Japanese alcohol flushers and nonflushers. *Hum. Genet.* 82: 14-16.

Smith, M., Hopkinson, D.A. and Harris, H. (1971) Developmental changes and polymorphism in human alcohol dehydrogenase. *Ann. Hum. Genet.* 34: 251-271.

Thomasson, H.R., Edenberg, H.J., Crabb, D.W., Mai, X.-L., Jerome, R.E., Li, T.-K., Wang, S.-P., Lin, Y.-T., Lu, R.-B. and Yin, S.-J. (1991) Alcohol and aldehyde dehydrogenase genotypes and alcoholism in Chinese men. *Am. J. Hum. Genet.* 48: 677-681.

Vallee, B.L. and Bazzone (1983) Isozymes of human liver alcohol dehydrogenase. *Curr. Top. Biol. Med. Res.* 8: 219-244.

van Opheim, P.W., Van Beeumen, J. and Duine, J.A. (1992) NAD-linked, factor-dependent formaldehyde dehydrogenase or trimeric, zinc-containing, long-chain alcohol dehydrogenase from *Amycolatopsis methanolica*. *Eur. J. Biochem.* 206: 511-518.

Varughese, K.I., Skinner, M.M., Whiteley, J.M., Matthews, D.A. and Xuong, N.H. (1992) Crystal structure of rat liver dihydropteridine reductase. *Proc. Natl. Acad. Sci. USA* 89: 6080-6084.

Williamson, V.M. and Paquin, C.E. (1987) Homology of *Saccharomyces cerevisiae ADH 4* to an iron-activated alcohol dehydrogenase from *Zymomonas mobilis*. *Mol. Gen. Genet.* 209: 374-381.

Yasunami, M., Chen, C.-S. and Yoshida, A. (1991) A human alcohol dehydrogenase gene (*ADH6*) encoding an additional class of isozyme. *Proc. Natl. Acad. Sci. USA* 88: 7610-7614.

Toward a Molecular Basis of Alcohol Use and Abuse
ed. by B. Jansson, H. Jörnvall, U. Rydberg, L. Terenius & B. L. Vallee
© 1994 Birkhäuser Verlag Basel/Switzerland

Drug metabolism and signal transduction: Possible role of Ah receptor and arachidonic acid cascade in protection from ethanol toxicity

Daniel W. Nebert

Department of Environmental Health, University of Cincinnati Medical Center, Cincinnati, Ohio 45267-0056

Summary

It has become increasingly clear that the so-called 'drug-metabolizing enzymes,' and the receptors controlling them, play a key role in regulating the steady-state levels of ligands important in the transcription of genes involved in cell division, differentiation, apoptosis, homeostasis, and neuroendocrine functions. In this Chapter is presented an example of the possible role of the high-affinity Ah receptor in protection against ethanol-induced inflammation by way of the arachidonic acid cascade. Ethanol consumption is associated with cancer, inflammation, birth defects, and other toxic effects. Ethanol-induced toxicity appears to be mediated by many of the same oxidative stress signal transduction pathways used by other tumor promoters, teratogens, mitogens and comitogens.

Introduction

For more than two decades, this laboratory has studied the aromatic hydrocarbon-responsive [*Ah*] gene battery in the mouse. This work has led to a greater understanding of fundamental mechanisms underlying toxicity and cancer induced by such chemicals as 2,3,7,8-tetrachlorodibenzo-*p*-dioxin (TCDD; dioxin) and benzo[a]pyrene. These experiments have helped us to appreciate that drug-metabolizing enzymes and their receptors play an important role in maintaining the steady-state levels of endogenous ligands that effect ligand-modulated transcription of genes involved in signal transduction pathways, as well as the levels of metabolites involved in second-messenger pathways. These studies have also led to the recent isolation and sequencing of genes encoding DNA-binding and ligand-binding moieties of the Ah receptor complex. How the Ah receptor might play a role in protection against ethanol toxicity is the subject of this Chapter.

First, I briefly review the possible role of drug-metabolizing enzymes in signal transduction pathways. Second, I describe our current understanding of the Ah receptor. Third, I discuss previous work from this laboratory, in which the murine high-affinity Ah receptor was shown to protect against ethanol toxicity. Lastly, I focus on the likely mechanism by which the arachidonic acid cascade plays a role in this ethanol-induced toxicity.

232

Proposed role of drug-metabolizing enzymes (DMEs) in controlling steady-state levels of endogenous ligands

Enzymes that metabolize drugs, carcinogens, and other environmental pollutants have classically been called "drug-metabolizing enzymes" (DMEs). The total number of substrates for the DMEs includes virtually everything in the Merck Index. Non-peptide ligands that bind to endogenous receptors appear to be involved in ligand-modulated transcription of genes that participate in various signal transduction pathways effecting growth, morphogenesis, apoptosis, homeostasis, neuroendocrine functions, and cell type-specific proliferation including tumor promotion (Fig. 1). It was proposed (Nebert, 1991) that DME genes might have existed on this planet more than 2 billion years before the presence of plants, animals, and drugs. An early role for these enzymes in prokaryotes probably included energy substrate utilization: splitting of the molecule, or insertion of oxygen, starting with various inaccessible carbon and other food sources, thereby rendering them accessible to further metabolism. Another early role for these DMEs in prokaryotes and early eukaryotes was likely related to their metabolic ability to control the steady-state levels of the ligands that modulate cell division, growth, morphogenesis, and mating. It is very probable that these roles diversified into a very large number of additional complex signal transduction pathways that exist today in all eukaryotes (Nebert, 1990; 1991; 1994).

The hypothesis summarized in Fig. 1 is based on numerous experimental observations. (i) Endogenous non-peptide ligands for receptors involved in growth regulation and homeostasis are

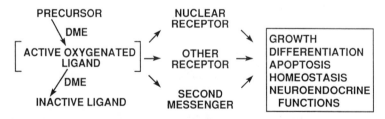

Fig. 1. Illustrated summary of the probable relationship between drug-metabolizing enzymes (DMEs) and signal transduction pathways effecting critical functions of the cell. Examples of a "nuclear receptor" would be the glucocorticoid receptor and the vitamin D_3 receptor (O'Malley, 1990). Examples of "other receptors" would include the recently discovered Per/ARNT/Sim (PAS) superfamily of transcription factors (Swanson and Bradfield, 1993) and the odorant receptor superfamily that appears specific to the olfactory epithelium (Buck and Axel, 1991). An example of "second messenger" would be the arachidonic acid cascade (Capdevila *et al.*, 1990). [Modified and reproduced, with permission from Nebert, 1991].

small organic molecules ($M_r = 250 \pm 200$). (ii) Synthesis and degradation of these molecules always involve DMEs. (iii) Binding of the natural ligand to the receptor always appears to be associated with cell- and/or developmental-specific increases (or decreases) in particular subsets of DMEs (Nebert and Gonzalez, 1987). (iv) Foreign chemicals that induce these enzymes also appear to bind to endogenous receptors, acting as either agonists or antagonists of receptor function. This hypothesis would also suggest that, when an "inducer of drug metabolism" is administered in large quantities to a cell or the intact animal, the cell or organism "senses" an abnormally elevated concentration of what is perceived to be an endogenous effector molecule; the cellular response thus would include the turning on of DME genes in order to degrade the excessive amounts of this chemical signal (Nebert, 1991; 1994).

P450-mediated arachidonic acid metabolites elicit many physiologic and subcellular changes. Relevant to protection against ethanol toxicity, one important component of these DME-mediated signal transduction pathways includes the arachidonic acid cascade. Virtually all enzymes in the arachidonic acid cascade may be defined as DMEs, including at least six different P450 enzymes (Nebert, 1991; 1994). Although generally regarded as manifesting pronounced effects in the kidney and lung (Capdevila *et al.*, 1992), arachidonic acid metabolites exist in virtually all mammalian cells and extracellular spaces. P450 metabolism of arachidonic acid leads to

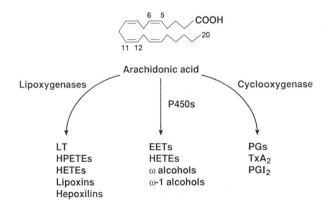

Fig. 2. Formation of arachidonic acid metabolites by lipoxygenases (left), cyclooxygenase (right), and cytochromes P450 (center). Although at least six P450 enzymes have been shown to be capable of forming EETs, CYP1A1 is by far the most effective to date (Capdevila *et al.*, 1992). LT, leukotriene. HPETEs, hydroperoxy-eicosatetraenoic acids. HETEs, hydroxyeicosatetraenoic acids. EETs, epoxyeicosatrienoic acids. PGs, prostaglandins. TxA$_2$, thromboxane A$_2$. PGI$_2$, prostaglandin I$_2$.

Table I. Physiologic and subcellular effects attributed to P450-mediated metabolites of arachidonic acid*

Effects of EETs:
  Bronchodilation, renal vasoconstriction, intestinal vasodilation, inhibition of cyclooxygenase, mitogenesis, inhibition of platelet aggregation, modulation of ion transport, enhanced peptide hormone secretion, mobilization of intracellular Ca^{++}

Effects of HETEs:
  Inhibition of Na,K ATPase, vasodilation, chemotaxis of neutrophils

Effects of ω- and ω-1 alcohols:
  Stimulation of Na,K ATPase, vasoconstriction, bronchoconstriction

---

*EETs, epoxyeicosatrienoic acids. HETEs, hydroxyeicosatetraenoic acids.

formation of EETs, HETEs, and ω and ω-1 alcohols (Fig. 2). As summarized in Table I, these metabolites are known to exert a broad spectrum of physiologic effects on growth, differentiation and homeostasis.

Enhanced availability of arachidonic acid usually requires increases in membrane-bound phospholipid biosynthesis (Burgoyne and Morgan, 1990). Endogenous pools of EETs are important as esters of cellular glycerolipids (Fig. 3). Approximately 92% of total liver EETs,

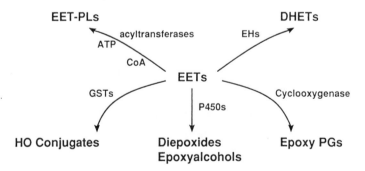

Fig. 3. EETs formed by P450 metabolism of arachidonic acid are then subject to five subsequent pathways, one of which is also mediated by P450 enzymes. EET-PLs, epoxyeicosatrienoic-phospholipids. CoA, acetyl Coenzyme A. GSTs, glutathione transferases. HO, hydroxylated. EHs, epoxide hydrolases. DHETs, dihydroxyeicosatetraenoic acids. PGs, prostaglandins.

for example, are esterified to phospholipids (Karara *et al.*, 1991). Tumor promoters are well known to enhance phospholipid biosynthesis. In the case of TCDD, this induction process is mediated by the Ah receptor (reviewed in Nebert, 1989).

Current understanding of the Ah receptor

The murine aromatic hydrocarbon-responsive [*Ah*] polymorphism was described more than 20 years ago (reviewed in Nebert and Gonzalez, 1987; Nebert, 1989). A particular P450-mediated activity [aryl hydrocarbon (benzo[a]pyrene) hydroxylase; AHH, now named the CYP1A1 enzyme] was found to be highly inducible by polycyclic aromatic hydrocarbons in some inbred mouse strains but not others. Mammalian CYP1A1 metabolizes polycyclic hydrocarbons such as benzo[a]pyrene. The lack of CYP1A1 induction behaves as an autosomal recessive trait between C57BL/6 (B6) and DBA/2 (D2) mice. The *Ah* terminology has recently been changed to *Ahr* (Eppig, 1993) because this regulatory gene is known to encode the Ah receptor.

The Ah receptor complex is a heterodimer which comprises a constitutive DNA-binding moiety, the ARNT protein, and the ligand-binding moiety encoded by the *Ahr* gene (reviewed by Swanson and Bradfield, 1993). The TCDD-bound Ah receptor complex interacts with DNA motifs, known as aromatic hydrocarbon-response elements (AhREs), which are present in upstream regulatory regions of the murine *Cyp1a1* and other genes that are up-regulated by TCDD (reviewed by Nebert *et al.*, 1993). The murine Ah receptor (*Ahr*) cDNA has been cloned (Burbach *et al.*, 1992). A single nucleotide difference between the high-affinity B6 *Ahr*[b-1] allele and the low-affinity D2 *Ahr*[d] allele, resulting in as little as a single amino acid difference, might account for the 15- to 20-fold difference in TCDD affinity between these two mouse strains (Chang *et al.*, 1993). An analogous polymorphism exists in the human population (discussed in Chang *et al.*, 1993), with about one-tenth of Caucasians having the high-affinity allele (*AHR*[H]). Cigarette smokers with the *AHR*[H] allele appear to have considerably greater risk of bronchogenic carcinoma than *AHR*[L] individuals (Kouri *et al.*, 1982; reviewed in Nebert et al., 1991).

High-affinity Ah receptor is associated with protection against ethanol toxicity

Following a selective breeding program of heterogenic stock (HS) mice for more than 30 generations, *SS* ("short sleep") and *LS* ("long sleep") lines were developed on the basis of their sleep times when challenged with an intraperitoneal dose of ethanol (McClearn and Kakihana,

1973). Work from this laboratory has shown that the *SS* mouse line exhibits markedly elevated hepatic levels of the high-affinity Ah receptor, while the *LS* mouse line contains the low-affinity Ah receptor (Bigelow *et al.*, 1989). This selective advantage occurring over 30 generations could not be explained on the basis of differences in central nervous system sensitivity or on the basis of changes in hepatic alcohol dehydrogenase or aldehyde dehydrogenase activities. In addition, comparison of (B6D2)F$_1$ x D2 backcross progeny, as well as comparison of the D2.B6-*Ahr^{b-1}* and B6.D2-*Ahr^d* congenic lines, revealed that mice having the high-affinity Ah receptor were 2-3 times more resistant to intraperitoneal ethanol toxicity than mice having the low-affinity Ah receptor (Bigelow *et al.*, 1989), providing further evidence of a direct relationship between the high-affinity Ah receptor and protection against ethanol toxicity.

Intraperitoneal ethanol is known to cause an acute chemical inflammation of the peritoneal cavity; the animal's response is mesenteric vasodilation which causes a marked lowering of body temperature (associated with "sleep time"). Hypothermia would affect absorption of ethanol from the peritoneal cavity, resulting in a slower rate of ethanol elimination. This laboratory showed that *LS* mice, when compared with *SS* mice, have a decreased rate of ethanol elimination; *Ahr^d/Ahr^d* mice were also shown to have a slower rate of ethanol elimination than *Ahr^{b-1}/Ahr^d* mice (Bigelow *et al.*, 1989), consistent with greater ethanol-induced inflammation in *LS* than *SS* mice. Interestingly, benoxaprofen, a cyclooxygenase inhibitor and anti-inflammatory agent, was reported to decrease Ah receptor-mediated halogenated hydrocarbon toxicity, suggesting that the high-affinity Ah receptor might play a role in arachidonic acid metabolism (Rifkind and Muschick, 1983). The selection program of the *SS* mouse line over two decades therefore appears to represent the selection of an animal more resistant to ethanol-induced hypothermia and peritonitis involving the arachidonic acid cascade.

Speculation: Formation of EETs occurs more readily in the high-affinity Ah receptor mouse
Fig. 4 summarizes a mechanism to explain why the concurrent segregation of *SS* mice and the high-affinity Ah receptor might have occurred. It is theoretically possible that ethanol could exert its effects through a cell surface receptor (Fig. 4), since even a molecule as small as glycine is known to act through a cell surface receptor (Vandenberg *et al.*, 1992). Numerous foreign small organic molecules--while not having the same high affinity as the "true" endogenous ligand--are able to bind to endogenous receptors and act as agonists or antagonists

(Nebert, 1994). Ethanol is known to enhance arachidonic acid release in peritoneal macrophages (Balsinde, 1993), and the peritoneal macrophages from *LS* mice have more lipoxygenase metabolites than those from *SS* mice (Fradin *et al.*, 1987). The Ah receptor up-regulates CYP1A1, and the high-affinity AhR is 15-20 times more effective than the low-affinity Ahr (Nebert, 1989). Although six P450 enzymes are capable of forming EETs from arachidonic acid, CYP1A1 is by far the best (Capdevila *et al.*, 1990). EETs are responsible for vasodilation (Capdevila *et al.*, 1992), which would enhance ethanol elimination and protect against toxicity. We have found (by in situ hybridization) murine CYP1A1 mRNA in endothelial cells of blood vessels in virtually every organ of the body (Dey *et al.*, 1989; A. Dey and D.W. Nebert, in preparation), which would also be consistent with the location of an important enzyme involved in EET-mediated vasodilation and chemotaxis. Hence, over more than two decades "short sleep" mice were selected for by their resistance to ethanol-induced peritonitis by means of rapid ethanol elimination (McClearn and Kakihana, 1973). Interaction of an endogenous ligand with the high-affinity AhR^H (Fig. 4) would thus protect the *SS* mouse from lipoxygenase metabolites-mediated ethanol toxicity--by up-regulating CYP1A1 in order to form more EETs at lower ethanol threshold concentrations, as compared with that in the *LS* mouse having the AhR^L.

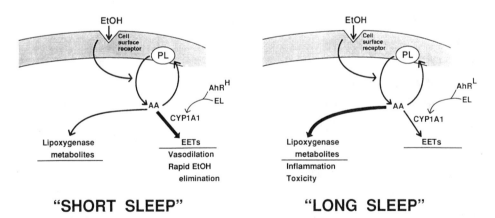

"SHORT SLEEP"   "LONG SLEEP"

Fig. 4. Simplified hypothesis of how the high-affinity Ah receptor (AhR^H) in *SS* mice might protect against ethanol-induced peritonitis more than the low-affinity AhR^L in *LS* mice. Different mechanisms for ethanol toxicity might occur in other tissues or cell types. EtOH, ethanol. Shaded area is the plasma membrane. PL, phospholipid. AA, arachidonic acid. EETs, epoxyeicosatrienoic acids. EL, unknown endogenous ligand. CYP1A1, the Ah receptor-controlled cytochrome P450 most capable of forming EETs from AA. It is possible (Kanetoshi *et al.*, 1992) that another enzyme in the CYP1 family is most responsible for forming EETs.

Conclusions

The Ah receptor is likely to play an important role in ethanol-induced inflammation. Although the endogenous ligand for the Ah receptor has not yet been identified, this study raises the possibility that a metabolite in the arachidonic acid pathway, or other factor activated by inflammation, might be the endogenous ligand for the Ah receptor. In this regard, it is interesting to note that when the peroxisome proliferator-activated receptor, PPAR, was cloned (Issemann and Green, 1990), no endogenous ligand was known. Recently, the synthetic arachidonic acid analogue 5,8,11,14-eicosatetraynoic acid was found to be the most potent ligand for the PPAR (Keller *et al.*, 1993). Arachidonic acid release occurs in all kinds of oxidative stress, including that from heavy metals (Käfer *et al.*, 1992) and ozone (McKinnon *et al.*, 1993). In this laboratory, arachidonic acid release, controlled by a repressor gene on mouse chromosome 7 and involved with the oxidative stress response, was shown to be associated with the expression of Ah receptor-inducible genes (reviewed in Nebert *et al.*, 1993; Nebert, 1994).

It is well known that excessive ethanol consumption is associated with inflammation and increases in certain types of cancer and birth defects (fetal alcohol syndrome). Ethanol, therefore, should be regarded as a tumor promoter and teratogen, although it is not mutagenic in the Ames test. The same is true of TCDD and numerous other tumor promoters. The pathways by which ethanol causes toxicity (Fig. 4) are not unlike the signal transduction pathways used by a large number of oxidative stress signals, mitogens, comitogens, and tumor promoters (Grinstein *et al.*, 1989), yet the entry point into the transduction pathway for any given stimulus might differ.

The true function(s) of the Ah receptor remains to be determined, but will likely be critical to the organism. For reasons not clear, an association has been found between the murine high-affinity Ah receptor and fertility, fitness and longevity (Nebert *et al.*, 1984). We and others have found that the B6 mouse appears to prefer ethanol to water, whereas the D2 mouse steadfastly refuses to drink ethanol. In the human population, would the AHR^H individual more likely enjoy, or crave or become dependent on, alcoholic beverages as compared with the AHR^L individual? And, would the AHR^H individual be more resistant to ethanol-induced toxicity, as compared with the AHR^L individual?

Acknowledgements

I thank Kathleen Dixon and Alvaro Puga for valuable discussions and critical reading of this manuscript. Supported by NIH Grants R01 AG09235, R01 ES06321, and P30 ES06096.

References

Balsinde, J. (1993) Mechanism of arachidonic acid liberation in ethanol-treated mouse peri toneal macrophages. *Biochim. Biophys. Acta* 1169: 54-58.

Bigelow, S.W., Collins, A.C. and Nebert, D.W. (1989) Selective mouse breeding for short ethanol sleep time has led to high levels of the hepatic aromatic hydrocarbon (Ah) receptor. *Biochem. Pharmacol.* 38: 3565-3572.

Buck, L. and Axel, R. (1991) A novel multigene family may encode odorant receptors: A molecular basis for odor recognition. *Cell* 65:175-187.

Burbach, K.M., Poland, A. and Bradfield, C.A. (1992) Cloning of the Ah-receptor cDNA reveals a distinctive ligand-activated transcription factor. *Proc. Natl. Acad. Sci. U.S.A.* 89: 8185-8189.

Burgoyne, R.D. and Morgan, A. (1990) The control of free arachidonic acid levels. *Trends Biol. Sci.* 15: 365-366.

Capdevila, J.H., Karara, A., Waxman, D.J., Martin, M.V., Falck, J.R. and Guengerich, F.P. (1990) Cytochrome P-450 enzyme-specific control of the regio- and enantiofacial selectivity of the microsomal arachidonic acid epoxygenase. *J. Biol. Chem.* 265: 10865-10871.

Capdevila, J.H., Falck, J.R. and Estabrook, R.W. (1992) Cytochrome P450 and the arachi donate cascade. *FASEB J.* 6: 731-736.

Chang, C.-Y., Smith, D.R., Prasad, V.S., Sidman, C.L., Nebert, D.W. and Puga, A. (1993) Ten nucleotide differences, five of which cause amino acid changes, are associated with the Ah receptor locus polymorphism of C57BL/6 and DBA/2 mice. *Pharmacogenetics* 3, in press.

Dey, A., Westphal, H. and Nebert, D.W. (1989) Cell type-specific expression of mouse inducible *Cyp1a1* mRNA during development. *Proc. Natl. Acad. Sci. U.S.A.* 86: 7446-7450.

Eppig, J.T. (1993) New changed gene symbols. *Mouse Genome* 91: 8-9.

Fradin, A., Henson, P.M. and Murphy, R.C. (1987) The effect of ethanol on arachidonic acid metabolism in the murine peritoneal macrophage. *Prostaglandins* 33: 579-589.

Grinstein, S., Rotin, D. and Mason, M.J. (1989) Na^+/H^+ exchange and growth factor-induced cytosolic pH changes. Role in cellular proliferation. *Biochim. Biophys. Acta* 988: 73-97.

Issemann, I. and Green, S. (1990) A member of the steroid hormone receptor superfamily activated by peroxisome proliferators. *Nature* 347: 645-649.

Käfer, A., Zöltzer, H. and Krug, H.F. (1992) The stimulation of arachidonic acid metabo lism by organic lead and tin compounds in human HL-60 leukemia cells. *Toxicol. Appl. Pharmacol.* 116: 125-132.

Kanetoshi, A., Ward, A.M., May, B.K. and Rifkind, A.B. (1992) Immunochemical identity of the 2,3,7,8-tetrachlorodibenzo-*p*-dioxin- and β-naphthoflavone-induced cytochrome P-450 arachidonic acid epoxygenases in chick embryo liver. *Mol. Pharmacol.* 42: 1020-1026.

Karara, A., Dishman, E., Falck, J.R. and Capdevila, J.H. (1991) Endogenous epoxyeicosatrienoyl-phospholipids. A novel class of cellular glycerolipids containing epoxidized arachidonate moieties. *J. Biol. Chem.* 266: 7561-7569.

Keller, H., Dreyer, C., Medin, J., Mahfoudi, A., Ozato, K. and Wahli, W. (1993) Fatty acids and retinoids control lipid metabolism through activation of peroxisome proliferator-activator receptor-retinoid X receptor heterodimers. *Proc Natl Acad Sci USA* 90: 2160-2164.

Knoll, A.H. (1992) The early evolution of eukaryotes: A geological perspective. *Science* 256: 622-627.

Kouri, R.E., McKinney, C.E., Slomiany, D.J., Snodgrass, D.R., Wray, N.P. and McLemore, T.L. (1982) Positive correlation between high aryl hydrocarbon hydroxylase activity and primary lung cancer as analyzed in cryopreserved lymphocytes. *Cancer Res.* 42: 5030-5037.

McClearn, G.E. and Kakihana, R. (1973) Selective breeding for ethanol sensitivity in mice. *Behav. Genet.* 3: 409-410.

McKinnon, K.P., Madden, M.C., Noah, T.L. and Devlin, R.B. (1993) *In vitro* ozone exposure increases release of arachidonic acid products from a human bronchial epithelial cell line. *Toxicol. Appl. Pharmacol.* 118: 215-223.

Nebert, D.W. (1989) The **Ah** locus: Genetic differences in toxicity, cancer, mutation and birth defects. *Crit. Rev. Toxicol* 20: 153-174.

Nebert, D.W. (1990) Growth signal pathways. *Nature* 347: 709-710.

Nebert, D.W. (1991) Proposed role of drug-metabolizing enzymes: Regulation of steady state levels of the ligands that effect growth, homeostasis, differentiation, and neuroendocrine functions. *Mol. Endocrinol.* 5: 1203-1214.

Nebert, D.W. (1994) Drug-metabolizing enzymes in ligand-modulated transcription. *Biochem. Pharmacol,* in press.

Nebert, D.W., Brown, D.D., Towne, D.W. and Eisen, H.J. (1984) Association of fertility, fitness and longevity with the murine *Ah* locus among (C57BL/6N)(C3H/HeN) recombinant inbred lines. *Biol. Reprod.* 30: 363-373.

Nebert, D.W. and Gonzalez, F.J. (1987) P450 genes: Structure, evolution and regulation. *Annu. Rev. Biochem.* 56: 945-993.

Nebert, D.W., Petersen, D.D. and Puga, A. (1991) Human *AH* locus polymorphism and cancer: Inducibility of *CYP1A1* and other genes by combustion products and dioxin. *Pharmacogenetics* 1, 68-78.

Nebert, D.W., Puga, A. and Vasiliou, V. (1993) Role of the Ah receptor and the dioxin-inducible [*Ah*] gene battery in toxicity, cancer and in signal transduction. *Ann. N.Y. Acad. Sci.* 685: 624-640.

O'Malley, B. (1990) The steroid receptor superfamily: More excitement predicted for the future. *Mol. Endocrinol.* 4: 363-369.

Rifkind, A.B. and Muschick, H. (1983) Benoxaprofen suppression of polychlorinated biphenyl toxicity without alteration of mixed function oxidase function. *Nature* 303: 524-526.

Rivera, M.C. and Lake, J.A. (1992) Evidence that eukaryotes and eocyte prokaryotes are immediate relatives. *Science* 257: 74-76.

Swanson, H.I. and Bradfield, C.A. (1993) The Ah receptor: Genetics, structure and function. *Pharmacogenetics* 3: 213-230.

Vandenberg, R.J., French, C.R., Barry, P.H., Shine, J. and Schofield, P.R. (1992) Antagonism of ligand-gated ion channel receptors: Two domains of the glycine receptor alpha subunit form the strychnine-binding site. *Proc. Natl. Acad. Sci. U.S.A.* 89: 1765-1769.

Toward a Molecular Basis of Alcohol Use and Abuse
ed. by B. Jansson, H. Jörnvall, U. Rydberg, L. Terenius & B. L. Vallee
© 1994 Birkhäuser Verlag Basel/Switzerland

Recruitment of enzymes and stress proteins as lens crystallins

Joram Piatigorsky, Marc Kantorow, Rashmi Gopal-Srivastava and Stanislav I. Tomarev

Laboratory of Molecular and Developmental Biology, National Eye Institute, National Institutes of Health, Bethesda, Maryland 20892

Summary

The major water-soluble proteins - or crystallins - of the eye lens are either identical to or derived from proteins with non-refractive functions in numerous tissues. In general, the recruitment of crystallins has come from metabolic enzymes (usually with detoxification functions) or stress proteins. Some crystallins have been recruited without duplication of the original gene (i.e., lactate dehydrogenase B and α-enolase), while others have incurred one (i.e., argininosuccinate lyase and a small heat shock protein) or several (i.e., glutathione S-transferase) gene duplications. Enzyme (or stress protein)-crystallins often maintain their non-refractive function in the lens and/or other tissues as well as their refractive role, a process we call gene sharing. α-Crystallin/small heat shock protein/molecular chaperone is of special interest since it is the major crystallin of humans. There are two α-crystallin genes (αA and αB), with αB retaining the full functions of a small heat shock protein. Here we describe recent evidence indicating that αA and αB have kinase activity, which would make them members of the enzyme-crystallins. We also describe various regulatory elements of the mouse α-crystallin genes responsible for their expression in the lens and, for αB, in skeletal muscle. Delineating the control elements for gene expression of these multifunctional protective proteins provides the foundations for their eventual use in gene therapy. Finally, comparison of the mouse and chicken αA-crystallin genes reveals similarities and differences in their functional cis-acting elements, indicative of evolution at the level of gene regulation.

Introduction

Since the role of the transparent eye lens is to focus light onto the retina it has been assumed that its characteristic crystallins, which account for approximately 90% of the soluble proteins of the lens, are highly specialized for this refractive function. However, the crystallins are a surprisingly diverse, developmentally regulated group of proteins which often vary quantitatively and qualitatively among species in a taxon-specific manner (see Wistow and Piatigorsky, 1988). Moreover, many of the crystallins are present in low concentrations in numerous non-lens tissues, consistent with their having non-refractive functions. Sequence data have shown, unexpectedly, that a large number of the crystallins are related to or identical with metabolic enzymes, and may retain enzymatic activity in the lens (see Piatigorsky and Wistow, 1989; de Jong et al., 1989; Piatigorsky, 1992; Wistow, 1993). In addition, the α-crystallins, which are present in all vertebrate lenses, are derived from the small heat shock proteins (sHSPs), with αB-crystallin continuing to serve as a sHSP and both αA and αB being able to act as a molecular chaperone (Horwitz, 1992; see de Jong et al., 1993).

The dual use of a metabolic enzyme or stress protein for refraction in the lens without loss of its original function in other tissues has been called gene sharing (see Piatigorsky and Wistow, 1989). The pragmatic use of the same protein for entirely different roles in different tissues implies that the innovation of new functions for a protein is associated with changes in the regulation of its gene (see Piatigorsky and Wistow, 1991; Piatigorsky, 1992). We have thus been studying the molecular basis for the recruitment of enzymes and stress proteins as crystallins at the level of gene regulation. Greater understanding of the mechanisms generating high expression of these genes in different tissues is clearly relevant to their eventual utilization for different medical purposes through induction and gene therapy.

Results and Discussion

The taxon-specific enzyme-crystallins

Many different enzymes have been recruited as lens crystallins (Fig. 1). Although these so-called enzyme-crystallins are not related by any single activity or metabolic pathway, many have de-toxification functions. The recruitment of detoxification enzymes as crystallins is not limited to vertebrates, but has also occurred among the invertebrates which have convergently evolved

TAXON-SPECIFIC CRYSTALLINS

Distribution	Crystallin	(Related) or Identical	Activity
Most birds and reptiles	δ1	(argininosuccinate lyase)	−
	δ2	argininosuccinate lyase	+
Duck, crocodile	ε	lactate dehydrogenase B	+
Guinea pig, camel, degus, rock cavy, llama	ζ	(alcohol dehydrogenase)	−
		quinone oxido reductase (novel)	+
Elephant shrews	η	cytoplasmic aldehyde dehydrogenase	−
Rabbits, hares	λ	(hydroxyl CoA dehydrogenase)	−
		(vat-1 [membrane protein, syn. vesicle])	
Australian marsupials	μ	(ornithine cyclodeaminase)	−
Frog	ρ	(NADPH-aldo-keto reductases)	−
		PGH₂ 9, 11-endoperoxide reductase	+/−
Many vertebrates	τ	α-enolase	+
Cephalopods	S	(glutathione S-transferases)	−
		glutathione S-transferase	+/−
	Ω/L	(aldehyde dehydrogenase)	−
Gastropods (Aplysia)	NN⁺	novel proteins (80, 63 kDa)	?
Jellyfish	J1	novel proteins (35, 20, 19 kDa)	?

⁺NN, not named.

Fig. 1. The phylogenetic distribution and activity of the taxon-specific crystallins.

complex eyes with cellular lenses. The best studied example is the cephalopod (squid and octopus) eye whose major crystallins (S-crystallins) are related to glutathione S-transferase (Tomarev
et al., 1992). Cephalopods (especially octopus) also have a minor crystallin (Ω-crystallin)
derived from aldehyde dehydrogenase (Zinovieva *et al.*, 1993). It is interesting to note that cytoplasmic class I aldehyde dehydrogenase has also been recruited as a crystallin (η-crystallin) in
the elephant shrew (Wistow and Kim, 1991) and class III aldehyde dehydrogenase may serve
a refractive function in the transparent corneal epithelial cells of mammals, where it comprises
as much as 40% of the soluble protein (Cuthbertson *et al.*, 1992). The derivations and functions,
if any, of the *Aplysia* and jellyfish crystallins are not known (see Piatigorsky, 1992).

Recruitment of crystallins from enzymes and stress proteins

Three different mechanisms have been used to recruit lens crystallins at the gene level (see
Piatigorsky, 1992 and Wistow, 1993 for reviews) (Fig. 2). The first dispenses with gene
duplication, using the identical gene encoding the metabolic enzyme as a crystallin. ϵ-
Crystallin/lactate dehydrogenase B (LDHB) and τ-crystallin/α-enolase are two examples of
recruitment without gene duplication. In these cases, the recruitment of the enzyme as a
crystallin depends upon changes in gene regulation leading to high expression in the lens. Any
changes in enzyme activity depend on posttranscriptional modifications in the lens, since the
structural gene must continue to encode an active enzyme for use in different tissues. The second
mechanism involves one duplication of the original enzyme or stress protein, with δ-
crystallin/argininosuccinate lyase (ASL) and α-crystallin/small heat shock protein (sHSP)
representing these cases (Fig. 2). In both of these examples, one of the duplicated genes has
specialized as a crystallin, although it is still expressed in some non-lens cells; the other
duplicated gene is also expressed most highly in the lens but has retained all of its original non-
refractive functions in numerous tissues. Finally, the third mechanism used to recruit lens
crystallins involves multiple duplications of the original enzyme. For example, there are at least
10 S-crystallin genes in the squid that were derived from the glutathione S-transferase (GST)
gene (Tomarev *et al.*, 1993). The S-crystallin genes are expressed only in the lens and cornea
and, except for very slight GST activity of one of the S-crystallins (SL11) (Tomarev, Chung and
Piatigorsky, unpublished), they lack enzymatic function (Tomarev *et al.*, 1992). The S-crystallin
genes encoding the inactive enzyme derivatives have acquired an additional exon which probably
contributes to the loss of enzyme activity of the crystallin.

RECRUITMENT OF CRYSTALLIN GENES

Original Gene		Crystallin Genes	Enzyme or Non-refractive Activity	Lens Expression	Examples
Lactate dehydrogenase B	No Duplication	ε/LDH$_B$	+	Preferred	Duck
α-Enolase	No Duplication	τ/α-enolase	+	Preferred	Duck
Argininosuccinate lyase	One Duplication	δ1 + δ2/ASL	δ1 −⎫ δ2 +⎭	Preferred	Chicken, Duck
Small heat shock protein	One Duplication	αA + αB/sHSP	αA +?⎫ αB + ⎭	Preferred	Mouse, Rat
Glutathione S-transferase	Several Duplications	SL11+SL20-1+SL20-3 +Other Genes	SL11 +/−⎫ Others −⎭	Specific	Squid

Fig. 2. Crystallin recruitment from enzymes or stress proteins has occurred without gene duplication (ε-crystallin/LDHB and τ-crystallin/α-enolase), with one gene duplication (δ-crystallin/ASL and α-crystallin/sHSP) and with several gene duplications (S-crystallin/GST).

The α-crystallins

The two α-crystallins (αA and αB) are particularly interesting because they are present in all vertebrates and are the major crystallins of the human lens. αA is expressed highly only in the lens, however it has also been reported to be present a low concentrations in other tissues, especially the spleen and thymus (Kato *et al.*, 1991; Srinivasan *et al.*, 1992). αB continues to be an active member of the sHSP family and is expressed constitutively in many different tissues as well as being inducible by physiological stress (Dubin *et al.*, 1989; Bhat and Nagineni, 1989; Dasgupta *et al.*, 1992; Klemenz *et al.*, 1991b). In general, the accumulation of αB-crystallin outside of the lens has been correlated with cells of high oxidative activity (see Sax and Piatigorsky, 1993 and de Jong *et al.*, 1993 for reviews). The structure and expression pattern of the two α-crystallin genes are shown in Fig. 3.

Stimuli resulting in the induction of αB-crystallin include oncogene (Ha-ras, v-mos) expression, osmotic stress, heat shock, dexamethasone, cadmium, sodium arsenite, and stretching of skeletal muscle. αB-crystallin is also overexpressed in numerous diseases, especially neuro (and other) degenerative diseases, such as scrapie-infection in hamster brains, and Alexander's disease, Creutzfeldt-Jacob disease, Lewy body disease, amyotrophic lateral sclerosis, Werner's syndrome, and Parkinson's disease in humans (see Sax and Piatigorsky, 1993, and Piatigorsky, 1993 for more complete discussion and references). An especially rele-

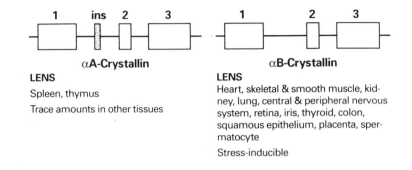

Fig. 3. Diagrammatic structures of the αA- and αB-crystallin genes. The boxes represent exons and the lines introns. Both are expressed principally in the lens and to a lesser extent in the other tissues listed. The stippled box in the αA-crystallin gene is the insert exon whose RNA is alternatively spliced into the mRNA in rodents and some other mammals to create the αA^{ins}-crystallin polypeptide (see Wistow and Piatigorsky, 1988).

vant case for the present symposium is the accumulation of αB-crystallin in ubiquitinated inclusion bodies (Mallory bodies) of alcoholic liver disease (Lowe *et al.*, 1992). It seems likely that protection against protein denaturation is at least one of the functions of αB/sHSP in these stressed cells in view its chaperone activity (Horwitz, 1992).

Is α-crystallin a kinase?

A proportion of both α-crystallin polypeptides are phosphorylated in the lens, and previous experiments in cell-free extracts have implicated a cyclic AMP-dependent kinase in this process (Spector *et al.*, 1985; Voorter *et al.*, 1986). Recently we have found that purified native α-crystallin or its individual polypeptides from bovine lenses (a kind gift of Dr. Joseph Horwitz, UCLA School of Medicine) become phosphorylated in a magnesium-dependent, cyclic AMP-independent manner in the presence of ^{32}P-γ-ATP (Kantorow and Piatigorsky, in preparation) (Fig. 4). While we cannot yet be certain that a separate kinase does not contaminate the α-crystallin preparations, the data suggest that α-crystallin posseses kinase activity, in particular for autophosphorylation. If α-crystallin is a kinase it joins the family of enzyme-crystallins and makes enzyme-crystallins present in all vertebrates, including humans. Moreover, this would be the first indication that small heat shock proteins have kinase activity and would fit with sequence data suggesting distant similarities between functional domains of small and large heat

246

shock proteins of eukaryotes and prokaryotes (Lee *et al.*, 1993). Finally, kinase activity by α-crystallin raises the possibility that this refractive protein/sHSP/molecular chaperone participates in one or more signal transduction pathways, providing new avenues of research for its high expression in growing tissues such as NIH 3T3 cells expressing oncogenes (Klemenz *et al.*, 1991a), astrocytic tumors of neuroectodermal origin (Iwaki *et al.*, 1991) and benign harmatomas (Iwaki and Tateishi, 1991). It is of interest in this connection that other sHSPs have been linked to cell growth and differentiation (see de Jong *et al.*, 1993). Thus, if α-crystallin is indeed a kinase, it would be a trifunctional protein with the capacity to serve a number of structural and metabolic roles.

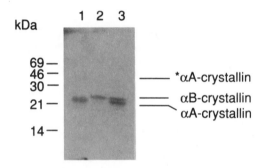

Fig. 4. Autoradiogram of a sodium dodecyl sulfate-polyacrylamide gel of purified αA-crystallin (lane 1), αB-crystallin (lane 2), or mixture of αA and αB polypeptides (lane 3) after incubation with ^{32}P-γ-ATP for 60 minutes at 37°C. The band labeled *αA is a dimer.

Molecular basis for αB-crystallin gene expression

Initial transfection and transgenic mouse experiments using a mouse αB-crystallin minigene lacking introns (Dubin *et al.*, 1989) and 5' flanking sequences from the αB gene fused to the bacterial chloramphenicol acetyltransferase (CAT) reporter gene (Dubin *et al.*, 1991) established that expression of the αB gene in skeletal muscle, heart and lens is regulated by transcription. Additional transfection experiments using the human growth hormone gene driven by the Herpes simplex thymidine kinase promoter fused to 5' flanking sequences of the mouse αB-crystallin gene revealed the existence of an enhancer at positons -426 to -257 of the αB gene which leads to high expression in the mouse C2C12 muscle cell line (especially after myotube formation in culture) and weaker expression in the mouse αTN4-1 lens cell line (which is transformed with the simian virus 40 large T-antigen) and the rabbit N/N1003A lens cell line (which is not transformed) (Dubin *et al.*, 1991). Recently we have identified by DNase I footprinting, site-specific mutagenesis and transfection assays three regulatory sites (αBE-1, αBE-2 and αBE-3) within the αB enhancer which function in both the muscle and lens cells and one regulatory site (MRF) which functions specifically in the muscle cells (Gopal-Srivastava and Piatigorsky, 1993). The αBE-2 site contains an AP-2-like sequence (TTCCCCTGG) and the MRF site contains an E box consensus sequence (CAGCTG); the region between the αBE-1 and αBE-2 sites contains a heat shock sequence (CCTAGGAAGATTCC). The results of immunological tests and co-transfection experiments in NIH 3T3 cells are consistent with one or more members of the MyoD/myogenin family of transcription factors being functional components of the MRF transcription complex. Finally, transgenic mice carrying 5' flanking sequences of the αB-crystallin gene (-426/+44) fused to the CAT gene confirmed that the sequence containing the αB enhancer leads to CAT expression specifically in skeletal muscle, heart and lens; by contrast, DNase I footprinting and transgenic mice carrying a shorter 5' flanking sequence of the αB gene (-164/+44) fused to the CAT gene identified a lens-specific regulatory sequence (-147/-118) (Gopal-Srivastava and Piatigorsky, in preparation). Thus, the selective expression of the αB-crystallin gene in different tissues is controlled in part by the utilization of different regulatory sequences.

Molecular basis for αA-crystallin gene expression

Although not absolutely lens-specific, the αA-crystallin gene is more specialized for lens expression than is the αB gene and, in contrast to the αB gene, does not appear to be inducible by physiological stress. Comparison of the 5' flanking regions of the αA-crystallin genes of mice, chicken and humans shows three highly conserved sequences (Jaworski *et al.*, 1991). These have been called DE1 (-111/-94), αA-CRYBP1 (-67/-40) and TATA/PE-1 (-35/+12) in the mouse. Relatively little sequence similarity exists further upstream than position -111 of these three species. Transfection experiments involving site-specific mutations, electrophoretic mobility shift assays and cloning of a putative transcription factor called αA-CRYBP1 (Nakamura *et al.*, 1990; Kantorow *et al.*, 1993) and transgenic mice (Sax *et al.*, 1993) have pointed to the functional importance of the αA-CRYBP1 site in the mouse gene.

Surprisingly, however, experiments using the αA-crystallin gene of the chicken have indicated that different regulatory mechanisms underlie the function of this orthologous gene. Initial transfection experiments showed that the poorly conserved sequence between position -162 and -121 is essential for activity of the chicken promoter in transfected lens cells (Klement *et al.*, 1989). Moreover, transfection experiments using a series of site-specific linker-scanning mutations indicated that the αA-CRYBP1 site is not necessary for activity of the chicken promoter (Klement *et al.*, 1993). This result fits with a previous demonstration that an oligonucleotide containing the core sequence of the chicken αA-CRYBP1 site (GAGAAATCCC) attached to the viral thymidine kinase promoter was unable to drive the CAT gene in transfected lens cells, while the core of the αA-CRYBP1 site of the mouse, which differs from the chicken sequence in its second nucleotide (GGGAAATCCC), was functional in the transfected cells (Sax *et al.*, 1990). Further analysis has shown that numerous cis-acting elements not present in the mouse gene are critical for promoter activity of chicken αA-crystallin gene (Matsuo and Yasuda, 1992; Klement *et al.*, 1993; see Sax and Piatigorsky, 1993). Nonetheless, a transgene containing 242 base pairs of 5' flanking sequence of the chicken αA-crystallin gene fused to the CAT gene functions with lens-specificity in transgenic mice (Klement *et al.*, 1989), indicating that some common features must exist for regulation of the chicken and the mouse αA-crystallin genes.

Conclusions

A wealth of sequence and functional data has shown that numerous metabolic enzymes (often with detoxification roles) and a sHSP have been recruited to become crystallins and contribute to the refractive properties of the transparent cellular lens of vertebrates and invertebrates. In some cases crystallin recruitment has involved one or several gene duplications, while in other cases it has occurred without gene duplication. Often at least one crystallin gene has retained some or all of its original non-refractive function and may continue to be used in non-lens tissues for its non-refractive role, a process we call gene sharing. Thus, changes in the regulation of gene activity lies at the heart of the recruitment process. Differences in the control elements of the mouse and chicken αA-crystallin gene provide a glimpse of the dynamism of the evolutionary process at the level of gene regulation. Gene sharing leaves open the possibility that virtually any gene may have two or more unrelated roles (see Piatigorsky and Wistow, 1991), raising concerns for gene therapy. The αB-crystallin gene is of special interest for eventual gene therapy against physiological stress in view of its multiple control elements, its powerful ability to protect proteins from denaturation, and its possible role in cellular growth and differentiation.

References

Bhat, S.P. and Nagineni, C.N. (1989) αB subunit of lens-specific α-crystallin in other ocular and non-ocular tissues. *Biochem. Biophys. Res. Comm.* 158: 319-325.

Cuthbertson, R.A., Tomarev, S.I. and Piatigorsky, J. (1992) Taxon-specific recruitment of enzymes as major soluble proteins in the corneal epithelium of three mammals, chicken, and squid. *Proc. Natl. Acad. Sci. USA* 89: 4004-4008.

Dasgupta, S., Hohman, T.C. and Carper, D. (1992) Hypertonic stress induces αB-crystallin expression. *Exp. Eye Res.* 54: 461-470.

de Jong, W.W., Hendriks, W., Mulders, J.W.M. and Bloemendal, H. (1989) Evolution of eye lens crystallins: the stress connection. *TIBS* 14: 365-368.

de Jong, W.W., Leunissen, J.A.M., and Voorter, C.E.M. (1993) Evolution of the α-crystallin/small heat shock protein family. *Mol. Cell. Biol.* 10: 103-126.

Dubin, R.A., Gopal-Srivastava, R., Wawrousek, E.F. and Piatigorsky, J. (1991) Expression of the murine αB-crystallin gene in lens and skeletal muscle: identification of a muscle-preferred enhancer. *Mol. Cell. Biol.* 11: 4340-4349.

Dubin, R.A., Wawrousek, E.F. and Piatigorsky, J. (1989) Expression of the murine αB-crystallin gene is not restricted to the lens. *Mol. Cell. Biol.* 9: 1083-1091.

Gopal-Srivastava, R. and Piatigorsky, J. (1993) The murine αB-crystallin/small heat shock protein enhancer: identification of αBE-1, αBE-2, αBE-3, and MRF control elements. *Mol. Cell. Biol.* 13: 7144-7152.

Horwitz, J. (1992) α-crystallin can function as a molecular chaperone. *Proc. Natl. Acad. Sci. USA* 89: 10449-10453.

Iwaki, T., Iwaki, A., Miyazono, M. and Goldman, J.E. (1991) Preferential expression of αB-crystallin in astrocytic elements of neuroectodermal tumors. *Cancer* 68: 2230-2240.

Iwaki, T. and Tateishi, J. (1991) Immunohistochemical demonstration of alphaB-crystallin in harmatomas of tuberous sclerosis. *Am. J. Path.* 139: 1303-1308.

Jaworski, C.J., Chepelinsky, A.B. and Piatigorsky, J. (1991) The αA-crystallin gene: conserved features of the 5' flanking regions in human, mouse, and chicken. *J. Mol. Biol.* 33: 495-505.

Kantorow, M., Becker, K., Sax, C.M., Ozato, K. and Piatigorsky, J. (1993) Binding of tissue-specific forms of αA-CRYBP1 to its regulatory sequence in the mouse αA-crystallin gene: double-label immunoblotting of UV-crosslinked complexes. *Gene* 131: 159-165.

Kato, K., Shinohara, H., Kurobe, N., Goto, S., Inaguma, Y. and Ohshima, K. (1991) Immuno-reactive αA-crystallin in rat non-lenticular tissues detected with a sensitive immunoassay method. *Biochim. Biophys. Acta* 1080: 173-180.

Klement, J.F., Wawrousek, E.F. and Piatigorsky, J. (1989) Tissue-specific expression of the chicken αA-crystallin gene in cultured lens epithelia and transgenic mice. *J. Biol. Chem.* 264: 19837-19844.

Klement, J.F., Cvekl, A. and Piatigorsky, J. (1993) Functional elements DE2A, DE2B, and DE1A and the TATA box are required for activity of the chicken αA-crystallin gene in transfected lens epithelial cells. *J. Biol. Chem.* 268: 6777-6784.

Klemenz, R., Frohli, E., Aoyama, A., Hoffmann, S., Simpson, R.J., Moritz, R.L., and Schafer, R. (1991a) αB-crystallin accumulation is a specific response to Ha-ras and v-mos oncogene expression in mouse NIH 3T3 fibroblasts. *Mol. Cell. Biol.* 11: 803-812.

Klemenz, R., Frohli, E., Steiger, R.H., Schafer, R. and Aoyama, A. (1991b) αB-crystallin is a small heat shock protein. *Proc. Natl. Acad. Sci. USA* 88: 3652-3656.

Lee, D.C., Kim, R.Y. and Wistow, G.J. (1993) An avian αB-crystallin: non-lens expression and sequence similarities with both small (HSP27) and large (HSP70) heat shock proteins. *J. Mol. Biol.* 232: 1221-1226.

Matsuo, I. and Yasuda, K. (1992) The cooperative interaction between two motifs of an enhancer element of the chicken αA-crystallin gene, αCE1 and αCE2, confers lens-specific expression. *Nucl. Acids Res.* 20: 3701-3712.

Nakamura, T., Donovan, D.M., Hamada, K., Sax, C.M., Norman, B., Flanagan, J.R., Ozato, K., Westphal, H. and Piatigorsky, J. (1990) Regulation of the mouse αA-crystallin gene: isolation of a cDNA encoding a protein that binds to a cis sequence motif shared with the major histocompatibility complex class I gene and other genes. *Mol. Cell. Biol.* 10: 3700-3708.

Piatigorsky, J. (1992) Lens crystallins. Innovation associated with changes in gene regulation. *J. Biol. Chem.* 267: 4277-4280.

Piatigorsky, J. and Wistow, G.J. (1989) Enzyme/crystallins: gene sharing as an evolutionary strategy. *Cell* 57: 197-199.

Piatigorsky, J. and Wistow, G.J. (1991) The recruitment of crystallins: new functions precede gene duplication. *Science* 252: 1078-1079.

Sax, C.M., Ilagan, J.G. and Piatigorsky, J. (1993) Functional reduncancy of the DE-1 and αA-CRYBP1 regulatory sites of the mouse αA-crystallin promoter. *Nucl. Acids Res.* 21: 2633-2640.

Sax, C.M., Klement, J.F. and Piatigorsky, J. (1990) Species-specific lens activation of the thymidine kinase promoter by a single copy of the mouse αA-CRYBP1 site and loss of tissue specificity by multimerization. *Mol. Cell. Biol.* 10: 6813-6816.

Sax, C.M. and Piatigorsky, J. (1993) Expression of the α-crystallin/small heat shock protein/molecular chaperone genes in the lens and other tissues. *Adv. Enzymol.*, in press.

Srinivasan, A.N., Nagineni, C.N. and Bhat, S.P. (1992) αA-crystallin is expressed in non-ocular tissues. *J. Biol. Chem.* 267: 23337-23341.

Spector, A., Chiesa, R., Sredy, J. and Garner, W. (1985) cAMP-dependent phosphorylation of bovine lens α-crystallin. *Proc. Natl. Acad. Sci. USA* 82: 4712-4716.

Tomarev, S.I., Zinovieva, R.D., Guo, K. and Piatigorsky, J. (1993) Squid glutathione S-transferase. Relationships with other glutathione S-transferases and S-crystallins of cephalopods. *J. Biol. Chem.* 268: 4534-4542.

Tomarev, S.I., Zinovieva, R.D. and Piatigorsky, J. (1991) Crystallins of the octopus lens. Recruitment from detoxification enzymes. *J. Biol. Chem.* 266: 24226-24231.

Tomarev, S.I., Zinovieva, R.D. and Piatigorsky, J. (1992) Characterization of squid crystallin genes. Comparison with mammalian glutathione S-transferase genes. *J. Biol. Chem.* 267: 8604-8612.

Voorter, C.E.M., Mulders, J.W.M., Bloemendal, H. and de Jong, W.W. (1986) Some aspects of the phosphorylation of α-crystallin A. *Eur. J. Biochem.* 160: 203-210.

Wistow, G. (1993) Lens crystallins: gene recruitment and evolutionary dynamism. *TIBS* 18: 301-306.

Wistow, G.J. and Kim, H. (1991) Lens protein expression in mammals: taxon-specificity and the recruitment of crystallins. *J. Mol. Evol.* 32: 262-269.

Wistow, G.J. and Piatigorsky, J. (1987) Recruitment of enzymes as lens structural proteins. *Science* 236: 1554-1556.

Wistow, G.J. and Piatigorsky, J. (1988) Lens crystallins: The evolution and expression of proteins for a highly specialized tissue. *Annu. Rev. Biochem.* 57: 479-504.

Zinovieva, R.D., Tomarev, S.I. and Piatigorsky, J. (1993) Aldehyde dehydrogenase-derived Ω-crystallins of squid and octopus. Specialization for lens expression. *J. Biol. Chem.* 268: 11449-11455.

Toward a Molecular Basis of Alcohol Use and Abuse
ed. by B. Jansson, H. Jörnvall, U. Rydberg, L. Terenius & B. L. Vallee
© 1994 Birkhäuser Verlag Basel/Switzerland

X-ray structure of PQQ-dependent methanol dehydrogenase

M. Ghosh,[1] A. Avezoux,[3] C. Anthony,[3] K. Harlos,[1,2] and C.C.F. Blake[1,2]

[1]Laboratory of Molecular Biophysics, [2]Oxford Centre for Molecular Sciences, University of Oxford, UK and [3]SERC Centre for Molecular Recognition, University of Southampton, UK

Summary

The three-dimensional structure of the PQQ-dependent quinoprotein, methanol dehydrogenase from *Methylobacterium extorquens* AM1, has been determined at 3Å resolution. The a_2b_2 tetrameric enzyme has a large a-chain of almost spherical form with a chain fold in which eight 4-stranded antiparallel b-sheets segments are arranged radially around a pseudo 8-fold molecular symmetry axis. The much smaller b-chain is surprisingly not globular, but has an extended conformation running across the surface of the α-subunit.

The PQQ prosthetic group is buried within the large a-subunit located on the pseudo 8-fold molecular symmetry axis. It is surrounded by protein side-chains but not covalently bound. Associated with the PQQ are two unexpected features: a vicinal disulphide bridge formed between Cys103 and Cys104, and a calcium ion bound between the protein and the PQQ. Vicinal disulphide bridges forming highly distorted structures containing a *cis* peptide bond, have been proposed to be present in one or two enzymes but have not previously been available for detailed structural investigation. Activity studies have indicated that the ability of the enzyme to transfer electrons derived from the reduction of the alcohols to the specific cytochrome C_L receptor is lost when the vicinal disulphide bridge is reduced. The roles of the calcium ions and the b-chain in the enzyme's activity remain to be determined.

Introduction

Methanol dehydrogenases are quinoproteins present in high concentrations in the periplasm of the methylotrophic gram negative bacteria. They couple the oxidation of primary alcohols to an electron transport chain whose first member is a unique c-type cytochrome, cytochrome C_L. The prosthetic group of the quinoproteins is pyrroloquinoline quinone (PQQ), see Fig. 1, first identified by Anthony and Zatman in 1967 in the methanol dehydrogenase of the bacterium now known as *Methylobacterium extorquens* AM1. It is the enzyme from this source whose three-dimensional structure has now been determined by X-ray diffraction.

Methanol dehydrogenase from *Methylobacterium extorquens* AM1 is an $\alpha_2\beta_2$ tetramer with a molecular mass of 148,000. The complete amino acid sequences of the two chains have been reported by Anderson *et al.* (1990) for an α-chain of 599 residues and by Nunn *et al.* (1989) for a much smaller β-chain of 74 residues. The two chains have not been derived by post-translational modification of a single chain, but are encoded on separate genes. Although not covalently linked the α- and β-chains cannot be separated without subjecting the enzyme to

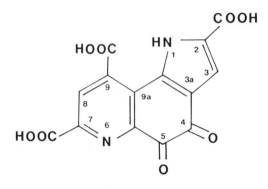

Pyrroloquinoline Quinone (PQQ)

Fig. 1. The prosthetic group of methanol dehydrogenase

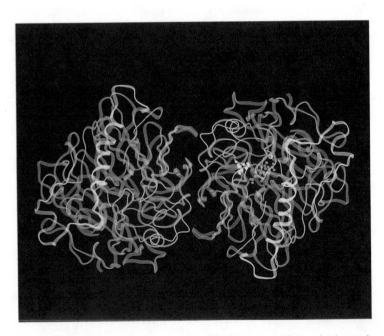

Fig. 2. Molecular structure of the $\alpha_2\beta_2$ tetramer of methanol dehydrogenase. The α-subunits are shown in pink and purple, both β-chains in yellow. The PQQ (red), vicinal disulphide bridge (yellow) and calcium (large sphere) are shown in the active site of one subunit.

denaturing conditions. One PQQ prosthetic group is non-covalently associated with each $\alpha\beta$ unit, and cannot be released without denaturing the protein. In the last two or three years it has become evident that methanol dehydrogenases contain calcium ions (Adachi *et al.*, 1990), at least one ion is bound per tetramer. The discovery of mutants where no calcium in bound and which cannot reduce substrate (Richardson and Anthony, 1992) suggests that the metal may either have a structural or a catalytic role.

Materials and Methods

Methanol dehydrogenases have been crystallized from a variety of sources. These include the enzyme from *Methylophilus methylotrophus* (Lim *et al.*, 1986), *Methylosinus trichosporium* (Parker *et al.*, 1987), bacterium W3A1 (Xia *et al.*, 1992). The methanol dehydrogenase from *Methylobacterium extorquens* AM1 was crystallized from 14.4% polyethelene gylycol 6000 at pH 9.0 using the hanging-drop technique (Ghosh *et al.*, 1992). The crystals belong to the orthorhombic system, space group $P2_12_12_1$ with cell dimensions $a = 66.8$ Å, $b = 108.9$ Å, $c = 188.9$ Å, with one $\alpha_2\beta_2$ tetramer of 148,000 molecular weight in the asymmetric unit. These crystals have been shown to diffract to 1.5 Å on the Daresbury Synchrotron Radiation Source.

The structure of the crystal was solved by multiple isomorphous replacement (MIR) coupled with electron density averaging around the molecular symmetry axis. Four isomorphous heavy atom derivatives were discovered: K_2PtCl_4, K_2PtCl_6, ethyl mercuric phosphate (EMP) and p-chloromercuribenzene sulphonate (PCMBS). Of these the two platinum derivatives had a very similar pattern of binding and the PCMBS derivative was the least isomorphous. All the major heavy atom binding sites with the exception of one of the five K_2PtCl_4 sites, could be seen to occur in pairs related by the same non-crystallographic two-fold rotation axis (Ghosh *et al.*, 1992). This axis was taken to be the symmetry axis relating the two $\alpha\beta$ units in the tetrameric protein molecule.

X-ray intensity data were collected using a Marresearch Image Plate detector mounted with a laboratory rotating anode X-ray tube. These data were restricted to 3 Å resolution because of the size of the image plate and the long crystallographic c-axis. Intensities were collected from the native and the derivative crystals. Phases of the native structure factors were determined by the MIR method and then improved by symmetry averaging about the non-crystallographic symmetry axis determined previously. The resulting electron density maps were interpreted on

254

an ESV Computer Graphics System using the programs 'O' (Jones, 1991) and Frodo (Jones et al., 1978), incorporating the known amino acid sequence of this methanol dehydrogenase. Intensity data for the native crystals have been collected beyond 2 Å resolution on the large Marresearch Image Plate mounted on the Daresbury Synchrotron Radiation Source, and will be used to extend the resolution of the current electron density map in the near future.

Results

The electron density maps showed that overall the methanol dehydrogenase molecule is composed of two almost spherical subunits giving it a dimension of 110Åx55Åx55Å. Within each spherical $\alpha\beta$ subunit, there is no obvious division into the constituent α- and β-chains. The

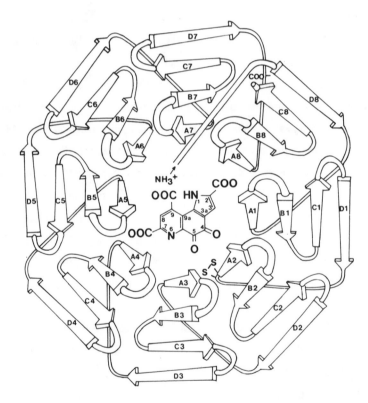

Fig. 3. A schematic drawing showing the overall chain folding in the α-subunit, and the relative position of the PQQ (not drawn to the scale). The arrows represent β-strands, labelled ABCD, in each motif which are numbered 1-8.

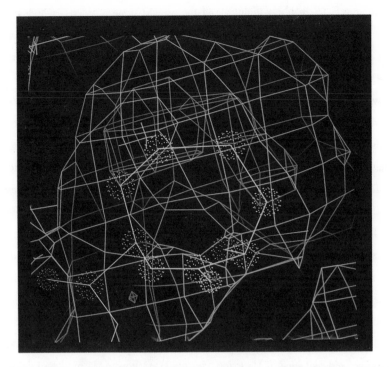

Fig. 4. The electron density around residues 103 and 104 at a contour level of 0.75σ (in blue) and a level five times as high (red). A model of a vicinal disulphide bridge is superimposed.

reason for this became clear when tracing the course of the polypeptide chain, when the 74-residue β-chain was seen not to fold into the expected globular form, but to be in an extended form running across the surface of the much larger α-chain (see Fig. 2). The N-terminal half of the β-chain is irregular and includes one intrachain disulphide bridge, but the C-terminal half is largely in the form of a single long α-helix which lies on the surface of the α-chain.

The α-subunit is composed of a single chain of 599 amino acid residues. The tertiary structure is remarkably regular and symmetrical. The polypeptide chain is organized into eight 4-stranded antiparallel β sheet segments, which are arranged radially around a pseudo 8-fold axis of molecular symmetry. This structure is shown schematically in Fig. 3 where it can be seen that the normal twist of the β-sheet enables space to be efficiently packed in the sub-unit as the distance from the molecular pseudo-symmetry axis increases. This molecular architecture allows the large polypeptide chain to be folded in a very compact form without any of the expected structural domains. There is no 'hole' along the pseudo-symmetry axis, which is filled with amino acid side chains from the eight A-strands of the β-sheets which are however more hydrophilic than those on particularly the B- and C-strands of each of the β-sheet motifs.

The PQQ moiety, a relatively small prosthetic group in such a large enzyme is entirely associated with the α-chain. It is buried quite deeply in the interior of the subunit and is located on the pseudo 8-fold molecular symmetry axis and is surrounded in a non-covalent association with a number of side-chains mostly from the A-strands of the eight β-sheet motifs (see Fig. 3). A convenient description of this active site region is of an internal 'chamber' in which the PQQ is located. This chamber communicates to the exterior through a funnel-shaped depression in the surface of the subunit which is quite narrow where it meets the chamber. The 'floor' of the chamber is formed by Trp243 whose indole group is parallel to, and in contact with, the planar PQQ ring system (see Fig. 5). Other side-chains, including Arg109, surround the PQQ interacting mostly with the carboxylate groups of the prosthetic factor forming hydrogen bonds or ionic interactions.

The 'ceiling' of the chamber is formed by a most unexpected structure. At the beginning of the A-strand of the second β-sheet motif, there are two neighbouring cysteine residues, 103 and 104 in the amino acid sequence of *Methylobacterium extorquens* AM1 methanol dehydrogenase. We were surprised to find a ring-shaped density feature which could only be interpreted as showing that the two neighbouring cysteine residues formed a disulphide bridge. Such a bridge, known as a vicinal disulphide bridge, has been considered theoretically but never previously discovered as an active feature in a protein structure. A vicinal disulphide was illustrated though not described, as an artefact in the active site of an inactive, oxidized form of mercuric ion reductase (Schiering *et al.*, 1991). The theoretical studies of Chandrasekharan and Balasubramanian (1969) have indicated that the vicinal disulphide bridge has a highly distorted structure with a *cis* - and possible non-planar conformation for the linking peptide group. The *cis*-peptide conformation has been confirmed in the X-ray determination of the L-Cys-L-Cys dipeptide (Mez, 1974; Capasso *et al.*, 1977). We used these two crystal structures, which differ largely at the dihedral angle of the sulphur-sulphur bonds, to fit the electron density. The conformation described by Mez and shown in Fig. 4 was the better fit to the protein feature.

A second unexpected observation was an isolated density feature in the active site which was too dense to be a water molecule. Since this feature was surrounded by oxygen atoms from protein

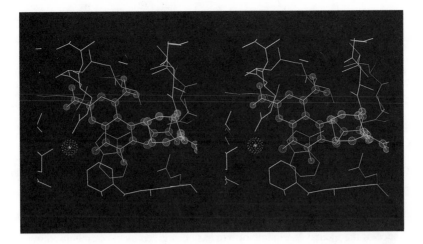

Fig. 5. A stereo drawing of the active site. The PQQ is shown in red, the vicinal disulphide bridge in yellow and the calcium ion as a large sphere.

side-chains and the prosthetic factor, we have tentatively identified this as the calcium ion (Richardson and Anthony, 1992). This ion as shown in Fig. 5 is situated on the 'wall' of the active site chamber, co-ordinated by the side-chains of Glu177, Asn261 and Asp303 and one carboxylate and one quinone oxygen of the PQQ.

Discussion

The X-ray study of the methanol dehydrogenase from *Methylobacterium extorquens* AM1 has shown that the enzyme has a number of interesting features. As the first PQQ-dependent enzyme to be studied in this way, it is too early to know if these features are generally present in PQQ-containing quinoproteins.

The overall structure of the α-subunit makes it a member of the family which is composed of multiply repeating 4-stranded antiparallel β-sheet motifs. The neuraminidase of the influenza virus has a 6-fold repetition of this motif (Varghese *et al.*, 1983); methylamine dehydrogenase, a non-PQQ-containing quinoprotein (Vellieux *et al.*, 1989) and galactose oxidase (Ito *et al.*, 1991) have a 7-fold repetition; and the methanol dehydrogenase from *Methylobacterium extorquens* AM1 (this work) and from *Methylophilus methylotrophus* (Xia *et al.*, 1993) have an 8-fold repetition of the motif. The extended non-globular form of the β-subunit is most unusual but similar in both *Methylobacterium extorquens* AM1 and *Methylophilus methylotrophus* enzymes. At present the functional role of the β-subunit is unclear: it does not seem to be associated with the active site, but we cannot rule out a role connected with the binding of, or electron transport to, the cytochrome C_L.

The observed close association of the PQQ prosthetic group with the vicinal disulphide bridge and the calcium ion is quite new and unexpected. Much work needs to be done to determine the behaviour of this arrangement of quinone group, disulphide bridge and the metal ion, as it relates to the oxidation of alcohol substrate and subsequent electron transfer. One factor that may be easily resolved by the X-ray study is the possibility that the substrate specificity of the enzyme for primary alcohol with an unsubstituted C-2 atom is determined by steric factors associated with the size of the 'funnel' leading to the active site chamber. Some indication of the significance of the vicinal disulphide bridge has been given by experiments showing that the bridge is readily reduced with a consequent loss of the electron transport activity of the enzyme. A readily reducible vicinal disulphide bridge has been proposed to be present in the acetyl choline receptor (Kao and Karlin, 1986) where it may be involved in receptor activation against agonists. The close association between the sulphurs in the vicinal disulphide bridge and the quinone group of the PQQ would facilitate the transfer of electrons derived from the oxidation of the substrate by the PQQ.

The exact configuration and distortion within the vicinal disulphide bridge will need analysis at higher resolution then we have at present. As predicted by Chandrasekharan and Balasubramanian (1969) the peptide linking the two cysteine residues at positions 103 and 104 is in a *cis*-conformation, but we cannot determine whether the peptide is, in addition, non-planar. Although reduction of the disulphide bridge, if this occurs during the catalytic cycle, might be expected to be accompanied by the conversion of the peptide bond to the lower energy *trans*-conformation, this change would seem to be opposed by the tight, overall fold of the polypeptide chain in the α-chain. It is interesting to note that two other enzymes in this family of structures also have unusual side-chain conformations: methylamine dehydrogenase has a tryptophan tryptophylquinone (TQQ) group which is produced from a covalent tryptophan-tryptophan interaction (Vellieux *et al.*, 1991 McIntire *et al.*, 1991 Chen *et al.*, 1991) and galactose oxidase has a covalent cysteine-tyrosine grouping in its active site (Ito *et al.*, 1991).

Conclusion

X-ray investigation of the molecular structure of methanol dehydrogenase from *Methylobacterium extorquens* AM1 has demonstrated the following features:

1. The large α-chain is globular but the small β-chain is extended and runs across the surface of the α-subunit.

2. The active site of the enzyme is buried within the structure of the α-subunit, but distant from the β-subunit.

3. Within the active site the PQQ prosthetic group is associated with a vicinal disulphide bridge of a highly distorted configuration, and a calcium ion.

4. The way in which these various factors couple the oxidation of alcohols to electron transport to a specific cytochrome receptor remains to be determined.

Acknowledgements

This work has been financially supported by the Wellcome Trust and by the UK Science and Engineering Research Council.

References

Adachi, O., Matsushita, K., Shinagawa, E. and Ameyama, M. (1988) Enzymatic determination of pyrroloquinoline quinone by a quinoprotein glycerol dehydrogenase. *Agric. Biol. Chem.* 52: 2081-2082.

Anderson, D.J., Morris, C.J., Nunn, D.N., Anthony, C. and Lidstrom, M.E. (1990) Nucleotide sequence of the *Methylobacterium extorquens* AM1 moxF and moxJ genes involved in methanol oxidation. *Gene* 90: 173-176.

Anthony, C. and Zatman, L.J. (1967) The microbial oxidation of methanol: the prosthetic group of alcohol dehydrogenase of *Pseudomonas* sp. M27; a new oxidoreductase prosthetic group. *Biochem. J.* 104: 960-969.

Capasso, S., Mattia, C. and Mazzarella, L. (1977) Structure of a cis-peptide unit: molecular conformation of the cyclic disulphide L-Cysteinyl-L-cysteine. *Acta Cryst.* B33: 2080-2083.

Chandrasekhar, R. and Balasubramaniun, R. (1969) Stereochemical studies of cyclic peptides, energy calculation on the cyclic disulphide Cysteinyl-cysteine. *Biochim. Biophys Acta.* 188: 1-9.

Chen, L.Y., Mathews, F.S., Davidson, V.L., Huizinga, E.G., Vellieux, F.M.D., Duine, J.A. and Hol, W.G.J. (1991) Crystallographic investigations of the tryptophan-derived cofactor in the quinoprtein methylamine dehydrogenase. *FEBS Lett.* 287: 163-166.

Ghosh, M., Harlos, K., Blake, C.C.F., Richardson, I. and Anthony, C. (1992) Crystallization and preliminary crystallographic investigation of methanol dehydrogenase *Methylobacterium extorquens* AM1. *J. Mol. Biol.* 228: 302-305.

Ito, N., Phillips S.E.V., Stevens C., Ogel, Z.B., McPherson M.J., Keen, J.N., Yadav, K.D.S. and Knowles P.F. (1991) Novel thioether bond revealed by a 1.7 Å crystal structure of galactose oxidase. Nature, 350: 87-90.

Jones, T.A. (1978) A graphics model building and refinement system for macromolecules *J. Appl. Crystallogr.* 11: 268-274.

Jones, T.A., Zuo, J.Y., Cowan, S.W. and Kjeldgaard, M. (1991) Improved methods for building protein models in electron density maps and the location of errors in these models. *Acta Crystallogr.* A47: 110-119.

Kao, P.N. and Karlin, A. (1986) Acetylcholine receptor binding site contains a disulphide cross-link between adjacent half-cysteinyl residue. *J. Biol. Chem.* 263: 1017-1022.

Lim, L.W., Xia, Z., Mathews, F.S. and Davidson, V.L. (1986) Preliminary X-ray crystallo graphic study of methanol dehydrogenase from *Methylophilus methylotrophus. J. Mol. Biol.* 191: 141-142.

McIntire, W.S., Wemmer, D.E., Christoserdov, A. and Lidstrom, M.E. (1991) A new cofactor in a prokaryotic enzyme - tryptophan tryptophylquinone as the redox prosthetic group in methylamine dehydrogenase. *Science* 252: 817- 824.

Mez, H. (1974) Cyclo-L-cystine acetic acid. *Cryst. Struct. Comm.* 3: 657-660.

Nunn, D., Day, D.J. and Anthony, C. (1989) The second sub-unit of methanol dehydrogenase of *Methylobacterium extorquens* AM1. *Biochem. J.* 260: 857-862.

Parker, M.W., Cornish, A., Gossain, V. and Best, D.J. (1987) Purification, crystallization and preliminary X-ray diffraction characterisation of methanol dehydrogenase from *Methylosinus trichosporium* OB3b. *Eur. J. Biochem.* 164: 223-227.

Richardson, I.W. and Anthony, C. (1992) Characterization of mutant forms of the quinoprotein methanol dehydrogenase lacking an essential calcium ion. *Biochem. J.* 287: 709-715.

Schiering, N., Kabsch, W., Moore, M.J., Distefano, M.D., Walsh, C. and Pai, E.F. (1991) Structure of the Detoxification catalyst mercuric ion reductase from Bacillus sp. strain RC607. *Nature* 352: 168-171.

Varghese, J.N., Laver, W.G. and Colman, P.M. (1983) Structure of the influenza virus glyco-protein antigen neuraminidase at 2.9 Å resolution. *Nature* 303: 35-40.

Vellieoux, F.M.D., Huitema, F., Groendijk, H., Kalk, K.H., Frank, J., Jongejan, J.A., Duine, J.A., Petratos, K., Drenth, J. and Hol, W.G.J. (1989) Structure of quinoprotein methylamine dehydrogenase at 2.25 Å resolution. *EMBO J.* 8: 2171-2178.

Xia, Z., Dai, W., Xiong, J., Hao, Z., Davidson, V.L., White, S. and Mathews, F.S. (1992) The three-dimensional structures of methanol dehydrogenase from two methylotrophic bacteria at 2.6 Å resolution. *J. Biol. Chem.* 267: 22289-22297.

Toward a Molecular Basis of Alcohol Use and Abuse
ed. by B. Jansson, H. Jörnvall, U. Rydberg, L. Terenius & B. L. Vallee
© 1994 Birkhäuser Verlag Basel/Switzerland

NMR, alcohols, protein solvation and protein denaturation

K. Wüthrich

Institut für Molekularbiologie und Biophysik, Eidgenössische Technische Hochschule-Hönggerberg, CH-8093 Zürich, Switzerland

Summary

The amino acid side chains of Ser, Thr and Tyr contain the hydroxyl functional group that is characteristic of alcohols, and a variety of different alcohols are used either as structure-inducing or denaturing co-solvents in physical-chemical studies of proteins. This article surveys selected aspects of the influence of protein-intrinsic and -extrinsic hydroxyl groups on studies of the solvation of peptides and proteins by nuclear magnetic resonance (NMR) spectroscopy.

Introduction

The following titles, which were selected arbitrarily from publications of the past 12 months, show that there is no scarcity of literature on the effects of alcohols on proteins: "Helix propagation in trifluoroethanol solutions" (Storrs *et al.*, 1992); "Hydrogen exchange in native and alcohol forms of ubiquitin" (Pan and Briggs, 1992); "A partially folded state of hen egg white lysozyme in trifluoroethanol: structural characterization and implications for protein folding" (Buck *et al.*, 1993). The phenomena alluded to in these titles are due to polypeptide solvation by the alcohol, to effects of the alcohol on polypeptide hydration, to changes of the solvent structure due to the addition of the alcohol, or to the combined impact of these different factors (e.g., Timasheff, 1992). Here, we focus on the use of NMR experiments for studies of polypeptide solvation.

NMR techniques have been introduced that enable studies of protein hydration in aqueous solution (Otting and Wüthrich, 1989; Otting *et al.*, 1991c; Wüthrich *et al.*, 1992; Wüthrich, 1993). These experiments can also be used for studies of polypeptide solvation in mixed solvents (Otting *et al.*, 1992), including mixtures with alcohols. Some fundamental aspects of such measurements will be reviewed.

The protein-intrinsic alcohol functional groups of Ser, Thr and Tyr interfere with the interpretation of NMR data on protein hydration (Liepinsh *et al.*, 1992a). This will be discussed in the second part of the paper.

NMR observations on protein hydration

Identification of individual molecules of hydration water by NMR in solution relies on the observation of NOEs (nuclear Overhauser enhancement) between hydrogen atoms of the polypeptide chain and water protons. These NOEs are due to time-dependent dipole-dipole coupling between nearby protons. The intensity of the NOE between two hydrogen atoms i and j is proportional to d_{ij}^{-6}, where d_{ij} is the distance between the two protons. Because of the strong distance dependence, NOEs can be observed only between spatially close protons, i.e., for d_{ij} < 4.0 Å. The NOE intensity is further related to a correlation function describing the stochastic modulation of the dipole-dipole coupling between the two protons. For studies of hydration it is important that this correlation function may be governed either by the Brownian rotational tumbling of the hydrated protein molecule, or by interruption of the dipolar interaction through translational diffusion of the interacting spins, whichever is faster (Otting et al., 1991a).

The NOEs can be measured either in the laboratory frame of reference by NOESY (nuclear Overhauser enhancement spectroscopy), or in the "rotating frame" by ROESY (Bothner-By et al., 1984). In suitably executed experiments, the measured NOE intensities reflect directly the cross relaxation rates in the laboratory frame, σ^{NOE}, or in the rotating frame, σ^{ROE}, respectively. The two rates differ in their functional dependence on the spectral densities in such a way that studies of the sign and value of the ratio $\sigma^{NOE}/\sigma^{ROE}$ can be used for investigations of the rate processes that determine the effective correlation time for modulation of the dipole-dipole coupling. As mentioned above, such rate processes include the Brownian rotational tumbling of the molecules considered, and translational diffusion of the two interacting spins, for example, upon rapid dissociation of a bimolecular complex. As an illustration, the plots of σ^{NOE} and σ^{ROE} versus the effective rotational correlation time (Fig. 1) show that positive values of $\sigma^{NOE}/\sigma^{ROE}$ are expected only for very short correlation times, i.e., shorter than about 0.3 nanoseconds at a proton frequency of 600 MHz (for modulation by translational diffusion, the sign change occurs at about 0.5 nanoseconds), and that the NOE and ROE intensities will increase rapidly for longer correlation times.

NOEs with hydration water have been observed using homonuclear 2D (two-dimensional) NOESY and 2D ROESY (Otting and Wüthrich, 1989), homonuclear 3D (three-dimensional)

TOCSY (total correlation spectroscopy)-relayed NOESY and ROESY (Otting *et al.*, 1991c), and heteronuclear 3D ^{15}N- or ^{13}C-correlated NOESY and ROESY (Wüthrich and Otting, 1992; Qian *et al.*, 1993a). These NMR experiments have to be performed in H_2O solution, and the intense solvent resonance can be suppressed only at the very end of the NOESY or ROESY pulse sequence. As an illustration, Fig. 2 shows cross sections from 2D NOESY and 2D ROESY experiments, which were taken at the chemical shift of the solvent water along the frequency axis ω_2 and contain NOE cross peaks between the polypeptide chain and hydration water molecules at the chemical shifts of the polypeptide protons along ω_2. Nearly all the resonances have opposite sign in the two experiments, showing that the system is in the "slow motional regime", with effective correlation times $\gg$ 0.5 nanoseconds (Fig. 1).

Based on the aforementioned principles, NMR can also be used to investigate how solvation by a given solvent is influenced through the addition of a cosolvent. A variety of different experimental approaches can be envisaged, as illustrated by the following examples: (i) All but one of two or several components of a mixed solvent are perdeuterated, so that they are not

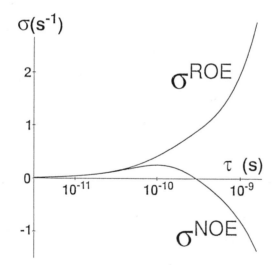

Fig. 1. Plot of the cross relaxation rates σ^{NOE} and σ^{ROE} between two protons *versus* the effective correlation time τ for modulation of the dipole-dipole coupling by Brownian rotational tumbling. The curves were calculated for a ^1H frequency of 600 MHz and a ^1H-^1H distance of 2.0 Å. (Reproduced with permission from Wüthrich, 1993).

observable in ¹H NMR spectra. Comparison of 2D NOESY or 2D ROESY spectra (Fig. 2) in the presence and absence of the deuterated cosolvents can then provide information on the competitive displacement of the hydrogen-bearing solvent molecules from the solute by the cosolvents. The hydration of the polypeptide hormone oxytocin in acetone-water mixtures was thus studied, which showed that addition of up to 40% v/v of acetone did not noticeably affect the hydration of the polypeptide chain (Otting *et al.*, 1992). If the solvent mixture includes more than one protic component this approach is not *a priori* straightforward, since magnetization can then be transferred by chemical exchange between different solvent components. (ii) In solvent mixtures without chemical exchange of protons, or with very slow exchange, such as chloroform and methanol, benzene and chloroform, or ethanol and acetone, solvation by all the different components can in principle be observed in a single experiment, for example, in 2D NOESY or 2D ROESY through inspection of cross sections at the chemical shifts of the different individual solvent components along ω_1 (Fig. 2). (iii) In solvent mixtures with proton exchange, which includes all alcohol-water mixtures, NOEs between the solute and all solvent components

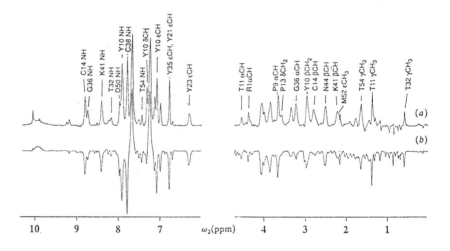

Fig. 2. ¹H NMR spectra containing NOE cross peaks between polypeptide hydrogen atoms of the protein BPTI and hydration water protons. Cross sections along ω_2 are shown, which were taken at the ω_1 chemical shift of the water resonance through a 2D NOESY (*a*) and a 2D ROESY (*b*) spectrum ("soft-NOESY" and "soft-ROESY", respectively; ¹H frequency = 600 MHz, mixing time = 50 ms, BPTI concentration = 20 mM, solvent = 90% H_2O/10% D_2O, pH = 3.5, T = 4°C). The peaks which were identified as NOEs with the H_2O resonance are identified above the NOESY cross section with the one-letter amino acid symbol, the sequence number in the polypeptide chain, and the proton type. (Reproduced with permission from Wüthrich, 1993).

may be observed at the same chemical shift along ω_1 (Fig. 2), so that solvation by the different individual components cannot *a priori* be distinguished. This situation is largely analogous to the interference of protein-intrinsic labile protons with studies of protein hydration (see below).

My laboratory has been involved in a variety of NMR investigations with peptides and proteins in mixed solvents, which also includes studies with alcohols. Unfortunately, except for BPTI the proteins chosen so far for these physico-chemical studies turned out to have a pronounced tendency to denature even upon addition of small amounts of alcohols. The interpretation of data collected with oligopeptides and BPTI is still going on.

Interference of the alcohol functions of Ser, Thr and Tyr with NMR studies of protein hydration

The experimental difficulties alluded to in the title of this section and in the above discussion on solvent mixtures with proton exchange are due to the fact that the hydrogen atoms bound to non-carbon heavy atoms of the polypeptide chain undergo chemical exchange with protic solvents, such as water or alcohol (Fig. 3). In general, these exchange reactions are acid- and base-catalyzed, and depending on the conditions of pH, temperature and ionic strength, two limiting situations may be encountered: (i) The exchange is fast on the chemical shift time scale so that the resonances of the solvent water and the labile polypeptide protons are merged into a single line. The 2D NOESY cross section at the water chemical shift (Fig. 2) can then contain cross peaks manifesting, for example, NOEs between γCH3 and γOH of Thr in addition to the polypeptide-water NOEs. (ii) The exchange is sufficiently slow so that two separate resonance lines can be observed. NOEs with the solvent and with labile polypeptide protons can then be observed in different cross sections along ω_1, but one must still consider possible pitfalls that might arise from "slow" chemical exchange of NOE intensity between these separate cross sections (Liepinsh *et al.*, 1992a). For the different types of labile polypeptide protons (Fig. 3) the desired slow-exchange situation is obtained with different experimental conditions, in particular at different pH values (Wüthrich, 1986). Ambiguities can thus often be resolved by repetition of the measurements at different conditions. Overall, no interference with studies of protein hydration comes from carboxylate groups (Liepinsh *et al.*, 1993), but potential interference must be carefully checked for all other functional groups of Fig. 3, in particular for the -OH groups of Ser, Thr and Tyr.

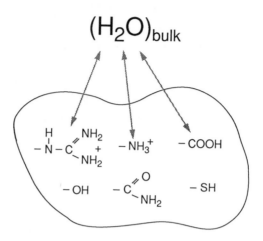

Fig. 3. Schematic drawing indicating the chemical exchange of labile polypeptide protons with protons of the solvent water. Similar situations are encountered in alcoholic solvents.

Two qualitatively different types of hydration water have so far been distinguished by NMR: (i) Globular proteins tend to contain a small number of "interior" hydration water molecules, which are an integral part of the molecular architecture and are not accessible to the bulk water. In several proteins so far these have been found in identical locations in X-ray crystal and NMR solution structures (Clore *et al.*, 1990; Forman-Kay *et al.*, 1991; Otting and Wüthrich, 1989). For these water molecules the life times with respect to exchange with the bulk solvent are long on the time scale of the Brownian rotational motions of the protein (Otting *et al.*, 1991b), so that the latter determine the effective correlation time for the NOEs and strong NOE and ROE cross peaks with opposite signs are observed (Figs. 1 and 2). (ii) Surface hydration waters exchange in and out of the hydration sites with very short residence times of about 20-500 picoseconds (Otting *et al.*, 1991a; 1992; Brunne *et al.*, 1993), giving rise to very weak NOE and ROE cross peaks with equal sign (Fig. 1). From the NMR viewpoint, surface hydration thus causes a "background" of weak NOE signals and the NOEs from longer-lived water molecules clearly stand out (Figs. 1 and 2).

Long-lived surface hydration sites could be expected to have special structural or functional roles, and observation and detailed investigation of such "specific" surface hydration water molecules would clearly be of special interest. However, identification of long-lived surface hydration waters in proteins will have to be based on careful exclusion of the protein-intrinsic alcohol functions as the cause of the corresponding NOEs, since the exchange rates of hydroxyl protons are in the same time regime as those of interior hydration water molecules (Liepinsh *et al.*, 1992a). A systematic study of the hydroxyl groups of Ser, Thr and Tyr has shown that differentiation of NOEs with solvation water from those with polypeptide alcohols can be achieved by investigations at different pH values in the range 3-7, and as an extra bonus the spatial orientation of the hydroxyl O-H bonds can often be precisely determined by such experiments (Liepinsh *et al.*, 1992a). "Specific" solvent-accessible hydration waters with long residence times have so far unambiguously been identified in the minor groove of A=T tracts in DNA duplexes in aqueous solution (Kubinec and Wemmer, 1992; Liepinsh *et al.*, 1992b), and "interior-type" water was identified in the protein-DNA interface of a homeodomain-operator-DNA complex (Qian *et al.*, 1993).

Conclusions

Alcohols in the broad sense of the chemist's definition as hydroxyl-bearing molecules have an important role as denaturing as well as structure-inducing solvents for physical-chemical studies of peptides and proteins. Currently emerging, novel NMR approaches promise to yield more detailed insights into the structural basis of these special features of alcoholic solvents.

Acknowledgements

In my laboratory, studies of biopolymer solvation have been pursued by Drs. E. Liepinsh, G. Otting and G. Wider, and by V. Dötsch. I thank R. Marani for the typing of the manuscript, and the Schweizerischer Nationalfonds for financial support (project 31.32033.91).

References

Bothner-By, A.A., Stephens, R.L., Lee, J., Warren, C.D. and Jeanloz, R.W. (1984) Structure determination of a tetrasaccharide: transient nuclear Overhauser effects in the rotating frame. *J. Am. Chem. Soc.* 106: 811-813.

Brunne, R.M., Liepinsh, E., Otting, G., Wüthrich, K. and van Gunsteren, W.F. (1993) Hydration of proteins. A comparison of experimental residence times of water molecules solvating the bovine pancreatic trypsin inhibitor with theoretical model calculations. *J. Mol. Biol.* 231: 1040-1048.

Buck, M., Radford, S.E. and Dobson, C.M. (1993) A partially folded state of hen egg white lysozyme in trifluoroethanol: structural characterization and implications for protein folding.

268

Biochemistry 32: 669-678.

Clore, G.M., Bax, A., Wingfield, P.T. and Gronenborn, A.M. (1990) Identification and localization of bound internal water in the solution structure of Interleukin 1b by heteronuclear three-dimensional ^{1}H rotating-frame Overhauser ^{15}N-^{1}H multiple quantum coherence NMR spectroscopy. *Biochemistry* 29: 5671-5676.

Forman-Kay, J.D., Gronenborn, A.M., Wingfield, P.T. and Clore, G.M. (1991) Determination of the positions of bound water molecules in the solution structure of reduced human thioredoxin by heteronuclear three-dimensional nuclear magnetic resonance spectroscopy. *J. Mol. Biol.* 220: 209-216.

Kubinec, M.G. and Wemmer, D.E. (1992) NMR evidence for DNA-bound water in solution. *J. Amer. Chem. Soc.* 114: 8739-8740.

Liepinsh, E., Otting, G. and Wüthrich, K. (1992a) NMR spectroscopy of hydroxyl protons in aqueous solutions of peptides and proteins. *J. Biomol. NMR* 2: 447-465.

Liepinsh, E., Otting, G. and Wüthrich, K. (1992b) NMR observation of individual molecules of hydration water bound to DNA duplexes: direct evidence for a spine of hydration water present in aqueous solution. *Nucl. Acids Res.* 20: 6549-6553.

Liepinsh, E., Rink, H., Otting, G. and Wüthrich, K. (1993) Contribution from hydration of carboxylate groups to the spectrum of water-polypeptide proton-proton Overhauser effects in aqueous solution. *J. Biomol. NMR* 3: 253-257.

Otting, G. and Wüthrich, K. (1989) Studies of protein hydration in aqueous solution by direct NMR observation of individual protein-bound water molecules. *J. Amer. Chem. Soc.* 111: 1871-1875.

Otting, G., Liepinsh, E. and Wüthrich, K. (1991a) Protein hydration in aqueous solution. *Science* 254: 974-980.

Otting, G., Liepinsh, E. and Wüthrich, K. (1991b) Proton exchange with internal water molecules in the protein BPTI in aqueous solution. *J. Amer. Chem. Soc.* 113: 4363-4364.

Otting, G., Liepinsh, E., Farmer II, B.T. and Wüthrich, K. (1991c) Protein hydration studied with homonuclear 3D ^{1}H NMR experiments. *J. Biomol. NMR* 1: 209-215.

Otting, G., Liepinsh , E. and Wüthrich, K. (1992) Polypeptide hydration in mixed solvents at low temperatures. *J. Amer. Chem. Soc.* 114: 7093-7095.

Pan, Y. and Briggs, M.S. (1992) Hydrogen exchange in native and alcohol forms of ubiquitin. *Biochemistry* 31: 11405-11412.

Qian, Y.Q., Otting, G. and Wüthrich, K. (1993) NMR detection of hydration water in the intermolecular interface of a protein-DNA complex. *J. Amer. Chem. Soc.* 115: 1189-1190.

Storrs, R.W., Truckses, D. and Wemmer, D.E. (1992) Helix propagation in trifluoroethanol solutions. *Biopolymers* 32: 1695-1702.

Timasheff, S.N. (1992) Water as ligand: preferential binding and exclusion of denaturants in protein unfolding. *Biochemistry* 31: 9857-9864.

Wüthrich, K. (1986) *NMR of proteins and nucleic acids*. Wiley, New York.

Wüthrich, K. (1993) Hydration of biological macromolecules in solution, surface structure and molecular recognition. In: *DNA and chromosomes*. 58th Cold Spring Harbor Laboratory Symposium on Quantitative Biology, in press.

Wüthrich, K. and Otting, G. (1992) Studies of protein hydration in aqueous solution by high-resolution nuclear magnetic resonance spectroscopy. *Int. J. Quant. Chem.* 42: 1553-1561.

Wüthrich, K., Otting, G. and Liepinsh, E. (1992) Protein hydration in aqueous solution. *Faraday Discuss.* 93: 35-45.

Toward a Molecular Basis of Alcohol Use and Abuse
ed. by B. Jansson, H. Jörnvall, U. Rydberg, L. Terenius & B. L. Vallee
© 1994 Birkhäuser Verlag Basel/Switzerland

Crystallographic investigations of alcohol dehydrogenases

Hans Eklund[1], S. Ramaswamy[1], Bryce V. Plapp[2], Mustafa El-Ahmad[3], Olle Danielsson[3], Jan-Olov Höög[3] and Hans Jörnvall[3]

[1]*Department of Molecular Biology, Swedish University of Agricultural Sciences, Uppsala, Sweden;* [2] *Department of Biochemistry, University of Iowa, Iowa City, USA;* [3] *Department of Medical Biochemistry and Biophysics, Karolinska Institutet, Stockholm, Sweden*

Summary

The structures of horse liver alcohol dehydrogenase class I in its apoenzyme form and in different ternary complexes have been determined at high resolution. The complex with NAD^+ and the substrate analogue pentafluorobenzyl alcohol gives a detailed picture of the interactions in an enzyme-substrate complex. The alcohol is bound to the zinc and positioned so that the hydrogen atom can be directly transferred to the C4 atom of the nicotinamide ring. The structure of cod liver alcohol dehydrogenase with hybrid properties (functionally of class I but structurally overall closer to class III) has been determined by molecular replacement methods to 3 Å resolution. Yeast alcohol dehydrogenase has been crystallized, and native data have been collected to 3 Å resolution.

Introduction

During the years 1965 - 1985 the structure of horse liver alcohol dehydrogenase of class I was thoroughly investigated in Carl-Ivar Brändén's laboratory using crystallographic methods (Eklund and Brändén, 1987). These studies provided one of the bases for our understanding of the enzyme. For some years thereafter, crystallographic investigations of the enzyme were quiescent, but in recent years crystallography of alcohol dehydrogenases has once again become intense. The horse liver enzyme has been refined in different crystal forms to high resolution, and the structure of the class I human $\beta_1\beta_1$ (Hurley *et al.*, 1991) and cod liver alcohol dehydrogenases have now been solved by molecular replacement methods. A number of inhibitor and substrate complexes have been investigated to a resolution of 2.1 - 2.4 Å and crystals of yeast alcohol dehydrogenase suitable for crystallographic investigations have been obtained. These studies provide a basis for understanding the specificities and physiological functions of the family of alcohol dehydrogenases.

Horse liver alcohol dehydrogenase

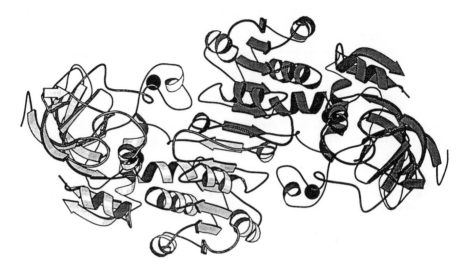

Fig. 1. A dimeric alcohol dehydrogenase molecule with one darker subunit to the right and a lighter subunit to the left. The two zinc atoms per subunit are shown as spheres.

General structural features

The liver alcohol dehydrogenases are dimeric molecules (Fig. 1). Each subunit is distinctly divided into two domains, the coenzyme binding domain, which contains most residues responsible for NAD binding, and the catalytic domain which contains most residues responsible for substrate binding including an active site zinc ion. The two coenzyme binding domains bind together to form the central part of the dimer. The two catalytic domains are located on each side of the two coenzyme binding domains, forming an elongated molecule about 100 Å long. The catalytic domain is built up essentially by antiparallel pleated sheets and a few helices. The coenzyme binding domain has the classical α/β fold with six parallel strands and surrounding helices. There is a deep cleft between the domains where NAD and substrate binds. When NAD binds to the enzyme, the catalytic domains rotate 10° relative to the coenzyme binding domain, thereby tightening the interactions with NAD and excluding water from the central part of the active site.

Coenzyme binding

NAD is bound at the carboxyl end of the strands of the pleated sheet (Fig. 2) (Eklund *et al.*,

1984). The adenine makes non-specific van der Waals interactions. The adenosine ribose is bound by Asp223, which is the main determinant of the specificity for NAD as compared to NADP, and also makes a hydrogen bond to Lys228. The pyrophosphate binding site has positively charged residues Arg47 and Arg369 and main chain amino groups at the amino end of αB and α1, some of which are hydrogen bonded via water molecules. The NMN ribose is bound by hydrogen bonds to Ser48, His51 and a main chain carbonyl. The nicotinamide ring is firmly positioned in the center of the molecule with three hydrogen bonds between the carboxyamide group and main chain atoms.

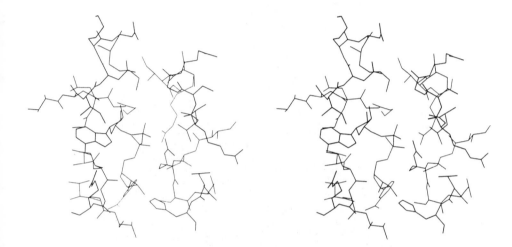

Fig. 2. Stereo drawing of the coenzyme bound to horse liver alcohol dehydrogenase. Dots represent water molecules.

The catalytic domain

The framework of the catalytic domain is a network of antiparallel pleated sheet strands (Fig. 3), containing 45% of the residues of the domain (Jones *et al.*, unpublished). Only about 25% of the residues of the domain are in helices. Two of the helices are of the 3_{10} type, one of which is 10 residues long and forms three full turns of 3_{10} helix. This helix is at the center of the rotation axis for the catalytic domain in the conformational transition upon NAD binding. The second hinge is at a turn of an α_L-helix.

Three regions of the catalytic domain have little secondary structure. One is the loop formed by residues 93-115 which binds the structural zinc ion by four cysteines. Two of these zinc ligands are located in a short helix. The sulfur atoms of the liganding cysteine residues accept two hydrogen bonds each, usually from main chain N-H groups. Of the two cysteine ligands to the active site zinc ion, only Cys46 has one hydrogen bond of this type, to 48 N. The side chain of the third zinc ligand, His67, is hydrogen bonded to the side chain of Asp49, which is stabilized by hydrogen bonds to 66 N and an internal water molecule. The fourth zinc ligand is a water molecule which is hydrogen bonded to Ser48.

Fig. 3. The catalytic domain with the structural zinc loop at the top and the active site zinc atom to the left.

Active site

The active site zinc atom is coordinated by Cys46, His67 and Cys174. The fourth coordination is a water molecule in the absence of substrate or inhibitors. In the complexes of substrate analogues, the oxygen atoms corresponding to the alcohol or aldehyde oxygen bind to the active site zinc atom in a distorted tetrahedral arrangement. Ser48 forms a hydrogen bond to this oxygen atom. Other than the groups mentioned, the active site of this class I enzyme is highly hydrophobic, and devoid of water molecules at the reaction center.

Specificity pocket

The substrate binding site is formed partly by the presence of the coenzyme. The full binding cleft for substrates is formed by positioning of the nicotinamide ring and the conformational change of the enzyme when NAD binds. Especially for small substrates these interactions explain the ordered mechanism. The specificity pocket is very different between the classes of alcohol dehydrogenases, as indicated by model building studies (Eklund et al., 1990). This is consistent with the differences in substrate specificity between the different classes of enzyme.

The structure of the enzyme has been examined in view of the different human isozymes and classes as well as their mutants and other vertebrate variants (Eklund et al., 1987, 1990; Danielsson et al., 1992; Höög et al., 1992). Class II and III subunits have 8 and 10 replacements respectively out of 11 residues in the substrate specificity pocket. The inner part of the substrate cleft in the class II and III enzymes is smaller than in the horse class I enzyme because both Ser48 and Phe93 are replaced by Thr and Tyr respectively. In class II, half of the residues in the specificity pocket are larger, Phe and Tyr, which makes the pocket smaller. In class III the outer part of the pocket is considerably wider and more polar, consistent with the glutathione-linked formaldehyde dehydrogenase activity of this enzyme.

Mechanism of action

The zinc atom polarizes the aldehyde substrates such that electrons are delocalized to the oxygen atom and a partial positive charge is created on the C1 atom which can then accept a hydride from NADH. For alcohol substrates, binding to the zinc and the positively charged NAD^+ change the pK_a of the alcohol so that an alcoholate ion is produced. This facilitates the hydride transfer from alcohol to NAD^+. The hydroxyl group of Ser48 is probably important for the transfer of the proton from the hydroxyl group of the substrate to the surface of the protein via a proton relay system involving also the 2'-hydroxyl of NMN-ribose and the imidazole of His51.

Unresolved problems

The crystallographic investigations of alcohol dehydrogenase up to the mid 1980's were only done on the horse class I enzyme and in many cases only carried out to about 2.9 Å resolution with refinement to R-values of 20-25%. Data were collected to higher resolution since some crystals diffracted to better than 2 Å resolution, but these studies were not finished. The accuracy of the structures was thus not as high as desired. The availability of alcohol dehydrogenases from other classes and other species was also limited. Some of the major differences with the class I horse enzyme were highlighted by model building studies (Eklund *et al.*, 1990), which give general aspects on the structures but are uncertain on details.

The situation has changed during recent years. Due to the improvements in data collection facilities it has become much easier to collect higher resolution data sets. Improved purification methods have also allowed the preparation of alcohol dehydrogenase from different species in quantities sufficient for crystallographic investigations.

Crystallographic refinement

Horse liver alcohol dehydrogenase was partly refined in four different crystal forms in the beginning of the 1980's; the orthorhombic apoenzyme crystal form, the ternary complexes in triclinic and monoclinic crystal forms, and a monoclinic crystal form of isonicotinimidylated enzyme (Plapp *et al.*, 1983). The refinement of the three first crystal forms have been continued independently by us and by others (Al-Karadaghi, 1993).

The apoenzyme was the first to be refined, to an R-factor of about 20% at 2.4 Å and was deposited at the Brookhaven data bank. These coordinates have since then been the basis for further refinement.

Horse liver alcohol dehydrogenase crystallizes in orthorhombic crystals of space group $C222_1$ in absence of the coenzyme (cell dimensions: a=56.0 Å, b=75.2 Å, c=181.7 Å). The structure of this crystal form has been refined with various techniques to an R-factor of 16.7% for all reflections in the resolution range 7.0-2.4 Å. The final model has deviations of 0.011 Å from standard bond lengths and 2.7° from standard bond angles. All 374 amino acid residues, two zinc ions and 175 water molecules have been included in the refinement.

Several complexes crystallize in both triclinic and monoclinic crystals and no general rules for obtaining one or the other are obvious. We have therefore collected data from both types of crystals and have several complexes in both crystal forms. The triclinic crystals (cell dimensions: a=52.0 Å, b=44.6 Å, c=94.4 Å, α= 104.4°, β=101.9°, γ=70.7°) were first investigated (Eklund et al., 1981) and the LADH-NADH-DMSO complex was partly refined at 2.9 Å resolution (Eklund et al., 1984). These coordinates were refined further for the NAD$^+$-pentafluoro-benzyl alcohol complex. Presently the R-factor at 2.4 Å resolution is R = 18.7% for this complex.

Data for the same complex have been collected also in the monoclinic P2$_1$crystal form (cell dimensions: a=51.1 Å, b=180.0 Å, c=44.2 Å, β=108.0°). The structure of the dimer from the triclinic complex was positioned in the monoclinic cell using the rotation function, Patterson correlation, and translation functions in X-plor (Brünger et al., 1983). Only one significant solution was obtained. This structure was then refined first as a rigid body and then by simulated annealing using the X-plor program. At present an R-factor of 18.3 % at 2.1 Å resolution has been obtained.

The refined protein coordinates have then been used at the start of the refinement of the other complexes in the triclinic and monoclinic crystal forms.

The position of substrate analogues

In our studies on the *p*-bromobenzyl alcohol complex, there was a question about the oxidation state of the coenzyme and the substrate (Eklund et al., 1982). We have continued this line of research to clarify the remaining questions. Fluorine derivatives of alcohols are inhibitors, due to electron withdrawal by the fluorine atoms. The complex with NAD$^+$ and a fluoroalcohol that does not oxidize should be close to a complex that is posed to reacting with a depronated hydroxyl group. We have crystallized several fluorinated substrates and data have been collected at 2.4 Å resolution of the 2,2,2-trifluoroethanol complex and the 2.1 Å data of a complex with 2,3,4,5,6-pentafluorobenzyl alcohol. The positions of the substrate analogues are very well defined in the electron density map due to the fluorines (Fig. 4). The oxygen is bound to the zinc ion and the O1 - C1 bond is roughly parallel to the C4 - C5 bond in the nicotinamide ring. This position is very close to that suggested to be optimal for hydride transfer (Eklund et al., 1982). The distance from the hydrogen on C1 to be transferred to C4 on the nicotinamide ring is 2.5 Å. It appears that this position of the substrate would be ideal for hydride transfer if the electronic properties of the ring were more favorable.

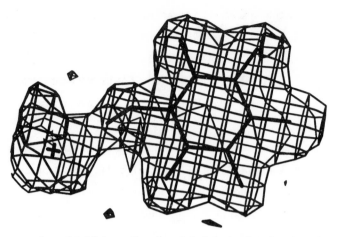

Fig. 4. The substrate analogue 2,3,4,5,6-pentafluorobenzyl alcohol when bound to the active site of horse liver alcohol dehydrogenase. The zinc atom is represented by the cross to the left.

Cod Liver Alcohol Dehydrogenase

Cod liver alcohol dehydrogenase of "hybrid" properties (Danielsson *et al.*, 1992), with class I type activity but overall residue similarities to class III has been crystallized in a NADH complex in the monoclinic space group $P2_1$ with cell dimensions a=103.3 Å, b=47.4 Å, c=81.0 Å, β=105.4. The position of the molecule in these crystals was determined by molecular replacement methods at 3.0 Å resolution. The successful search object was a model based on the horse enzyme where the side chains were replaced according to the amino acid sequence. The rotation and translation solutions were obtained by the X-plor program (Brünger *et al*, 1983). The structure of the cod enzyme was refined using X-plor. After the refinement, the correct solution was verified from interpretable density for the NADH molecule at plausible positions, although no NAD molecule was included in the refinement.

Yeast Alcohol Dehydrogenase

The recombinant cytoplasmic yeast alcohol dehydrogenase isozyme I has been crystallized as a complex with NAD^+ and 2,2,2-trifluoroethanol in the hexagonal space group P622 (Ramaswamy *et al.*, 1994). The cell dimensions are a=b=149 Å c=69 Å. A native data set to 3.2 Å resolution has been collected. There is one subunit of the tetrameric enzyme per asymmetric unit. Molecular replacement programs did not give a plausible solution, suggesting that the structure is somewhat different from the liver enzymes. We have started the search for heavy atom derivatives.

References

Al-Karadaghi, S. (1993) Refined crystal structures of Zn(II)- and Cu(II)- alcohol dehydrogenase. A comparative study. PhD thesis, Stockholm University.

Brünger, A.T., Kuriyan, J. and Karplus, M. (1987) Crystallographic R factor refinement by molecular dynamics. *Science* 235: 458-460.

Danielsson, O., Eklund, H. and Jörnvall, H. (1992) The major piscine liver alcohol dehydro genase has class-mixed properties in relation to mammalian alcohol dehydrogenase of class I and III. *Biochemistry* 31: 3751-3759.

Eklund, H., Samama, J.-P., Wallén, L., Brändén, C.-I., Åkeson, Å. and Jones, T.A. (1981) Structure of a triclinic ternary complex of horse liver alcohol dehydrogenase at 2.9 Å resolution. *J. Mol. Biol.* 146: 561-587.

Eklund, H., Plapp, B.V., Samama, J.-P. and Brändén, C.-I. (1982) Binding of substrate in a ternary complex of horse liver alcohol dehydrogenase. *J. Biol. Chem.* 257: 14349-14358.

Eklund, H., Samama, J.-P. and Jones, T.A. (1984) Crystallographic investigations of nicotinamide adenine dinucleotide binding to horse liver alcohol dehydrogenase. *Biochemistry* 23: 5982-5996.

Eklund, H. and Brändén, C.-I. (1987) Active Sites of Enzymes. In: Jurnak, F. and McPherson, A. (eds): *Alcohol dehydrogenase. Biological Molecules and Assemblies 3.* Wiley and Sons, New York, pp. 73-142.

Eklund, H., Horjales, E., Vallee, B.L. and Jörnvall, H. (1987) Computer-graphics inter- pretations of residue exchanges between the α, β and γ subunits of human-liver alcohol dehydrogenase class I isoenzymes. *Eur. J. Biochem.* 167: 185-193.

Eklund, H., Müller-Wille, P., Horjales, E., Levina, O., Holmquist, B., Vallee, B.L., Höög, J.-O., Kaiser, R. and Jörnvall, H. (1990) Comparison of three classes of human liver alcohol dehydrogenase. *Eur. J. Biochem.* 193: 303-310.

Hurley, T.D., Bosron, W.F., Hamilton, J.A. and Amzel, L.M. (1991) Structure of human $\beta_1\beta_1$ alcohol dehydrogenase: catalytic effects of non-active site substitutions. *Proc. Natl. Acad. Sci. USA,* 88: 8149-8153.

Plapp, B.V., Eklund, H., Jones, T.A. and Brändén, C.-I. (1983) Three-dimensional structure of isonicotinimidylated liver alcohol dehydrogenase. *J. Biol. Chem.* 258: 5537-5547.

Ramaswamy, S., Kratzer, D.A., Hershey, A.D., Rogers, P.H., Arnone, A., Eklund, H. and Plapp, B.V. (1994) Crystallization and preliminary crystallographic studies of *Saccharomyces cervisiae* alcohol dehydrogenase I. *J. Mol. Biol.,* in press.

Toward a Molecular Basis of Alcohol Use and Abuse
ed. by B. Jansson, H. Jörnvall, U. Rydberg, L. Terenius & B. L. Vallee
© 1994 Birkhäuser Verlag Basel/Switzerland

Retinoids and the alcohol dehydrogenase gene family

Gregg Duester

La Jolla Cancer Research Foundation, 10901 North Torrey Pines Road, La Jolla, California 92037 USA

Summary

Alcohol dehydrogenase (ADH) is best known as the enzyme which catalyzes the reversible oxidation/reduction of ethanol/acetaldehyde. However, mammalian ADH has also been shown to function *in vitro* as a retinol dehydrogenase in the conversion of retinol (vitamin A alcohol) to retinoic acid, a hormone which regulates gene expression at the transcriptional level. It is clear that retinol must be converted to more active retinoid forms in order to fulfill its roles in growth, development, and cellular differentiation. An important unsolved issue in retinoid research is the control of retinoic acid synthesis from retinol during differentiation. Several enzymes which participate in the conversion of retinol to retinoic acid *in vitro* have been isolated, but more information on their relative importance is needed.

Human ADH exists as a family of isozymes encoded by seven genes which are differentially expressed in adult and fetal mammalian tissues, being found preferentially in the epithelial cells which are known to synthesize and respond to retinoic acid. Retinoic acid is also known to play a role in neural tube development in vertebrate embryos. Excessive doses of retinoic acid or ethanol are both teratogenic for neural tube development. A relationship may exist between these two types of teratogenesis due to the role of ADH in both retinol and ethanol metabolism and the ability of ethanol to competitively inhibit retinol oxidation. There is a lack of information on the expression patterns of ADH genes in early embryos, but transgenic mouse studies are presented here which show that the human *ADH3* gene can be expressed in several mouse embryonic tissues including the neural tube. Thus, ethanol-induced neural tube defects seen in cases of fetal alcohol syndrome may be due to ethanol inhibition of retinol oxidation catalyzed by an embryonic ADH. This could potentially lower retinoic acid levels in the neural tube to the extent that gene expression is not properly regulated, resulting in morphological defects.

Introduction

The vitamin A metabolite retinoic acid is known to act as a potent regulator of gene expression by serving as a regulatory ligand for a family of retinoic acid receptors known to function in the control of transcription in vertebrate animals (Petkovich *et al.*, 1987). Retinoic acid is known to play a role in cellular differentiation (Roberts and Sporn, 1984) and plays a role in the growth and development of the embryonic central nervous system (Hunter *et al.*, 1991). The neural tube has been reported to possess the enzymatic machinery to convert retinol to retinoic acid (Wagner *et al.*, 1990), and retinoic acid has been identified as an endogenous molecule in this tissue (Hunter *et al.*, 1991). When exogenously introduced in large amounts, retinoic acid acts as a teratogen for human neural tube development, particularly of the hindbrain and cephalic neural crest (Lammer *et al.*, 1985). This indicates that the synthesis and accumulation of endogenous

280

retinoic acid must normally be limited to very small amounts in discrete embryonic locations.

Studies on the enzymatic pathway controlling the synthesis of retinoic acid from its vitamin A precursor retinol have shown that the enzyme alcohol dehydrogenase (ADH) from humans and other mammals functions as a cytosolic retinol dehydrogenase in the conversion of retinol to retinoic acid *in vitro* (Mezey and Holt, 1971; Kim *et al.*, 1992; Yang *et al.*, 1993). Mammalian ADH consists of a family of isozymes that function as multifunctional enzymes involved in the conversion of many alcohols to their corresponding aldehydes including the oxidation of ethanol and retinol (Pietruszko, 1979). A microsomal retinol dehydrogenase activity distinct from ADH which oxidizes retinol to retinal has been identified, but the enzyme responsible has not been purified or characterized extensively (Posch *et al.*, 1991; Kim *et al.*, 1992). Retinal can be further oxidized to retinoic acid irreversibly by an isozyme of aldehyde dehydrogenase (Lee *et al.*, 1991). Mammalian ADH has recently been shown to possess a dismutase activity that also irreversibly oxidizes aldehydes to carboxylic acids (Henehan and Oppenheimer, 1993) including retinal to retinoic acid (Pocker and Li, 1993). Thus, there may exist alternative pathways for synthesis of retinoic acid from retinol (Fig. 1).

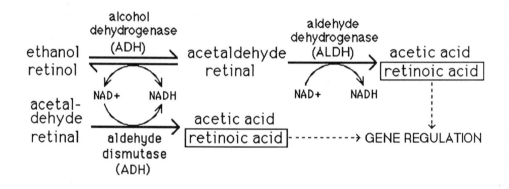

Fig. 1. Reactions catalyzed by alcohol dehydrogenase. See text for details.

Mammalian ADH has been best studied in the human where it exists as a family of proteins encoded by seven genes falling into five classes (Duester *et al.*, 1986; Yasunami *et al.*, 1991; Parés *et al.*, 1992). Class I ADH in humans is active as a retinol dehydrogenase with a Km value for retinol near 20 μM (Mezey and Holt, 1971; Yang *et al.*, 1993). Purified human class I, II, III, and IV ADHs have been compared with respect to retinol oxidation, and it was shown that class IV and class II oxidized retinol more efficiently than class I ADH, with class III inactive (Yang *et al.*, 1993).

The human class I ADH sub-family has been studied extensively at the gene level. Humans possess three closely related class I ADH genes (*ADH1*, *ADH2*, and *ADH3*) and several transcription factors regulating promoter activity have been identified (Van Ooij *et al.*, 1992). It has been demonstrated that the *ADH3* gene is inducible by retinoic acid in tissue culture cells (Shean and Duester, 1992) and that the *ADH3* promoter possesses a retinoic acid response element consisting of a direct repeat of the sequence AGGTCA spaced by five nucleotides able to bind the retinoic acid receptor (Duester *et al.*, 1991). This led to the hypothesis that the retinol dehydrogenase function of ADH may be feedback regulated by retinoic acid. Thus, studies on both the enzymology and molecular biology of human ADH have provided evidence that this enzyme may participate in retinoic acid synthesis.

The distribution of human ADH in adult tissues is suggestive of a function in retinoic acid synthesis. ADH has been localized primarily in the epithelial tissues (Bühler *et al.*, 1983). Since epithelial cells have a requirement for retinoic acid to regulate their state of differentiation (Roberts and Sporn, 1984) it is possible that ADH expression in epithelial cells contributes to retinoic acid synthesis. Retinoic acid is also involved in the early stages of post-implantation embryonic development, particularly neurogenesis (Hunter *et al.*, 1991), but there have been no reports on the distribution of any retinoic acid synthetic enzymes in early vertebrate embryos at either the protein or gene expression level. In order to explore the potential role of human ADH as a retinoic acid synthetic enzyme in embryonic central nervous system tissue it is necessary to determine if ADH gene expression occurs there. Due to the inaccessibility of early human embryos undergoing neurogenesis, i.e. embryos of 3-4 weeks gestation (Rugh, 1990), it is difficult to analyze expression of the human ADH gene family during this time. In order to study the embryonic expression pattern of a well-characterized human ADH gene this laboratory has turned to the use of transgenic indicator mice to allow an analysis of human ADH promoter activity in mouse embryos. Analysis has centered on the expression pattern of the human *ADH3* gene encoding the enzyme γ-ADH which has previously been detected in human fetal epithelial tissues (Smith *et al.*, 1971), and which acts as a retinol dehydrogenase (Yang *et al.*, 1993). The human *ADH3* promoter was fused to the *lacZ* indicator gene encoding β-galactosidase, and this construct was introduced into transgenic mice to enable a visualization of human *ADH3* promoter activity in early mouse embryos. The findings have implications for

ADH function during embryogenesis and for the mechanism of fetal alcohol syndrome.

Materials and Methods

Fusion of the ADH3 promoter to the lacZ gene

A 1.15 kilobase fragment containing the human *ADH3* promoter derived from the plasmid *ADH3-cat*(-1102) (Duester *et al.*, 1991) was excised with *Hind*III and ligated upstream of the *E. coli lacZ* gene encoding β-galactosidase derived from the plasmid pCH110 (Pharmacia, Inc.). In this gene fusion, *ADH3-lacZ(-1102)*, the AUG translation start codon for *lacZ* is the first such codon downstream of the *ADH3* transcription initation site at position +1 base pairs.

Introduction of the ADH3-lacZ fusion gene into transgenic mice

A 4.96 kilobase *Xho*I-*Bam*HI DNA fragment from plasmid *ADH3-lacZ(-1102)* was purified by agarose gel electrophoresis, then injected (2 ng/μl) into the male pronucleus of fertilized mouse eggs (FVB female x C57BL6 male) which were then transferred to pseudopregnant FVB females (Hogan *et al.*, 1986). Five offspring representing different integration sites were identified as carrying the transgene based upon Southern blot analysis of tail DNA.

Detection of β-galactosidase activity in mouse embryos

Embryos were staged by counting somites (Rugh, 1990). The embryo was removed from extraembryonic tissues, fixed, and stained for β-galactosidase activity as previously described using the chromophore X-gal (Mendelsohn *et al.*, 1991). β-Galactosidase activity was observed as a blue staining pattern in a light yellow background. In all cases, non-transgenic embryos of the same stage were negative for β-galactosidase activity when subjected to identical staining conditions (i.e. only the light yellow background was observed).

Results

Construction of an ADH3-lacZ transgene and introduction into transgenic mice

A region of the *ADH3* promoter was fused to the *E. coli lacZ* gene encoding β-galactosidase as shown in Fig. 2. This region of the *ADH3* gene has previously been shown to contain promoter and enhancer sequences which direct transcription upon transfection into tissue culture cells (Duester *et al.*, 1991; Van Ooij *et al.*, 1992).

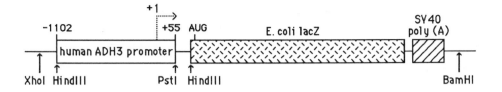

Fig. 2. Fusion of *ADH3* promoter to *lacZ*. The human *ADH3* promoter from position -1102 to +55 relative to the transcription initiation site is shown fused to the *E. coli lacZ* gene containing an AUG translation start codon. The SV40 polyadenylation site is located downstream of *lacZ*. For microinjection into fertilized eggs, an *XhoI/BamHI* DNA fragment was isolated after release from the plasmid.

In order to analyze expression of the human *ADH3* gene during early embryogenesis, we have produced lines of transgenic mice carrying *ADH3-lacZ* randomly integrated into their genomic DNA. Five independent transgenic founder mice were characterized, and two were identified which expressed the transgene in the same pattern. This was an indication that *ADH3* promoter expression in these lines was not effected by the site of transgene integration in mouse chromosomal DNA. The data presented here is from those two lines of mice.

Expression of the ADH3-lacZ transgene in mouse embryo kidney

In human fetuses *ADH3* has been demonstrated by starch gel analysis to be expressed in the kidney, but absent in liver and lung (Smith *et al.*, 1971). Upon analysis of transgenic mice carrying *ADH3-lacZ* it was noticed that this expression pattern was recapitulated in late mouse embryos [i.e. 13.5 days post-coitum (d.p.c.) to 18.5 d.p.c. near the time of birth] with β-galactosidase staining being observed in the kidney, but not in the liver or lung. This provides further evidence that the transgene is being appropriately expressed. It is known that the closely related human genes *ADH1* and *ADH2* are expressed much differently than *ADH3* in human fetuses (Smith *et al.*, 1971). *ADH1* is expressed in liver and *ADH2* is expressed in liver and lung as demonstrated by starch gel analysis which enables a separation of all three class I ADH species. Thus, in kidney, liver, and lung the *ADH3* transgene appears to have retained the unique pattern of human *ADH3* expression in a murine background.

Transgene expression in early embryos

Transgene expression was not observed in embryos at 7.5 or 8.5 d.p.c., but was detected from

Table I. Expression of *ADH3-lacZ* in embryonic tissues

Tissues examined	day 9.5	day 12.5	day 13.5	day 14.5
Central Nervous System				
neural tube	+	+	+	-
spinal cord	+	+	+	-
hindbrain	+	+	+	-
midbrain	+	+	-	-
forebrain	-	-	-	-
Head				
craniofacial region	+	+	+	+
otic vesicles (ears)	+	+	+	+
eyes +	+	+	+	
Limbs				
forelimb buds	+	+	+	+
hindlimb buds	+	+	+	+
Organs				
heart +	+	+	-	
kidney	-	-	+	+
liver -	-	-	-	
lung -	-	-	-	
digestive tract	-	-	-	+

(+), β-galactosidase staining observed; (-), no staining observed

9.5 d.p.c. onwards in the central nervous system, head, limbs, and several organs (Table I). Examination of embryos at 9.5 d.p.c. revealed staining in the central nervous system, especially along the ventral midline of the neural tube including the floor plate. Neural tube staining extended from its most anterior position in the midbrain to more posterior positions in the hindbrain and future spinal cord. No staining was observed in the forebrain. This pattern of staining continued in older embryos (12.5 and 13.5 d.p.c.), but no staining was observed in the central nervous system in embryos of 14.5 d.p.c. or older.

In the head, staining was observed in the first, second and third branchial arches of 9.5 d.p.c. embryos which contribute to craniofacial structures including the jaw and mouth. Craniofacial staining continued through 14.5 d.p.c. The otic vesicles (developing ears) and eyes stained from 9.5 to 14.5 d.p.c.

Dorsoventral Gradient of ADH3-lacZ Expression in Neural Tube

To more easily view the staining in the neural tube, the heads of whole 9.5 and 12.5 d.p.c. embryos were removed with a scalpel to analyze the pattern of *ADH3* transgene expression in the dorsoventral axis. In a 9.5 d.p.c. embryo cut at a level just anterior to the forelimb buds (the cervical region of the spinal cord), a dorsoventral gradient of *ADH3* expression was observed with highest expression ventrally. Intense staining was observed in the ventral midline corresponding to the floor plate, and staining tapered off about halfway to the dorsal roof plate which was unstained. In a 12.5 d.p.c. embryo cut at the level of the cervical spinal cord, the neural tube continued to show a dorsoventral gradient of *ADH3* transgene expression with the floor plate most intensely stained. Located laterally on both sides of the floor plate, the ventral horns of the neural tube which contain the motor neurons were stained. Dorsal to those regions there was no staining observed. This is summarized in Fig. 3.

Discussion

The oxidation of retinol leads to the synthesis of retinoic acid, the active form of vitamin A involved in regulating cellular differentiation, growth, and development. The human ADH gene family encodes several enzymes known to function as cytosolic retinol dehydrogenases oxidizing retinol to retinal which can then be oxidized to retinoic acid. One ADH gene, the human *ADH3* gene encoding a class I enzyme, has been demonstrated to be regulated transcriptionally by retinoic acid, further implicating ADH as a player in the control of retinoic acid synthesis. In human adults and fetuses the expression of most ADH genes correlates with tissues known to synthesize retinoic acid from retinol, particularly the epithelial cells of various organs. However, the expression pattern of human ADH in embryos has not previously been examined due to the inaccessibility of early human embryos for detailed studies. To address this problem we have analyzed the expression pattern of a human *ADH3* promoter in transgenic mouse embryos. Several transgenic founders expressed in a pattern consistent with what was previously known about expression of *ADH3* in human fetal organs; i.e. expression in late embryonic kidney, but not liver or lung. This suggests that the promoter fragment including 1.15 kb of 5'-flanking DNA was able to direct transcription accurately. Analysis of these transgenic mice has indicated several new sites of ADH expression. Expression of the *ADH3-lacZ* transgene was observed in the embryonic central nervous system, craniofacial region, heart, and limb buds suggesting that ADH may exist in these tissues and play a role in their development possibly as a retinoic acid synthetic enzyme.

Expression in the embryonic central nervous system was particularly interesting. The *ADH3-lacZ* transgene was expressed in the neural tube from 9.5 d.p.c. onwards with highest expression in the ventral floor plate, exhibiting a ventral to dorsal gradient of decreasing expression. Previous studies have shown that the neural tube possesses the enzymatic machinery to convert retinol to retinoic acid via an NAD+ linked reaction, and that the ventral floor plate is enriched in this activity compared to dorsal neural tube tissue (Wagner *et al.*, 1990). These studies are supported by an analysis of the amount of endogenous retinoic acid in various portions of the neural tube which has shown that a higher level (1.5-fold) exists in ventral tissue compared to dorsal (Wagner *et al.*, 1992). The dorsoventral gradient of *ADH3* transgene expression we noticed in the neural tube suggests that a gradient of ADH-catalyzed retinoic acid synthesis could be established with the high end located in the ventral floor plate. Our data does not address the role retinoic acid plays in central nervous system development, which is still unclear, but our data suggests that ADH is expressed in the correct location to serve as a retinoic acid synthetic enzyme for this system.

Retinol concentrations in whole mouse embryos are aproximately 0.4 μM at day-10 rising to 1.2 μM by day-16 (Satre *et al.*, 1992). However, this is whole embryos and there may be regions of the embryo where the local concentration of retinol is higher. Retinol concentrations in the adult livers of vitamin A adequate rats is in the range of 4-10 μM (Yost *et al.*, 1988). Thus, the concentration of retinol in embryo and liver tissue is close to the Km values for several ADHs which are in the low micromolar range (Pietruszko, 1979; Yang *et al.*, 1993). However, much of it may be bound to cellular retinol-binding protein type I (CRBP-I) in tissues where this protein is present (Yost *et al.*, 1988). CRBP-I participates in the oxidation of retinol by a microsomal retinol dehydrogenase, and it has been proposed that the sequestration of retinol by CRBP-I inhibits cytosolic retinol oxidation by ADH (Posch *et al.*, 1991). However, since only very small amounts of retinol need to be converted to retinoic acid to achieve the nanomolar quantities needed for retinoic acid receptor activation (Petkovich *et al.*, 1987), it is possible that the sequestration of retinol by CRBP-I may not totally hinder the ability of ADH to oxidize small amounts of retinol and serve as a retinoic acid synthetic enzyme. Also, CRBP-I has been found to be absent from the floor plate of the mouse embryonic neural tube (Maden *et al.*, 1990; Ruberte *et al.*, 1993), a site where our *ADH3* promoter transgene was expressed at high levels (Fig. 3). Thus, if CRBP-I has an inhibitory role on ADH retinol

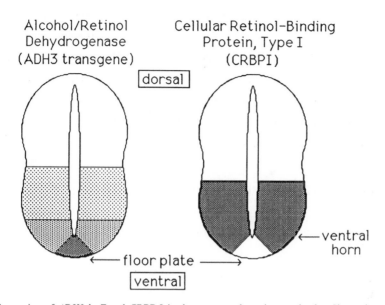

Fig. 3. Expression of *ADH3-lacZ* and CRBP-I in the mouse embryonic neural tube. Shown is a diagram of a transverse section through the neural tube at the region where the spinal cord meets the hindbrain. This shows the dorsoventral polarity of the neural tube with dorsal at the top and ventral at the bottom of the diagram. The floor plate is a group of cells located at the extreme ventral region of the neural tube, and the ventral horns are cells located dorsolaterally to the floor plate (Rugh, 1990). The expression pattern for the *ADH3* transgene is shown on the left with the intensity of shading corresponding to the observed intensity of β-galactosidase staining in embryos from 9.5-13.5 d.p.c. The pattern of expression of cellular retinol-binding protein type I (CRBP-I) is shown for comparison. This was determined by others using the techniques of immunohistochemistry (Maden *et al.*, 1990) and *in situ* hybridization (Ruberte *et al.*, 1993). Note the high level of *ADH3* expression in the ventral floor plate, but the lack of CRBP-I expression at this location.

oxidation, it will not be able to do this in the floor plate of the neural tube, allowing ADH (if present there) to act as a retinoic acid synthetic enzyme in that location.

Since ethanol and retinol are both substrates for ADH, it is possible for ethanol to competitively inhibit retinol oxidation. Human class I ADH isozymes exhibit a Km for ethanol of about 1.0 mM, and a Km for retinol of about 0.028 mM (Mezey and Holt, 1971). Ethanol is a competitive inhibitor of retinol oxidation catalyzed by human liver class I ADH with a Ki of 0.36 mM (Mezey and Holt, 1971). Thus, during alcohol intoxication when blood alcohol levels reach 20-100 mM (Lindblad and Olsson, 1976) most of the class I ADH activity will be tied up in ethanol oxidation and will not be available for retinol oxidation. Ethanol inhibition of rat class IV ADH has also been analyzed. The rat ADH-1 enzyme is a class IV enzyme exhibiting a very high Km for ethanol of 340 mM and a low Km for retinol of 0.02 mM (Julià *et al.,* 1986). Rat class IV ADH retinol oxidation is competitively inhibited by ethanol with a high Ki of 600 mM (Julià *et al.,* 1986). Thus, even during conditions of excessive ethanol consumption, class IV ADH should be only partially inhibited. However, since human class IV ADH has a Km for ethanol of 41 mM (Moreno and Parés, 1991) which is significantly lower than that of the rat enzyme, ethanol inhibition of retinol oxidation may occur at a lower Ki for human class IV ADH. The inhibitory effect of ethanol on retinol oxidation may result in a reduction in retinoic acid synthesis (Fig. 4).

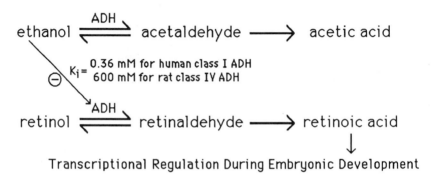

Fig. 4. Ethanol inhibition of retinoic acid synthesis. See text for details.

Since retinoic acid can regulate central nervous system development, we have hypothesized that the neuroteratogenic properties of ethanol observed in cases of fetal alcohol syndrome may be caused by a reduction in retinoic acid synthesis catalyzed by ADH (Duester, 1991). The expression of the human *ADH3-lacZ* transgene in mouse embryos presented here is the first indication that ADH may be expressed in embryonic central nervous system tissue. Interestingly, the major types of ethanol-induced damage in fetal alcohol syndrome, i.e. microcephaly, mental retardation, craniofacial defects, and heart defects (Jones and Smith, 1973) correlate with the major sites of *ADH3* transgene expression in early mouse embryos, i.e. the neural tube, brain, craniofacial region, and heart. Since we have now shown that an *ADH3* transgene can be expressed in the tissues sensitive to ethanol teratogenesis, it is more likely that retinoic acid synthesis in those tissues is dependent upon ADH and thus sensitive to ethanol inhibition.

Acknowledgements

M. Zgombic and M. Satre contributed to the analysis of the *ADH3-lacZ* transgenic mice. This work was supported by NIH grant AA07261.

References

Bühler, R., Pestalozzi, D., Hess, M. and von Wartburg, J.-P. (1983) Immunohistochemical localization of alcohol dehydrogenase in human kidney, endocrine organs and brain. *Pharmacol. Biochem. Behav.* 18: Suppl.1,55-59.

Duester, G. (1991) A hypothetical mechanism for fetal alcohol syndrome involving ethanol inhibition of retinoic acid synthesis at the alcohol dehydrogenase step. *Alcohol. Clin. Exp. Res.* 15: 568-572.

Duester, G., Shean, M.L., McBride, M.S. and Stewart, M.J. (1991) Retinoic acid response element in the human alcohol dehydrogenase gene *ADH3*: Implications for regulation of retinoic acid synthesis. *Mol. Cell. Biol.* 11: 1638-1646.

Duester, G., Smith, M., Bilanchone, V. and Hatfield, G.W. (1986) Molecular analysis of the human class I alcohol dehydrogenase gene family and nucleotide sequence of the gene encoding the β subunit. *J. Biol. Chem.* 261: 2027-2033.

Henehan, G.T.M. and Oppenheimer, N.J. (1993) Horse liver alcohol dehydrogenase-catalyzed oxidation of aldehydes: Dismutation precedes net production of reduced nicotinamide adenine dinucleotide. *Biochemistry* 32: 735-738.

Hogan, B., Costantini, F. and Lacy, E. (1986) *Manipulating the Mouse Embryo*. Cold Spring Harbor Laboratory, Cold Spring Harbor.

Hunter, K., Maden, M., Summerbell, D., Eriksson, U. and Holder, N. (1991) Retinoic acid stimulates neurite outgrowth in the amphibian spinal cord. *Proc. Natl. Acad. Sci. USA* 88: 3666-3670.

Jones, K.L. and Smith, D.W. (1973) Recognition of the fetal alcohol syndrome in early infancy. *Lancet* 2: 999-1001.

Julià, P., Farrés, J. and Parés, X. (1986) Ocular alcohol dehydrogenase in the rat: Regional distribution and kinetics of the ADH-1 isoenzyme with retinol and retinal. *Exp. Eye Res.* 42: 305-314.

Kim, C.-I., Leo, M.A. and Lieber, C.S. (1992) Retinol forms retinoic acid via retinal. *Arch. Biochem. Biophys.* 294: 388-393.

Lammer, G.J., Chen, D.T., Hoar, R.M., Agnish, N.D., Benke, P.J., Braun, J.T., Curry, C. J., Fernhoff, P.M., Grix, A.W., Lott, I.T., Richard, J.M. and Sun, S.C. (1985) Retinoic acid embryopathy. *N. Engl. J. Med.* 313: 837-841.

Lee, M.-O., Manthey, C.L. and Sladek, N.E. (1991) Identification of mouse liver aldehyde dehydrogenases that catalyze the oxidation of retinaldehyde to retinoic acid. *Biochem. Pharmacol.* 42: 1279-1285.

Lindblad, B. and Olsson, R. (1976) Unusually high levels of blood alcohol? *JAMA* 236: 1600-1602.

Maden, M., Ong, D.E. and Chytil, F. (1990) Retinoid-binding protein distribution in the developing mammalian nervous system. *Development* 109: 75-80.

Mendelsohn, C., Ruberte, E., LeMeur, M., Morriss-Kay, G. and Chambon, P. (1991) Developmental analysis of the retinoic acid-inducible RAR-b2 promoter in transgenic animals. *Development* 113: 723-734.

Mezey, E. and Holt, P.R. (1971) The inhibitory effect of ethanol on retinol oxidation by human liver and cattle retina. *Exp. Mol. Pathol.* 15: 148-156.

Moreno, A. and Parés, X. (1991) Purification and characterization of a new alcohol dehydrogenase from human stomach. *J. Biol. Chem.* 266: 1128-1133.

Parés, X., Cederlund, E., Moreno, A., Saubi, N., Höög, J.-O. and Jörnvall, H. (1992) Class IV alcohol dehydrogenase (the gastric enzyme): structural analysis of human $\sigma\sigma$-ADH reveals class IV to be variable and confirms the presence of a fifth mammalian alcohol dehydrogenase class. *FEBS Lett.* 303: 69-72.

Petkovich, M., Brand, N.J., Krust, A. and Chambon, P. (1987) A human retinoic acid receptor which belongs to the family of nuclear receptors. *Nature* 330: 444-450.

Pietruszko, R. (1979) Nonethanol substrates of alcohol dehydrogenase. In: Majchrowicz, E. and Noble, E.P. (eds.): *Biochemistry and Pharmacology of Ethanol,* Vol. 1, Plenum Press, New York, pp. 87-106.

Pocker, Y. and Li, H. (1993) The catalytic specificity of liver alcohol dehydrogenase: vitamin A alcohol and vitamin A aldehyde activities. *Adv. Exp. Med. Biol.* 328: 411-418.

Posch, K.C., Boerman, M.H.E.M., Burns, R.D. and Napoli, J.L. (1991) Holocellular retinol binding protein as a substrate for microsomal retinal synthesis. *Biochemistry* 30: 6224-6230.

Roberts, A.B. and Sporn, M.B. (1984) Cellular biology and biochemistry of the retinoids. In: Sporn, M.B., Roberts, A.B. and Goodman, D.S. (eds.): *The Retinoids Vol. 2,* Academic, Orlando, pp. 209-286.

Ruberte, E., Friederich, V., Chambon, P. and Morriss-Kay, G. (1993) Retinoic acid receptors and cellular retinoid binding proteins. III. Their differential transcript distribution during mouse nervous system development. *Development* 118: 267-282.

Rugh, R. (1990) *The Mouse: Its Reproduction and Development.* Oxford Univ. Press, New York.

Satre, M.A., Ugen, K.E. and Kochhar, D.M. (1992) Developmental changes in endogenous retinoids during pregnancy and embryogenesis in the mouse. *Biol. Reprod.* 46: 802-810.

Shean, M.L. and Duester, G. (1992) The role of alcohol dehydrogenase in retinoic acid homeostasis and fetal alcohol syndrome. *Alcohol and Alcoholism* 28, Suppl. 2: 51-56.

Smith, M., Hopkinson, D.A. and Harris, H. (1971) Developmental changes and polymorphism in human alcohol dehydrogenase. *Ann. Hum. Genet.* 34: 251-271.

Van Ooij, C., Snyder, R.C., Paeper, B.W. and Duester, G. (1992) Temporal expression of the human alcohol dehydrogenase gene family during liver development correlates with differential promoter activation by HNF1, C/EBPa, LAP, and DBP. *Mol. Cell. Biol.* 12: 3023-3031.

Wagner, M., Thaller, C., Jessell, T. and Eichele, G. (1990) Polarizing activity and retinoid synthesis in the floor plate of the neural tube. *Nature* 345: 819-822.

Wagner, M., Han, B. and Jessell, T.M. (1992) Regional differences in retinoid release from embryonic neural tissue detected by an in vitro reporter assay. *Development* 116: 55-66.

Yang, Z.N., Davis, G.J., Hurley, T.D., Stone, C.L., Li, T.-K. and Bosron, W.F. (1993) Catalytic efficiency of human alcohol dehydrogenases for retinol oxidation and retinal reduction. *Alcohol. Clin. Exp. Res.* 17: 496.

Yasunami, M., Chen, C.-S. and Yoshida, A. (1991) A human alcohol dehydrogenase gene (*ADH6*) encoding an additional class of isozyme. *Proc. Natl. Acad. Sci. USA* 88: 7610-7614.

Yost, R.W., Harrison, E.H. and Ross, A.C. (1988) Esterification by rat liver microsomes of retinol bound to cellular retinol-binding protein. *J. Biol. Chem.* 263: 18693-18701.

Toward a Molecular Basis of Alcohol Use and Abuse
ed. by B. Jansson, H. Jörnvall, U. Rydberg, L. Terenius & B. L. Vallee
© 1994 Birkhäuser Verlag Basel/Switzerland

Alcohol and acetaldehyde dehydrogenase gene polymorphism and alcoholism

David I.N. Sherman, Roberta J. Ward, Akira Yoshida[1] and Timothy J. Peters

Department of Clinical Biochemistry, King's College Hospital School of Medicine & Dentistry, London, U.K. and [1]Department of Biochemical Genetics, Beckman Research Institute, City of Hope National Medical Centre, Duarte, Los Angeles, U.S.A.

Summary

Inherited variations in alcohol and aldehyde dehydrogenases, the principal enzymes of ethanol metabolism, have been implicated in determining susceptibility to alcoholism and alcohol-related organ damage. An association between an RFLP for the alcohol dehydrogenase-2 (ADH2) gene and alcohol-induced liver damage was demonstrated in a Caucasian population. Genotyping studies revealed an increase in the ADH^3_2 allele in patients with alcohol-induced cirrhosis. PCR studies of the ALDH5 gene have demonstrated diverse polymorphism within a short segment of its coding region, with marked inter-racial variation in allele frequencies. In addition, the Caucasian alcohol-induced flushing reaction has been characterised and its relationship with phenotypic polymorphism of ALDH1 examined.

Introduction

Over the past twenty years there have been important advances in our understanding of the biochemistry and molecular genetics of alcoholism and its consequences. In Western societies alcohol misuse has a prevalence of between 10 to 15%, and yet within these patients there is striking variability in individual responses to the chronic effects of ethanol. Evidence from studies of concordance in monozygotic and dizygotic twins and adoption studies showing inheritance of alcoholism in sons of alcoholic fathers despite adoption into a non-alcoholic environment, have confirmed the importance of genetic influences, which may account for up to one quarter of this variation between individuals. It has also become clear that a simple Mendelian model is inappropriate for the genetics of alcoholism, as it is a heterogeneous disorder with a complex genetic background. Inherited predisposition is probably determined by a number of 'major' and 'minor' polymorphic gene loci, each with 2 or more alleles that exhibit varying degrees of penetrance. Racial and gender differences also play a part.

Variation in individual susceptibility is also seen in alcohol-related organ damage. Surveys of liver histology amongst chronic alcohol misusers reveal that between 90 and 100% have fatty liver, but only 30% have alcoholic hepatitis and 10-20% cirrhosis. A polygenic model is also applicable to the genetics of alcoholic liver disease, with major genes interacting with environmental influences such as alcohol consumption and nutritional status. Differential organ toxicity, for example the low reported incidence of cardiomyopathy in patients with cirrhosis, may be partly determined by polymorphic genes specific to the organ in question.

There is considerable evidence implicating the two principal enzymes of ethanol metabolism, alcohol dehydrogenase (ADH) and aldehyde dehydrogenase (ALDH), as candidate genes influencing susceptibility to alcohol-related disorders. Both enzymes account for at least 90% of hepatic ethanol metabolism, and determine the rate of formation and elimination of acetaldehyde, which is responsible for much ethanol-induced toxicity. Acetaldehyde may be involved in addiction to alcohol as well as in the pathogenesis of liver disease and damage to other organs. Conversely, it also mediates aversion to ethanol via the Oriental flushing reaction. Studies of ethanol metabolism in twins have shown that genetic factors determine most of the repeatable variation in ethanol metabolism between individuals (Martin *et al.*, 1984). In addition, dependent alcoholics undergoing detoxification show increased ethanol elimination rates and peak acetaldehyde levels in comparison with either misusers without signs of dependency or control subjects (Peters *et al.*, 1987). This and other work suggests a link between ethanol metabolism and susceptibility.

The biochemical properties of the ADH and ALDH isoenzymes have been well characterized, although the relationship between kinetic data obtained in vitro on different isoenzymes and ethanol metabolism *in vivo* is not straightforward. The recent cloning of cDNAs for the genes encoding the major isoenzymes of ADH and ALDH has allowed the study of these loci in populations. Our studies have focused on the polymorphic ADH gene loci, by investigating RFLP in the ADH_2 gene (Sherman *et al.*, 1993a) and direct genotyping of ADH_3 by PCR in Caucasian alcohol misusers. In addition, we have studied the relationship between variations in $ALDH_1$ activity and alcohol-induced flushing in Caucasians. Finally, we have examined the frequencies of common mutations of the recently cloned $ALDH_5$ gene in both Caucasians and Orientals.

Restriction fragment length polymorphism in the ADH_2 gene

We studied 46 unrelated alcohol misusers and 23 non-alcoholic Caucasian controls. All of the alcohol misusers had consumed at least 80g of ethanol daily for a minimum of 2 years, and had come to medical attention because of alcoholism or alcohol-related liver damage. Thirty eight had histologically proven alcoholic liver disease (6 fatty liver, 4 alcoholic hepatitis and fibrosis, 14 alcoholic hepatitis with cirrhosis, 14 cirrhosis). Information on quantity, duration and pattern of alcohol intake, severity of alcohol dependency and family history was obtained by interview and completion of a questionnaire incorporating DSM-III and SADQ criteria for alcoholism and dependency. Twenty one showed clinical features of, and satisfied questionnaire criteria for, alcohol dependency and 19 had a positive family history for alcoholism with at least one affected first degree relative. The control subjects, who were recruited voluntarily from laboratory staff, drank less than than an average of 24 g of ethanol daily. Gender distribution was similar in the two groups (56% male in the controls, 58% in the alcoholics). The average age of the controls was younger (36 versus 54 years).

Leucocyte DNA was prepared from peripheral blood by standard techniques. DNA was digested with PvuII restriction enzyme, the resulting fragments separated by agarose gel electrophoresis and transferred to nylon filters by Southern hybridisation ADH 36, a 1.3 Kb genomic DNA probe consisting of exon 3 of the ADH2 gene and segments of introns 2 and 3, was labelled with 32P by the random oligonucleotide technique and hybridised with the filters washed to high stringency and auto-radiographed.

Table I. Genotypes and allele frequencies for ADH36 RFLP in controls and alcohol misusers. Number of subjects shown between parentheses. $\chi^2 = 25.8$, p < 0.001 (2 degrees of freedom)

	pADH36 genotype			Allele Frequency (%)	
	AA	AB	BB	A	B
Controls (23)	18	3	2	85	15
Alcoholics (45)	7	19	19	37	63

Southern hybridisation of radiolabelled ADH36 probe revealed a two allele polymorphism. 'A' and 'B' alleles were denoted by 5.1 / 0.8 Kb and 3.1 / 2.9 Kb doublets, respectively. Allele frequencies for the whole sample conformed to Hardy-Weinberg equilibrium, and Mendelian inheritance of the A and B alleles was confirmed in 7 family pedigrees. The A allele appeared to be the 'wild type', with a frequency of 85% in non-alcoholic controls, with only two individuals homozygous for the B allele (Table I). In contrast, the frequency of the B allele in the alcohol misusers was 63% compared to 15% in the controls, a highly significant increase. With respect to liver disease, there was a clear association of the B allele with more severe liver histology (Table II). In patients with cirrhosis or alcoholic hepatitis with cirrhosis, the B allele frequencies were 86% and 61%, respectively, in comparison with much lower frequencies in small numbers of patients with fatty liver and alcoholic hepatitis with fibrosis.

Table II. Genotypes and allele frequencies of pADH36 RFLP in patients with alcoholic liver disease sub-divided according to histology. Number of subjects shown between parentheses. All cirrhosis vs non-cirrhotics: $\chi 2 = 9.9$, $p < 0.01$ (2 degrees of freedom).

	pADH36 Genotype			Allele Frequency (%)	
	AA	AB	BB	A	B
Fatty liver (6)	1	5	0	58	42
Alcoholic hepatitis and fibrosis (4)	2	2	0	75	25
Alcoholic hepatitis and cirrhosis (14)	2	7	5	39	61
Cirrhosis (14)	1	2	11	14	86

The strength of the association was upheld in the 21 alcoholics with alcohol dependency and in those with positive family history, who showed B allele frequencies of 64% and 61%, respectively ($p < 0.001$ in comparison with controls). Patients posessing the B allele had been abusing alcohol for a shorter period before the onset of liver disease ($p < 0.02$), but the age of onset of the liver disease did not differ significantly between ADH36 genotypes ($p > 0.4$).

Genotyping studies of ADH3 in alcohol-induced cirrhosis

Twenty-six patients with alcohol-induced cirrhosis and 16 control subjects were studied. Leucocyte DNA was amplified by PCR as described by Couzigou *et al.* (1990) using oligonucleotide primers specific for exon 8 of the ADH_3 gene. The product was digested with SSp 1 restriction enzyme and DNA fragments were separated by 10% polyacrilamide gel electrophoresis.

Table III. ADH3 genotypes in alcoholic liver disease patients and controls. Number of subjects shown between parentheses. $\chi2 = 5.9$, p = 0.05

| | ADH36 Genotypes | | | Allele Frequencies (%) | |
	$ADH_3^1\text{-}ADH_3^1$	$ADH_3^1\text{-}ADH_3^2$	$ADH_3^2\text{-}ADH_3^2$	ADH_3^1	ADH_3^2
Alcoholic Cirrhosis (26)	2	16	8	39	62
Controls (16)	5	10	1	63	38

Genotypes were identified after ethidium bromide staining. Analysis of ADH3 genotypes in the controls (Table III) showed that the ADH_3^1 allele was more common, with a frequency of 63%, which is in agreement with frequencies reported in two recent studies of different populations (Couzigou et al., 1990; Day et al., 1991). There was an increased frequency of the ADH_3^2 allele in the alcoholic cirrhotic patients compared with the controls. There was no correlation between the alleles of the ADH36 RFLP and ADH_3 genotype in the patients studied.

Studies of alcohol-induced flushing and possible $ALDH_1$ polymorphism in Caucasians

The physiological, biochemical and molecular basis for the Oriental alcohol-induced flushing reaction has been well characterised. In contrast, although alcohol-induced flushing is well documented in Caucasians, relatively little is known about its character or its relationship to alcohol-metabolising enzymes. The point mutation in the $ALDH_2$ gene that is responsible for the Oriental flushing phenomenon has not been identified in Caucasians. We have previously reported lowered erythrocyte aldehyde dehydrogenase ($ALDH_1$) activity and altered enzyme properties associated with alcohol-related flushing in a small number of Caucasian subjects (Yoshida et al., 1989). One family pedigree showed inheritance of these abnormalities. We have now extended these original observations by examining the incidence of this flushing reaction and characterising the changes in cutaneous blood flow that occur in affected individuals after oral alcohol or topical administration of ethanol or acetaldehyde, in addition to measuring erythrocyte $ALDH_1$ activity (Ward et al., 1993).

Two hundred Caucasian medical students aged 18-24 were asked to complete a detailed questionnaire containing questions on alcohol consumption, the presence or absence of alcohol-induced flushing, the timing, duration, extent and nature of any reaction, and the presence of a family history of alcohol-related flushing. Heparinised blood was obtained from 12 individuals with a flushing response and erythrocyte $ALDH_1$ activity assayed after isolation by DEAE-cellulose chromatography (Yoshida *et al.*, 1989). Changes in cutaneous blood flow in response to oral ethanol (0.4 g/kg) were investigated in six flushers with low erythrocyte $ALDH_1$ activity, six non-flushers with normal $ALDH_1$ activity and 2 Oriental flushers deficient in $ALDH_2$. The duration and intensity of the cutaneous flush were measured by laser Doppler velocimetry with the probe applied to the zygoma. Blood ethanol was determined enzymatically and acetaldehyde levels were assayed by HPLC after formation of a fluorescent acetaldehyde adduct (Peters *et al.*, 1987). Changes in forearm cutaneous blood flow were also measured after topical application of filter paper squares soaked in ethanol or acetaldehyde.

Over half (57%) of female medical students reported a flush after a small amount of alcohol, in contrast to only 9% of males. At least two-thirds of affected individuals reported a similar reaction in other family members. Interestingly, average weekly alcohol intake was not reduced in those students with a flushing response. Eighty percent of subjects with alcohol-related flushing had erythrocyte $ALDH_1$ activity of less than 3.6 Units/mg protein. In addition, two family pedigrees with a number of affected members showed an association between alcohol-related flushing and low $ALDH_1$ activity.

There were no significant differences in pulse rate, mean arterial pressure, blood ethanol or acetaldehyde levels, or rates of elimination of ethanol or acetaldehyde between the six Caucasian 'flushers' and six 'non-flushers' after an oral ethanol dose. Blood acetaldehyde levels were similar in both Caucasian groups, reaching maximum values of 1 μmol/L one hour after ethanol ingestion: in comparison to peak levels of 17 μmol/L in the $ALDH_2$-deficient subjects. Whereas laser Doppler velocimetry showed no changes in cutaneous blood flow after oral ethanol in the Caucasian non-flushers, the flushers showed a significant increase reaching a peak at 25 minutes after ethanol ingestion, after which there was a steady decline. $ALDH_2$-deficient individuals showed a 2-fold greater increase in blood flow than the Caucasian flushers, with a more rapid response reaching a sustained peak in 15 minutes. There were no changes in forearm blood flow in the Caucasian flushers after topical application of water, 100% ethanol or 1M acetaldehyde. However, there was visible erythema and increased blood flow within 5 minutes of application of 5M acetaldehyde.

Polymorphism of the ALDH5 gene

The ALDH$_5$ gene was recently cloned by Hsu and Chang (1991), and shown by Northern hybridisation to be expressed in human testis as well as liver. Comparisons of cDNA sequences showed considerable homology between the corresponding amino acid sequences of ALDH$_5$ and both ALDH$_1$ (64.6%) and ALDH$_2$ (72.5%). ALDH5 had an unusual genomic structure, with an intronless coding region containing the sequence for 517 amino acid residues. We have examined the allele frequencies of 3 separate mutations (at nucleotides 183, 257 and 320) occurring within a short (137 bp) segment of the ALDH$_5$ coding region in Caucasian and Oriental populations (Sherman *et al.*, 1993b). Two of these mutations result in amino acid substitutions. We have also obtained preliminary data on the relationship between this polymorphism and hypogonadism in a group of Caucasians with chronic alcoholic and non-alcoholic liver disease. Leucocyte genomic DNA obtained from 37 male British Caucasians (8 non-alcoholic controls, 22 with alcoholic cirrhosis and 7 with Hepatitis C cirrhosis) and 30 Japanese non-alcoholic males was studied. Patients with cirrhosis were assessed for hypogonadism by clinical examination including orchidometry, and measurement of serum concentrations of testosterone, sex hormone binding globulin, FSH and LH. The relevant coding sequences were amplified by PCR using 5 separate oligonucleotide primers. Nucleotide changes

Table IV. Genotypes and allele frequencies of ALDH$_5$ locus in Caucasian subjects. (Numbers of subjects shown in parentheses)

Nucleotide change and amino acid substitution	Genotypes			Allele Frequencies	
C <--> T at nt 183	C/C	C/T	T/T	C	T
Thr <--> *Thr* at 44	23	9	2	0.81	0.19
(34)					
C <--> T at nt 257	C/C	C/T	T/T	C	T
Ala <--> *Val* at 69	20	12	2	0.76	0.24
(34)					
G <--> T at nt 320	T/T	G/T	G/G	T	G
Arg <--> *Leu* at 90	15	17	2	0.69	0.31
(34)					

Table V. Genotypes and allele frequencies of ALDH5 locus in Japanese subjects. (Numbers of subjects shown in parentheses)

Nucleotide change and amino acid substitution	Genotypes			Allele Frequencies	
C <--> T at nt 183	C/C	C/T	T/T	C	T
Thr <--> Thr at 44	9	8	3	0.65	0.35
(20)					
C <--> T at nt 257	C/C	C/T	T/T	C	T
Ala <--> Val at 69	6	10	4	0.55	0.45
(20)					
G <--> T at nt 320	T/T	G/T	G/G	T	G
Arg <--> Leu at 90	2	9	13	0.27	0.73
(34)					

were identified by specific restriction endonuclease digestion, agarose gel electrophoresis and ethidium bromide staining. Mutagenesis was employed to create restriction sites in the PCR products for mutations at nucleotides 257 and 320 by including appropriate single-base mismatches in 2 primers.

The three polymorphic sites in the ALDH$_5$ coding region were shown to be dimorphic in both Caucasian and Oriental populations (Tables IV and V), each with 2 common alleles conforming to Hardy-Weinberg equilibrium. Significant differences in allele frequencies between the 2 populations were seen at nucleotides 257 and 320, but not at nucleotide 183. The presence of T183 correlated strongly with T257 in Japanese (P < 0.001), but not in Caucasians. No significant associations between ALDH$_5$ genotypes and testicular atrophy were found in Caucasian patients with chronic alcoholic liver disease. Patients with clinical hypogonadism showed lower frequencies of C183 than those without, but this difference was not significant (p < 0.1).

Discussion

Alcohol dehydrogenase polymorphism

These studies have shown an association between alcohol-induced liver damage and an RFLP for the ADH$_2$ gene using a genomic DNA probe. The high frequency of the B allele in the alcohol misusers contrasted markedly with that in the non-alcoholic controls. The increased B allele frequency in patients with more severe liver damage suggests that this polymorphism denotes susceptibility to liver damage in particular. It does not, however, exclude an association with alcohol dependency syndrome. Given that the frequency of the ADH_2^2 allele in Caucasian populations is less than 5% (Couzigou *et al.*, 1991), it is extremely unlikely that the B allele of the ADH36 RFLP corresponds to this genotype. As a genomic DNA probe was used, this RFLP may exist in linkage disequilibrium with either a coding or non-coding sequence, or alternatively an adjacent regulatory region.

The finding of an increased frequency of the ADH_3^2 allele of the ADH_3 gene in cirrhotic patients differs from the findings of Day *et al.* (1991), who found an increased frequency of the ADH_3^1 allele in cirrhotics from North-East England. This may reflect differences between the populations, although allele frequencies in both control populations were similar. There is, therefore, no clear evidence implicating the ADH_3 gene in susceptibility to cirrhosis in Caucasians.

Aldehyde dehydrogenase polymorphism

We have shown that alcohol-related flushing is common in Caucasians, particularly females. The increase in cutaneous blood flow that occurs during the Caucasian flush is less marked than that of Orientals, and is of faster onset and shorter duration. The absence of raised acetaldehyde levels confirms that this flush is not caused by an inactive $ALDH_2$ isoenzyme. Although the biological role of $ALDH_1$ is unclear, the possibility remains that the lowered erythrocyte ALDH activities seen in these subjects may correspond to a mutation in the $ALDH_1$ gene and that the flush is attributable to impaired metabolism of vasoactive amines.

Our studies of $ALDH_5$ gene polymorphism show clear differences in allele frequencies between Caucasians and Orientals. In addition, we have demonstrated remarkably diverse polymorphism within a short 138 bp segment of the $ALDH_5$ coding region. If the three mutation sites are taken together, twenty seven possible haplotypes can be distinguished. The properties of the $ALDH_5$ enzyme(s) are at present unknown, and await expression studies. However, it seems likely that the mutations at nucleotides 257 and 320 may encode isoenzymes with different functional properties. Further studies will be required to determine the role of $ALDH_5$ in susceptibility to alcohol-induced testicular or liver damage.

Conclusion

Recent advances in the molecular biology of alcohol metabolising enzymes have led to an increased understanding of the relationship between genetic polymorphism and phenotypes such as alcohol-induced liver damage, alcohol dependency and alcohol-induced flushing.

References

Couzigou, P., Fleury, B., Groppi, A., Cassaigne, A., Begueret, J., Iron, A. and The French Group for Research on Alcohol & Liver (1990) Genotyping study of alcohol dehydrogenase class I polymorphism in French patients with alcoholic cirrhosis. *Alcohol Alcohol.* 25: 623-626.

Day, C.P., Bashir, R., James, O.F.W., Bassendine, M., Crabb, D.W., Thomasson, H.R., Li, T.-K. and Edenberg, H. (1991) Investigation of the role of polymorphisms at the alcohol and aldehyde dehydrogenase loci in genetic predisposition to alcohol-related-end organ damage. *Hepatology* 14: 798-801.

Hsu, L.C. and Chang, W-C. (1991) Cloning and characterisation of a new functional human aldehyde dehydrogenase gene. *J. Biol. Chem.* 266: 12257-12265.

Martin, N.G., Perl, J., Oakeshott, J.G., Gibson, J.B., Starmer, G.A. and Wilks, A.V. (1984) A twin study of ethanol metabolism. *Behav. Gen.* 15: 93-109.

Peters, T.J., Ward, R.J., Rideout, J. and Lim C.K. (1987) Blood acetaldehyde and ethanol levels in alcoholism. *Progr. Clin. Biol. Res.* 241: 215-230.

Sherman, D.I.N., Ward, R.J., Warren-Perry, M., Williams, R. and Peters, T.J. (1993a) Association of an RFLP marker for the alcohol dehydrogenase-2 gene with alcohol-induced liver damage. *Brit. Med. J.* 307: 1388-1391.

Sherman, D.I.N., Dave, V., Hsu, L.C., Peters, T.J. and Yoshida, A. (1993b) Diverse polymorphism within a short coding region of the human aldehyde dehydrogenase-5 (ALDH5) gene. *Hum. Genet.*, in press.

Ward, R.J., Macpherson, A.J.S., Chow, C., Ealing, J., Sherman, D.I.N., Yoshida, A. and Peters, T.J. (1994) Identification and characterisation of alcohol-induced flushing in Caucasian subjects. *Alcohol Alcohol.*, in press.

Yoshida, A., Dave, V., Ward, R.J. and Peters, T.J. (1989) Cytosolic aldehyde dehydrogenase (ALDH$_1$) variants found in alcohol flushers. *Ann. Hum. Gen.* 53: 1-7.

Toward a Molecular Basis of Alcohol Use and Abuse
ed. by B. Jansson, H. Jörnvall, U. Rydberg, L. Terenius & B. L. Vallee
© 1994 Birkhäuser Verlag Basel/Switzerland

Site-directed mutagenesis and enzyme properties of mammalian alcohol dehydrogenases correlated with their tissue distribution

Jan-Olov Höög, Mats Estonius and Olle Danielsson

Department of Medical Biochemistry and Biophysics, Karolinska Institutet, S-171 77 Stockholm, Sweden

Summary

Site-directed mutagenesis of mammalian alcohol dehydrogenases has helped to explain functional differences between enzymes within the protein family and traced these characteristics to specific amino acid residues. A threonine/serine exchange at position 48 in the human β/γ subunits can explain sensitivity to testosterone inhibition, as well as steroid dehydrogenase activity. It is possible to correlate the glutathione-dependent formaldehyde dehydrogenase activity of class III alcohol dehydrogenase with an arginine at position 115.

Tissue distribution analysis of the three initially established classes of mammalian alcohol dehydrogenase show pronouncedly different patterns. Class I alcohol dehydrogenase is widespread but varies between the tissues, and exists in small amounts in the brain. The occurrence of class II is limited in contrast to the class III enzyme which is abundant in all tissues examined. The latter probably reflects the need for scavenging of formaldehyde in cytoprotection.

Additional enzyme forms of mammalian alcohol dehydrogenase have been detected and have to be investigated further, together with the enzymes characterized earlier, regarding their physiological role in alcohol metabolism.

Introduction

Five classes of mammalian alcohol dehydrogenases (ADH) differ markedly in function and tissue distribution (cf. Jörnvall, this volume). The class I ADH oxidizes ethanol at low substrate concentrations, and in humans is divided into isozymes that are built up of α, β and γ subunits with characteristic properties; e.g., only isozymes containing the γ subunit are sensitive to testosterone inhibition (Mårdh *et al.*, 1986) and have steroid dehydrogenase activity (McEvily *et al.*, 1988). The γ subunit is also the class I polypeptide that shows the highest similarity to class I ADHs from other species (Jörnvall *et al.*, 1989). The class III ADH is unique within the protein family in many respects, an even tissue distribution, conserved primary structure, and dual enzymatic activity. The class III ADH has ethanol dehydrogenase activity, but cannot be saturated, and prefers longer aliphatic alcohols (Beisswenger *et al.*, 1985). It is also identical to glutathione-dependent formaldehyde dehydrogenase (Koivusalo *et al.*, 1989).

Classes I to III have been investigated extensively by analyses of divergent natural variants (Persson *et al.*, 1993) and correlation of their replacements with the residues lining the substrate-binding pocket, where about ten positions are of major importance (Eklund *et al.*, 1987; 1990). Three-dimensional models for these ADHs have been constructed by computer-graphics using the structure of horse class I ADH.

Interpretation of the substrate binding in ADH has focused on the hydrogen-bonding network between the catalytic zinc, the substrate, the amino acid residue at position 48, the coenzyme and the residue at position 51 (Eklund *et al.*, 1987; 1990). All known zinc-containing ADHs, from any class or species, have Thr or Ser at postion 48 (Jörnvall *et al.*, 1987; Höög *et al.*, 1993; Persson *et al.*, 1993). Therefore, this position have been of interest for site-directed mutagenesis. Position 115 in class III ADH has been ascribed a critical role for the glutathione-dependent formaldehyde dehydrogenase activity (Holmquist *et al.*, 1993) and has therefore also been of interest to change to the alternative residues of other ADHs. In addition, the functional characteristics of the different ADH classes has been correlated with their tissue distributions (Estonius *et al.*, 1993).

Materials and Methods

cDNA-libraries have been screened, and isolated cDNAs subcloned as described (Höög *et al.*, 1993). Expression of recombinant mammalian ADHs was performed in *Escherichia coli* using an inducible expression plasmid, pKK223-3 (Pharmacia Biotechnology), with a tac promoter (Hurley *et al.*, 1990; Höög *et al.*, 1992). Recombinant proteins were purified to homogeneity with ion-exchange and affinity chromatography prior to kinetic characterizations. Plasmid modifications were carried through according to standard protocols, and site-directed mutagenesis was performed on single-stranded DNA with mismatched oligonucleotide primers (Höög *et al.*, 1992). Tissue distribution studies utilized hybridizations to class specific probes (Estonius *et al.*, 1993).

Results and Discussion

Enzyme properties

Kinetic measurements show that a mutated $\beta\beta$ enzyme, with the residue at position 48 changed from Thr to Ser, is inhibited by testosterone, has steroid dehydrogenase activity, and has several kinetic constants affected in the direction toward those characteristic of the $\gamma\gamma$ isozyme (Table I), while an alanine at position 48 creates an inactive enzyme (Höög *et al.*, 1992). This shows that Ser is essential at position 48 to maintain all typical characteristics, the hydrogen-bonding

Table I. Established differences between the human $\beta\beta$ and $\gamma\gamma$ alcohol dehydrogenases compared to the mutant enzyme with β48Ser subunits

	$\beta\beta$	$\gamma\gamma$	β48Ser
Km for methanol (mM)	6	74	26
Km for cyclohexanol (mM)	14.5	0.042	0.280
pH effect	+	-	(-)
Steroid dehydrogenase activity (k_{cat}/K_m)	-	450	15
Testosterone inhibition (I_{50}, μM)	-	20	100
Amino acid residue at position 48	Thr	Ser	Ser
Total number of residue differences β/γ		21	

network, and enough space for steroids. The same mutation has been performed in yeast ADH but with different results (Creaser et al., 1990). This discrepancy can be explained by the fact that yeast ADH deviates also in many further respects from the human enzymes. An investigation of rabbit class I ADH strengthened the hypothesis that a serine at position 48 is necessary, to obtain steroid dehydrogenase activity and sensitivity to testosterone inhibition (Höög et al., 1993). However, Ser is not the only requirement for steroid interaction (cf. horse E ADH with Ser at position 48, but without steroid dehydrogenase activity, Table II). The serine side-chain, like the threonine side-chain in the β subunit, hydrogen-bonds to the oxygen of the substrate and to the O_2' of the ribose of the coenzyme, which is further hydrogen-bonded to His51 (Eklund et al., 1987). Substitutions of the latter residue in yeast ADH with Gln or Glu (Plapp et al., 1991), and in the human $\beta\beta$ isozyme with Gln (Ehrig et al., 1991) drastically reduces the activity. In the other mammalian ADH classes, a different type of interaction must be present owing to the fact that they have Thr, Tyr and Lys instead of His at position 51 (Table II).

Position 47, another site of interest, interacts electrostatically with the pyrophosphate of the coenzyme. In most class I ADHs this position is occupied by Arg and by His in class II and III. A His at this position was also found in the "Oriental" β_2 allelic subunit of ADH (cf. Jörnvall et al., 1989), instead of the Arg of the β_1 subunit. In the fetal form position 47 is occupied by Gly and mutagenesis of this position in the $\beta\beta$ enzyme (Arg47→Gly) resulted in a decreased V_{max} and weaker coenzyme binding (Hurley et al., 1990). Another site that differs in the fetal form

Table II. Amino acid residues lining the substrate-binding cleft of different mammalian ADHs. Species and residues in italics indicate those class I enzymes that have been shown to have steroid dehydrogenase activity. Δ indicates a deletion. Data from Jörnvall *et al.* (1987), Kaiser *et al.* (1989), Yasunami *et al.* (1991), Höög *et al.* (1993), and Parés *et al.* (1993)

Position			inner part of binding cleft				middle and outer part of binding cleft						
	47	51	48	93	140	141	57	110	115	116	117	294	318
Horse E	Arg	His	Ser	Phe	Phe	Leu	Leu	Phe	Asp	Leu	Ser	Val	Ile
Horse S	*Arg*	*His*	*Ser*	*Phe*	*Phe*	*Leu*	*Leu*	*Leu*	*Δ*	*Leu*	*Ser*	*Val*	*Ile*
Human α	Gly	His	Thr	Ala	Phe	Leu	Met	Tyr	Asp	Val	Ser	Val	Ile
Human β	Arg	His	Thr	Phe	Phe	Leu	Leu	Tyr	Asp	Leu	Gly	Val	Val
Human γ	*Arg*	*His*	*Ser*	*Phe*	*Phe*	*Val*	*Leu*	*Tyr*	*Asp*	*Leu*	*Gly*	*Val*	*Ile*
Rabbit	*Arg*	*His*	*Ser*	*Phe*	*Phe*	*Ile*	*Ile*	*Phe*	*Asp*	*Leu*	*Gly*	*Val*	*Ile*
Rat	*Arg*	*His*	*Ser*	*Phe*	*Phe*	*Leu*	*Leu*	*Tyr*	*Asn*	*Leu*	*Thr*	*Val*	*Ile*
Human II	His	Thr	Thr	Tyr	Phe	Phe	Phe	Phe	Ser	Asn	Leu	Val	Phe
Horse III	His	Tyr	Thr	Tyr	Tyr	Met	Asp	Leu	Arg	Thr	Thr	Val	Ala
Human III	His	Tyr	Thr	Tyr	Tyr	Met	Asp	Leu	Arg	Val	Thr	Val	Ala
Rat III	His	Tyr	Thr	Tyr	Phe	Met	Asp	Leu	Arg	Val	Thr	Val	Ala
Rat IV	Gly	His	Thr	Phe	Phe	Met	Met	Leu	Asp	Leu	Thr	Ala	Val
Human V	Gly	Lys	Thr	Phe	Phe	Gly	His	Phe	Lys	Gln	Ser	Val	Val

is position 93 (Table II). This position has been mutated in the ββ enzyme (Phe93→Ala), and results in a hybrid enzyme with the substrate specificity towards secondary alcohols (Hurley and Bosron, 1992). The recently characterized ADHs, human class V and rat class IV, also have Gly at position 47 (Yasunami *et al.*, 1991; Parés *et al.*, 1993), but the interaction with the coenzyme for these enzymes is not known in detail.

In the substrate-binding pocket of the class III ADH unique amino acid residues are localized to a limited number of positions (Table II). Of these, Asp57 is one with a unique characteristic, changing the hydrophobic pocket of class I ADH into a hydrophilic one in class III (Kaiser et al., 1989; Eklund *et al.*, 1990). However, good substrates for class III ADH are ω-fatty acid alcohols (e.g. 12-hydroxy dodecanoate) and hydroxymethyl-glutathione (Holmquist *et al.*, 1993). It has also been shown that it is possible to activate class III ADH with fatty acids to obtain an ethanol oxidizing potential at a level comparable to that of other ADHs (Moulis *et al.*, 1992). All these substrates/activators include negatively charged carboxyl groups and, therefore, suggest a corresponding, positively charged residue in the substrate-binding pocket. The only possible candidate for this was at position 115, arginine in class III ADH. In class I ADHs this position is oppositely charged (Asp) and those enzymes do not share any of these class III characteristics.

Position 115 in class III has been investigated both by means of amino acid labelling techniques (Holmquist *et al.*, 1993) and site-directed mutagenesis (Engeland *et al.*, 1993). Changing Arg115 in the human class III ADH to the residue of class I ADH results in an enzyme that cannot be activated with fatty acids, and glutathione-dependent formaldehyde dehydrogenase activity is decreased by a factor of 1300, while the activity toward 12-hydroxy dodecanoate is decreased by a factor of 140 (Engeland *et al.*, 1993). All these features are in agreement with a charge interaction between the substrate and the protein chain. Without activation, the class III ADH participation in alcohol metabolism is insignificant because of a very high K_m (> 3 M) for ethanol, but with extreme alcohol intake the enzyme can contribute to a minor part of ethanol metabolism when other ADHs are saturated.

Tissue distribution

Tissue distributions of mammalian ADHs have been studied both at the protein and mRNA levels, and for rat these studies are fairly complete (Julià *et al.*, 1987; Boleda *et al.*, 1989; Estonius *et al.*, 1993). For class III ADH/glutathione-dependent formaldehyde dehydrogenase all investigations show the same result; this ADH is present in all tissues examined at a fairly even level (Boleda *et al.*, 1989; Ratnagiri *et al.*, 1989; Estonius *et al.*, 1993). However, for class I ADH the results differ between tissues. In kidney, stomach and spleen large amounts are detected at the mRNA level, while no or very low amounts are found at the protein level. Many tissues show no ADH activity at all, where small or minute amounts are detected at the mRNA level, e.g. brain, heart and muscle (Estonius *et al.*, 1993; Table III). If the class I ADH in rat brain contributes to ethanol metabolism is not known. In fact, the prescence of any other ADH in the central nervous system aside from class III has been debated for a long time (Beisswenger *et al.*, 1985; Rout, 1992). Tissue distributions for class II ADH have been reported only at the mRNA level for the human and rat (Engeland and Maret, 1993; Estonius *et al.*, 1993), but the results differ. The rat shows the presence of message only in a limited number of tissues (Table III), but in man this message seems to be present in almost every tissue (Engeland and Maret, 1993). Class IV ADH has only been screened for at the protein level in the rat (at that time referred to as a class II type enzyme). This enzyme has high activity in the stomach and the enzyme is often referred to as the stomach ADH, but a high amount is also found in the eye (Boleda *et al.*, 1989). The latter activity agrees well with results obtained with the baboon, where a class II type of enzyme is present in the eye (Holmes and VandeBerg, 1986). Finally, for class V, the mRNA has been found in liver and stomach (Yasunami *et al.*, 1991).

Additional types of mammalian ADH

Additional forms of mammalian alcohol dehydrogenases have been found in the rat. A class II cDNA clone was isolated from a liver cDNA-library (Parés *et al.*, 1990; Höög, 1991) showing

Table III. Tissue distribution of mammalian ADH, comparison of results obtained from different studies at mRNA (Estonius *et al.*, 1993) and protein levels (Boleda *et al.*, 1989).
The mRNA values represent the densities of autoradiographs after correction for the occurrence of glyceraldehyde-3-phosphate dehydrogenase in each tissue. The values obtained for glyceraldehyde-3-phosphate dehydrogenase in thymus were calibrated to one, allowing inter-tissue comparison of mRNA amounts within any of class I-III ADH. The protein values represent intensities in activity stained starch gels with 2-buten-1-ol as substrate. All numbers given are relative, and the the absolute numbers for mRNA and protein can thereby not be compared. * indicates the prescence of class III ADH, but no level is given. nd not determined

Tissue	Class I		Class II	Class III		Class IV
	mRNA	protein	mRNA	mRNA	protein	protein
Male liver	56	4	19	3	3	-
Female liver	100		9	4		
Kidney	31	1	1	3	1	-
Stomach	30	-	1	4	2	3
Spleen	38	-	<1	4	*	-
Thymus	2	-	-	4	*	-
Testis	12	3	<1	3	1	1
Brain	1	-	-	3	*	-
Heart	1	-	-	2	*	-
Muscle	1	-	-	2	*	-
Lung	10	1	-	3	1	2
Epididymis	11	1	-	4	1	1
Duodenum	21	nd	12	2	nd	nd
Colon	21	3	-	5	1	1
Ovary	7	-	-	2	1	1
Uterus	22	1	-	4	1	1
Small intestine	7	2	-	1	1	1
Adrenal	2	-	-	2	1	1
Eye	5	-	-	2	2	3

that this class is very divergent and that the rat variant in relation to the human counterpart differs significantly in the substrate binding pocket. A PCR-amplified cDNA coding for human class V ADH was used as a probe in screening for a corresponding rat ADH. A full-length cDNA including 5' and 3'non-coding regions has been isolated from a rat liver cDNA-library, also ascribing a fifth type of ADH to the rat (Höög, unpublished). However, the human class V ADH and the fifth rat type show only 65% positional identity at the protein level. This low identity may place the question whether it is a class V ADH, a new class, or a subtype of a class defined in the mammalian set of alcohol metabolizing enzymes.

The interaction between a large number of enzymes/isozymes of mammalian ADHs focuses on the importance of different activities in different tissues and raises the question of the physiological role for every member in this enzyme family. In higher mammals the class I ADH is duplicated into different subunits, but these isozymes (α and β types) lack steroid dehydrogenase activity and sensititvity to testosterone inhibition. So far, class I ADH of the human γ type is the only mammalian ADH of any class with the above characteristics that creates enough space in the active site pocket for steroids. In contrast, classes other than class III ADH have Arg at position 115, but it is not known if they have retained the glutathione-dependent formaldehyde dehydrogenase activity. In summary, the mammalian system of ADHs seems to expand even further.

Acknowledgements

Work from this department was supported by grants from the Swedish Medical Research Council and Karolinska Institutet. A fellowship to JOH from Procordia AB is gratefully acknowledged.

References

Beisswenger, T. B., Holmquist, B. and Vallee, B.L. (1985) χ-ADH is the sole alcohol dehydrogenase isozyme of mammalian brains: Implications and inferences. *Proc. Natl. Acad. Sci. USA* 82: 8369-83.

Boleda, D. M., Julià, P. and Parés, X. (1989) Role of extrahepatic alcohol dehydrogenase in rat ethanol metabolism. *Arch. Biochem. Biophys.* 274: 74-75.

Creaser E.H., Murali, C. and Britt, K.A. (1990) Protein engineering of alcohol dehydrogenases; effects of amino acid changes at position 93 and 48 of yeast ADH1. *Prot. Engng.* 3: 523-536.

Ehrig, T., Hurley, T.D., Edenberg, H.J. and Bosron, W.F. (1991) General base catalysis in a glutamine for histidine mutant at position 51 of human liver alcohol dehydrogenase. *Biochemistry* 30: 1062-1068.

Eklund, H., Horjales, E., Vallee, B.L. and Jörnvall, H. (1987) Computer-graphics interpretations of residue exchange between α, β and γ subunits of human-liver alcohol dehydrogenase class I isozymes. *Eur. J. Biochem.* 167: 185-193.

308

Eklund, H., Müller-Wille, P., Horjales, E., Futer, O., Holmquist B., Vallee, B.L., Höög, J.-O., Kaiser, R. and Jörnvall, H. (1990) Comparison of three classes of human liver alcohol dehydrogenase. Emphasis on different substrate binding pockets. *Eur. J. Biochem.* 193: 303-311.

Engeland, K. and Maret, W. (1993) Extrahepatic, differential expression of four classes of human alcohol dehydrogenase. *Biochem. Biophys. Res. Commun.* 193: 47-53.

Engeland, K., Höög, J.-O., Holmquist, B., Estonius, M., Jörnvall, H. and Vallee, B.L. (1993) Mutation of Arg-115 of human class III alcohol dehydrogenase: A binding site required for formaldehyde dehydrogenase activity and fatty acid activation. *Proc. Natl. Acad. Sci. USA* 90: 2491-2494.

Estonius, M., Danielsson, O., Karlsson, C., Persson, H., Jörnvall, H. and Höög, J.-O. (1993) Distribution of alcohol and sorbitol dehydrogenases: Assessment of mRNAs in rat tissues. *Eur. J. Biochem.* 215: 497-503.

Holmes, R.S. and VandeBerg, J.L. (1986) Ocular NAD-dependent alcohol dehydrogenase and aldehyde dehydrogenase in the baboon. *Exp. Eye Res.* 43: 383-396.

Holmquist, B., Moulis, J.-M., Engeland, K. and Vallee, B.L. (1993) Role of arginine 115 in fatty acid activation and formaldehyde dehydrogenase activity of human class III alcohol dehydrogenase. *Biochemistry* 32: 5139-5144.

Hurley, T.D. and Bosron, W.F. (1992) Human alcohol dehydrogenase: Dependence of secon dary alcohol oxidation on the amino acids at positions 93 and 94. *Biochem. Biophys. Res. Commun. 183,* 93-99.

Hurley, T.D., Edenberg, H.J. and Bosron, W.F. (1990) Expression and kinetic characterization of variants of human $\beta_1\beta_1$ alcohol dehydrogenase containing substitutions at amino acid 47. *J. Biol. Chem.* 265: 16366-16372.

Höög, J.-O. (1991) Mammalian class II alcohol dehydrogenase: Species and class comparisons at genomic and protein levels, In: Weiner, H., Wermuth, B. and Crabb, D.W. (eds.): *Enzymology and Molecular Biology of Carbonyl Metabolism 3*, Plenum Press, New York, pp. 285-286.

Höög, J.-O., Eklund, H. and Jörnvall, H. (1992) A single-residue exchange gives human recombinant $\beta\beta$ alcohol dehydrogense $\gamma\gamma$ isozyme properties. *Eur. J. Biochem.* 205: 519-526.

Höög, J.-O., Vagelopoulos, N., Yip, P.-K., Keung, W.M. and Jörnvall, H. (1993) Isozyme development in mammalian class I alcohol dehydrogenase. cDNA cloning, functional correlations, and lack of evidence for genetic isozymes in rabbit. *Eur. J. Biochem.* 213: 31-38.

Julià, P., Farrés, J. and Parés, X. (1987) Characterization of three isoenzymes of rat alcohol dehydrogenase. *Eur. J. Biochem.* 162: 179-180.

Jörnvall, H., Persson, B. and Jeffery, J. (1987) Characteristics of alcohol/polyol dehydrogenases. The zinc-containing long-chain alcohol dehydrogenases. *Eur. J. Biochem.* 167: 195-201.

Jörnvall, H., von Bahr-Lindström, H. and Höög, J.-O. (1989) Alcohol dehydrogenases - structure. In: Batt, R.D. and Crow, K.E. (eds.): *Human Metabolism of Alcohol*, vol. II, CRC Press, Boca Raton FL, pp. 43-64.

Kaiser, R., Holmquist, B., Vallee, B.L. and Jörnvall, H. (1989) Characteristics of mammalian class III alcohol dehydrogenases, an enzyme less variable than the traditional liver enzyme of class I. *Biochemistry* 28: 8432-8438.

Koivusalo, M., Baumann, M. and Uotila, L. (1989) Evidence for the identity of glutathione-dependent formaldehyde dehydrogenase and class III alcohol dehydrogenase. *FEBS*

Lett. 257: 105-106.

McEvily, A.J., Holmquist, B., Auld, D.S. and Vallee, B.L. (1988) 3β-Hydroxy-5β-steroid dehydrogenase activity of human liver alcohol dehydrogenase is specific to γ-subunits. *Biochemistry* 27: 4284-4288.

Moulis, J.-M., Holmquist, B. and Vallee, B.L. (1991) Hydrophobic activation of human liver χχ alcohol dehydrogenase. *Biochemistry* 30: 5743-5749.

Mårdh, G., Falchuk, K.H., Auld, D.S. and Vallee, B.L. (1986) Testosterone allosterically regulates ethanol oxidation by homo- and heterodimeric γ-subunit-containing isozymes of human alcohol dehydrogenase. *Proc. Natl. Acad. Sci. USA* 83: 2836-2840.

Parés, X., Moreno, A., Cederlund, E., Höög, J.-O. and Jörnvall, H. (1990) Class IV mammalian alcohol dehydrogenase. Structural data of the rat stomach enzyme reveal a new class well separated from those already characterized. *FEBS. Lett.* 277: 115-116.

Parés, X., Cederlund, E., Moreno, A., Hjelmqvist, L., Farrés, J. and Jörnvall, H. (1993). Mammalian class IV alcohol dehydrogenase (stomach ADH): Structure, origin and correlation with enzymology. *Proc. Natl. Acad. Sci. USA,* in press.

Persson, B., Bergman, T., Keung, W.M., Waldenström, U., Holmquist, B., Vallee, B.L. and Jörnvall, H. (1993) Basic features of class-I alcohol dehydrogenase: variable and constant segments coordinated by inter-class and intra-class variability. Conclusions from characterization of the alligator enzyme. *Eur. J. Biochem.* 216: 49-56.

Plapp, B.V., Ganzhorn, A.J., Gould, R.M., Green, D.W., Jacobi, T., Warth, E. and Kratzer, D.A. (1991) Catalysis by yeast alcohol dehydrogenase. In: Weiner, H., Wermuth, B. and Crabb, D.W. (eds.): *Enzymology and Molecular Biology of Carbonyl Metabolism 3*, Plenum Press, New York, pp. 241-251.

Rathnagiri, P., Linnoila, M., Blanche O'Neill, J. and Goldman, D. (1989) Distribution and possible metabolic role of class III alcohol dehydrogenase in the human brain. *Brain Res.* 481: 131-132.

Rout, U.K. (1992) Alcohol dehydrogenases in the brains of mice. *Alcoholism: Clin. Exp. Res.* 16: 286-287.

Yasunami, M., Chen, C.-S. and Yoshida, A. (1991) A human alcohol dehydrogenase gene (ADH 6) encoding an additional class of isozyme. *Proc. Natl. Acad. Sci. USA* 88: 7610-7611.

Toward a Molecular Basis of Alcohol Use and Abuse
ed. by B. Jansson, H. Jörnvall, U. Rydberg, L. Terenius & B. L. Vallee
© 1994 Birkhäuser Verlag Basel/Switzerland

Control of alcohol metabolism

Bryce V. Plapp

Department of Biochemistry, The University of Iowa, Iowa City, Iowa 52242 USA

Summary

The rate of alcohol metabolism is determined by the kinetic characteristics and concentrations of the alcohol and aldehyde dehydrogenases and by the rate of restoration of the redox state of the cell. Several potent competitive and uncompetitive inhibitors of the alcohol dehydrogenases can decrease the rate of alcohol metabolism; they may be useful for preventing the potentially deleterious effects of ethanol metabolism. Alcohol dehydrogenases have very broad specificity and can readily reduce a variety of carbonyl compounds by exchange reactions while ethanol is metabolized. Agents that increase the rate of metabolism need to be developed.

Introduction

Ethanol intoxicates, its metabolism affects physiology, and together they may cause alcoholism. Thus, detoxification of ethanol is a critical process, and the enzymes involved have been extensively studied. We believe that knowledge of the substrate specificities, kinetics and regulation of the alcohol and aldehyde dehydrogenases could be used to control the rates of alcohol metabolism and to develop therapeutic agents for acute and chronic alcoholism.

In this article, we selectively highlight several questions about the control of alcohol metabolism. What are the rate-limiting steps? How can the process be inhibited or accelerated? What are the consequences of alcohol metabolism?

Material and Methods

The rates of alcohol metabolism were determined in male, Sprague-Dawley rats, 180-280g, fed certified rat chow ad libitum. Inhibitors or potential activators at 0.1 M or lower in physiologic saline were administered ip, and ten min later, ethanol (3.3 M) was given ip at a dose of 20 mmol/kg of body weight. Blood samples were taken at timed intervals, deproteinized and analyzed by gas chromatography (Chadha *et al.*, 1983; Plapp *et al.*, 1984). The rate of ethanol metabolism was calculated according to Widmark's method (Kalant, 1971) in which the dose of ethanol given (in mmol/kg body weight) is divided by the time (hours) required to eliminate alcohol from the blood, as extrapolated from the linear elimination curve.

Results and Discussion

Kinetics of alcohol metabolism

The debate about which metabolic reactions are rate-limiting for ethanol metabolism has gone on for many years. Reviewing data on enzyme activities, effects of inhibitors and activators, and redox changes in the liver during alcohol oxidation, we concluded that alcohol and aldehyde dehydrogenases and reoxidation of NADH were at least partially rate-limiting (Plapp, 1975). Subsequent investigations have provided more supporting data, but quantitative analyses are still required. Recent advances in computing have facilitated the simulation of progress curves (Barshop et al., 1983) when the mechanism and kinetic parameters for the enzymes are known. Thus, we can now simulate ethanol elimination curves and steady-state levels of acetaldehyde and test the effects of varied parameters (Fig. 1).

The simulation is based on a set of reasonable rate constants and describes alcohol metabolism very well in terms of the enzymatic properties. Varying the simulation parameters shows that the rate of disappearance of ethanol is sensitive to the concentration of alcohol dehydrogenase. This agrees with previous correlations of enzyme activity and elimination rate (Plapp, 1975; Crabb et al., 1983). On the other hand, there was an "obvious discrepancy" in a study that showed that alcoholic men had 1.4-fold faster rates of ethanol metabolism and less than half of the alcohol dehydrogenase activity as compared to nonalcoholics (Nuutinen et al., 1983).

A feature of the simulation is that the steady-state level of acetaldehyde (from 0 to 6 h) is very sensitive to the concentration of aldehyde dehydrogenase. This may be significant because the levels of acetaldehyde in alcoholics and in individuals deficient in the mitochondrial, low-K_m, aldehyde dehydrogenase were significantly higher than in nonalcoholics (Korsten et al., 1975; Nuutinen et al., 1983; Mizoi et al., 1983). Furthermore, the simulation showed that the shape of the acetaldehyde disappearance curve was sensitive to the rate constants for binding and reaction of acetaldehyde. The shape of the curve in Fig. 1 is somewhat different than the shapes found in more recent studies, and the methods of determination of acetaldehyde in blood are critical. The concentration of acetaldehyde in the liver cells should be determined (Lindros, 1978).

This simulation is a first approximation in our attempts to understand the biochemical and pharmacokinetic parameters of alcohol metabolism. Future work should determine the kinetic

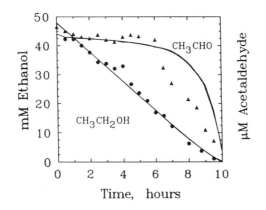

Fig. 1. Alcohol metabolism in an alcoholic man. The data points are taken from Figs. 1 and 2 of Korsten *et al.* (1975). The lines were simulated with KINSIM (Barshop *et al.*, 1983), for the metabolism of 48 mM ethanol, by coupling the reactions of alcohol and aldehyde dehydrogenases as described by the following mechanisms.

$$E \overset{k_1\ NAD^+}{\underset{k_{-1}}{=====}} E\text{-}NAD^+ \overset{k_2\ RCH_2OH}{\underset{k_{-2}}{=========}} E\text{-}NAD^+\text{-}RCH_2OH \overset{k_3}{\underset{k_{-3}\ RCHO}{========}} E\text{-}NADH \overset{k_4}{\underset{k_{-4}\ NADH}{======}} E$$

$$E2 \overset{k_5\ NAD^+}{\underset{k_{-5}}{=====}} E2\text{-}NAD^+ \overset{k_6\ RCHO}{\underset{k_{-6}}{========}} E2\text{-}NAD^+\text{-}RCHO \overset{k_7}{\dashrightarrow} E2\text{-}NADH \overset{k_8}{\underset{k_{-8}\ NADH}{======}} E2$$

association	dissociation	association	dissociation
$k_1 = 1.4 \times 10^6\ M^{-1}s^{-1}$	$k_{-1} = 150\ s^{-1}$	$k_5 = 1 \times 10^6\ M^{-1}s^{-1}$	$k_{-5} = 50\ s^{-1}$
$k_2 = 2.4 \times 10^5\ M^{-1}s^{-1}$	$k_{-2} = 560\ s^{-1}$	$k_6 = 1 \times 10^6\ M^{-1}s^{-1}$	$k_{-6} = 1\ s^{-1}$
$k_{-3} = 9.2 \times 10^4\ M^{-1}s^{-1}$	$k_3 = 64\ s^{-1}$	$k_{-7} = 0$	$k_7 = 100\ s^{-1}$
$k_{-4} = 2.1 \times 10^7\ M^{-1}s^{-1}$	$k_4 = 5.5\ s^{-1}$	$k_{-8} = 1 \times 10^6\ M^{-1}s^{-1}$	$k_8 = 20\ s^{-1}$

The rate constants were estimated from the data for horse liver alcohol dehydrogenase (Sekhar and Plapp, 1990; Shearer *et al.*, 1993) and for human mitochondrial E2 aldehyde dehydrogenase (Greenfield and Pietruszko, 1977; H. Weiner, personal communication). The concentration of alcohol dehydrogenase was 0.32 μN, calculated for the whole body mass and corresponding to about 1 mg enzyme/g liver, which is 2.3 % of body mass. The concentration of aldehyde dehydrogenase used was 0.125 μN. Fixed, steady-state levels of 500 μM NAD$^+$ and 2 μM NADH were used (Bücher *et al.*, 1972).

parameters for the multiple enzymes at pH 7 and 37°C (Bosron and Li, 1986; Ehrig *et al.*, 1990).

Inhibition of alcohol metabolism

Specific inhibitors of alcohol dehydrogenase can be used to assess the contribution of the

314

dehydrogenase to the overall rate of ethanol metabolism. They can also be used to differentiate between the direct effects of alcohol, formation of acetaldehyde, and changes in redox state in the pathogenesis of alcoholism. If metabolism leads to addictive compound (Lindros, 1978), inhibitors could be useful in therapy. Effective inhibitors also offer advantages over ethanol in the treatment of methanol or ethylene glycol poisoning.

Pyrazoles have been intensively investigated as inhibitors, as they bind tightly to alcohol dehydrogenases, and they are effective *in vivo* (Li and Theorell, 1969; Goldberg and Rydberg, 1969). 4-Methylpyrazole has been proposed for treatment of methanol poisoning (Blomstrand *et al.*, 1979). 4-Methylpyrazole is competitive against varied concentrations of ethanol (since it binds to the enzyme-NAD$^+$ complex) both with isolated alcohol dehydrogenase and in rats (Fig. 2A). Thus, inhibition is eliminated when the concentration of alcohol is saturating.

In contrast, a variety of amides, formamides, and tetramethylene sulfoxides (thiolane 1-oxides) are uncompetitive or noncompetitive inhibitors against varied concentrations of alcohols, since they bind most tightly to the enzyme-NADH complex. These compounds are potent inhibitors of isolated alcohol dehydrogenase and of alcohol metabolism (Winer and Theorell, 1960; Porter *et al.*, 1976; Chen *et al.*, 1981; Delmas *et al.*, 1983; Chadha *et al.*, 1983, 1985; Freudenreich *et al.*, 1984; Plapp *et al.*, 1984). As shown in Fig. 2B, inhibition is effective even at doses of ethanol that saturate the metabolic capacity.

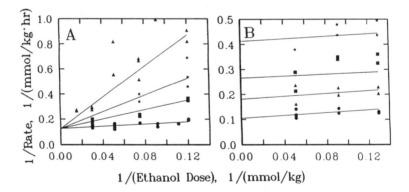

Fig. 2. Inhibition of ethanol metabolism in rats. (A) Competitive inhibition by 4-methylpyrazole. The rates of metabolism of the indicated doses of ethanol were studied with doses of 0 (●), 5 (■), 10 (★), and 20 (▲) μmol/kg body wt of inhibitor. The "rate" represents the observed rate corrected for the inhibitor-insensitive rate. Reproduced with permission from Plapp *et al.* (1984). (B) Uncompetitive inhibition by thiolane 1-oxide (tetramethylene sulfoxide). Rats were given doses of 0 (●), 0.25 (▲), 0.5 (■), and 1.0 (★) mmol/kg inhibitor. Reproduced with permission from Chadha *et al.* (1983).

The inhibition data in Table I show that the more potent inhibitors *in vitro* are also more effective *in vivo*. The K_i values *in vivo* are larger than those *in vitro*, most likely due to partitioning of the inhibitors between liver cytosol and other compartments, such as the lipid phase. The more hydrophobic compounds, such as 3-hexyl and 3-phenyl-thiolane 1-oxides, are particularly potent inhibitors of the enzymes, since they bind tightly to the hydrophobic substrate binding site. The inhibitors are also metabolized, and the effectiveness *in vivo* of isobutyramide and 4-methylpyrazole decreased with a half-time of about 2 h, whereas thiolane 1-oxide had a half-time of about 4 h (Chadha *et al.*, 1983; Plapp *et al.*, 1984). Inhibitor structure must be optimized for bioavailability and enzyme specificity. These new inhibitors should be considered further for experimental studies.

In these studies, it was noted that inhibition of the rate of elimination of ethanol was a linear function of low doses of inhibitor, consistent with the conclusion that the metabolic capacity was limited by the activity of alcohol dehydrogenase (Figs. 5 and 6 in Plapp *et al.*, 1984). However, higher doses of inhibitor did not completely stop the elimination of ethanol. The limiting rate observed with saturating levels of inhibitor showed a hyperbolic dependence upon the dose of

Table I. Inhibition of ethanol oxidation *in vitro* and *in vivo*[a]

Inhibitor	Liver alcohol dehydrogenase K_i, μM *vs* ethanol			Metabolism K_i, μmol/kg
	horse	monkey	rat	rat
Thiolane 1-oxide	19	1600	130	340
3-Methylthiolane 1-oxide	7.5	460	19	64
3-Butylthiolane 1-oxide	0.63	2.3	1.1	11
3-Hexylthiolane 1-oxide	0.19	0.35	0.13	65
3-Phenylthiolane 1-oxide	0.36	1.5	1.1	87
Isobutyramide	220	1500	300	1000
Isovaleramide	19	500	18	180
4-Methylpyrazole	0.013	0.084	0.11	1.4

[a]Enzyme isolated from monkey or rat liver was tested under approximately physiological conditions in 83 mM potassium phosphate and 40 mM KCl buffer, pH 7.3, and 37°C, with 0.5 mM NAD^+ and varied concentration of ethanol and inhibitor. The crystalline horse EE isoenzyme was tested at pH 7 and 25°C. Inhibition was uncompetitive for the amides and sulfoxides and K_i values are intercept inhibition constants, whereas 4-methylpyrazole was a competitive inhibitor. Inhibition *in vivo* by the four 3-substituted thiolane 1-oxides (tetramethylene sulfoxide) was determined with a single dose of 20 mmol of ethanol per kg body weight, and uncompetitive inhibition was assumed. Data are taken from Chadha *et al.* (1983, 1984) and Plapp *et al.* (1984).

ethanol, with an apparent K_m *in vivo* of 21 mmol/kg and a maximum, inhibitor-insensitive rate of elimination of 2.1 mmol/kg/h. In the absence of inhibitor, the maximal rate was 8.8 mmol/kg/h and the K_m was about 2.8 mmol/kg (Figs. 7 and 8 in Plapp *et al.*, 1984). Thus, up to 24% of the ethanol *can* be eliminated in rats by pathways that do not involve alcohol dehydrogenase. These other pathways can include excretion, catalase-H_2O_2, and the microsomal ethanol-oxidizing system. Since the kinetic characteristics of the systems differ, the relative contributions of the various pathways depend upon the dose of alcohol. A similar conclusion was reached from the analysis of the enzymatic activities in rat tissues (Boleda *et al.*, 1989).

Acceleration of alcohol metabolism

There have been many unsuccessful attempts to increase the rate of metabolism of ethanol in animals (reviewed in Plapp, 1975). Fructose may produce a significant, but quantitatively modest, effect of about 30% in humans (e.g., Nuutinen *et al.*, 1983). A similar increase was seen with 2,4-dinitrophenol in rats (Israel *et al.*, 1970). A rational attempt to use the coupled exchange reaction, in which the enzyme-NADH complex is reoxidized by a carbonyl compound

Table II. Survey of potential activators of ethanol metabolism in rats[a]

Moderate activation (1.1 to 1.3-fold increase in rate):
D-fructose, D-glucose, L-xylose, D-sorbitol, sodium pyruvate, fructose + dinitrophenol

No significant effect (0.95 to 1.05-fold effect):
D-ribose, D-mannitol, γ-D-galactonolactone, cycloheptanone, 1,4-cyclohexandione
D-fructose + methylene blue

Moderate inhibition (0.7 to 0.95 of normal rate):
cyclopentanone, cyclohexanone, γ-butyrolactone, β-butyrolactone, δ-valerolactone
γ-valerolactone, 2-deoxy-D-ribose, D-lyxose, D-ribonolactone
D-fructose + pentachlorophenol

Strong inhibition (0.2 to 0.7 of normal rate):
D-glucono-δ-lactone, fructose + pentachlorophenol + methylene blue
pentachlorophenol + methylene blue

---

[a]Compounds were tested at 10 mmol/kg body weight (0.1 mmol/kg for dinitrophenol or methylene blue, or 1 mmol/kg for pentachlorophenol), with a dose of ethanol of 20 mmol/kg. Ethanol alone was metabolized at a rate of 7.2 mmol/kg/h.

such as lactaldehyde or cyclohexanone, thereby bypassing the rate-limiting step of release of NADH in the alcohol dehydrogenase reaction, did not succeed (Gershman and Abeles, 1973). The coupled reaction should accelerate ethanol oxidation without altering the redox state as there is no net change in coenzymes. Ethanol metabolism could be resistant to acceleration if several activities of the metabolic pathway were partially rate-limiting. Then, a combination of compounds, which react with alcohol dehydrogenase, affect shuttle systems, and stimulate oxidative phosphorylation, might be required for a significant effect.

We explored additional sugars, carbonyl compounds (ketones and lactones) and combinations of compounds as potential activators of ethanol metabolism in rats (Table II). Fructose and some other compounds have significant effects alone, but multiple agents were not synergistic. In particular, fructose and dinitrophenol (an uncoupler of oxidative phosphorylation), which act on different steps in the pathway, were no better than fructose alone.

Substrate specificity of alcohol dehydrogenases and consequences of ethanol metabolism
Mammalian alcohol dehydrogenases are active on a broad range of nonpolar alcohols and carbonyl compounds, including primary, secondary, and cyclic substrates, steroids, lactones and hemiacetals. Aldehydes can be oxidized to acids (Shearer *et al.*, 1993). The enzymes have different specificities, which are related to the amino acid residues in the substrate binding pockets (Eklund *et al.*, 1987, 1990). We do not know and can not yet predict all of the substrates that might be physiologically significant. Of the seven enzymes identified in humans, at

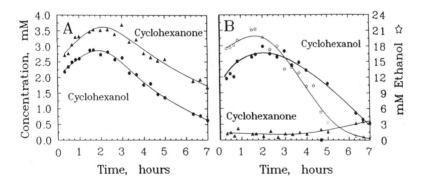

Fig. 3. Metabolism of cyclohexanol in rats and effects of ethanol. (A) The rat was given a dose of 5 mmol/kg of cyclohexanol i p, and the concentrations of cyclohexanol (●) and cyclohexanone (▲) in the blood were determined by gas chromatography. (B) The rat was given 5 mmol/kg of cyclohexanol and 20 mmol/kg of ethanol (☆).

least five have kinetic characteristics suited for metabolizing ethanol and other simple alcohols that might be ingested in beverages. These include the class I (α, β and γ), class II (π), and the stomach (σ) enzymes. The α-isoenzyme is especially active with secondary alcohols, including cyclohexanol (Stone et al., 1989; Light et al., 1992). The γ-isoenzyme is active on ethanol and 3-β-hydroxysteroids, as are the rat and horse SS liver enzymes (Cronholm et al., 1975; McEvily et al., 1988; Park and Plapp, 1992). The human π enzyme has low catalytic efficiency on ethanol and cyclohexanol, but high activity on long-chain primary alcohols, as do the σ-isoenzyme and the corresponding stomach enzymes from rat (Moreno and Parés, 1991; Juliá et al., 1987; Plapp et al., 1987). Given the variety of enzymes and their activities, it is clear that ethanol metabolism will affect the metabolism of various compounds. Ethanol should competitively inhibit oxidation of other alcohols, in proportion to the catalytic efficiencies, as shown for congeners in alcoholic beverages (Ehrig et al., 1988). Ethanol will also change the redox distribution of metabolites.

The experiment in Fig. 3A shows that cyclohexanol is rapidly converted to cyclohexanone in a rat, and an equilibrium is established while both compounds are slowly eliminated. When both ethanol and cyclohexanol are given (Fig. 3B), the concentration of cyclohexanone is greatly reduced until the ethanol is eliminated. These results demonstrate again the influence of ethanol on the redox state of the liver (Cronholm, 1987). Ethanol augments the normal reductive potential of the liver, which was also demonstrated when rats were given 5 mmol/kg of cyclohexanone with or without ethanol (experiments not shown). Cyclohexanone was reduced rapidly, so that within 20 to 60 min, the steady-state ratio of cyclohexanone to cyclohexanol shown in Fig. 3 was obtained.

The organism has a high capacity to reduce carbonyl compounds, especially in the presence of ethanol. Acetaldehyde concentrations are kept low, in part, due to the reduction catalyzed by alcohol dehydrogenase. The hydrogens from ethanol are readily incorporated into acetaldehyde and other metabolites, and the rate of exchange is higher than the net rate of elimination of ethanol (Cronholm et al., 1988). Human liver alcohol dehydrogenases readily reduce a variety of carbonyl compounds, such as the aldehydes produced during dopamine metabolism and lipid peroxidation, so that the fate of metabolites is affected by ethanol (Deetz et al., 1984; Mårdh and Vallee, 1986; Sellin et al., 1991; Tank and Weiner, 1979). Rat liver alcohol dehydrogenase can reduce dihydrotestosterone to 5α-androstane-3β,17β-diol, which is the predominant metabolite found in rat liver (Mezey and Potter, 1982; Van Doorn et al., 1975). During ethanol metabolism, the alcohol dehydrogenases could directly reduce other steroids, or the altered redox state could indirectly affect steroid metabolism (Murono and Fisher-Simpson, 1985; Andersson *et al.*, 1986; Chung, 1990: Orpana *et al.*, 1990). Fundamental biochemical studies are still required to determine the impact of alcohol metabolism on the pathology of alcoholism.

Acknowledgments

This work was supported by Grants AA00279 and AA06223 from the National Institute on Alcohol Abuse and Alcoholism, United States Public Health Service. The assistance of K. G. Leidal with the metabolism experiments was greatly appreciated.

References

Andersson, S., Cronholm, T. and Sjövall, J. (1986) Redox effects of ethanol on steroid metabolism. *Alcoholism Clin. Exp. Res.* 10: 55S-63S.

Barshop, B.A., Wrenn, R.F. and Frieden, C. (1983) Analysis of numerical methods for computer simulation of kinetic processes: Development of KINSIM - A flexible, portable system. *Anal. Biochem.* 130: 134-145.

Blomstrand, R., Östling-Wintzell, H., Löf, A., McMartin, K., Tolf, B.-R. and Hedström, K.-G. (1979) Pyrazoles as inhibitors of alcohol oxidation and as important tools in alcohol research: An approach to therapy against methanol poisoning. *Proc. Natl. Acad. Sci. U.S.A.* 76: 3499-3503.

Boleda, M.D., Julià, P., Moreno, A. and Parés, X. (1989) Role of extrahepatic alcohol dehydrogenase in rat ethanol metabolism. *Arch. Biochem. Biophys.* 274: 74-81.

Bosron, W.F. and Li, T.-K. (1986) Genetic polymorphism of human liver alcohol and aldehyde dehydrogenases, and their relationship to alcohol metabolism and alcoholism. *Hepatology* 6: 502-510.

Bücher, T., Brauser, B., Conze, A., Klein, F., Langguth, O. and Sies, H. (1972) State of oxidation-reduction and state of binding in the cytosolic NADH-system as disclosed by equilibration with extracellular lactate/pyruvate in hemoglobin-free perfused rat liver. *Eur. J. Biochem.* 27: 301-317.

Chadha, V.K., Leidal, K.G. and Plapp, B.V. (1983) Inhibition by carboxamides and sulfoxides of liver alcohol dehydrogenase and ethanol metabolism. *J. Med. Chem.* 26: 916-922.

Chadha, V.K., Leidal, K.G. and Plapp, B.V. (1985) Inhibition of liver alcohol dehydrogenase and ethanol metabolism by 3-substituted thiolane 1-oxides. *J. Med. Chem.* 28: 36-40.

Chen, W.-S., Bohlken, D.P. and Plapp, B.V. (1981) Inactivation of liver alcohol dehydrogenases and inhibition of ethanol metabolism by ambivalent active-site-directed reagents. *J. Med. Chem.* 24: 190-193.

Chung, K.W. (1990) Effects of chronic ethanol intake on aromatization of androgens and concentration of estrogen and androgen receptors in rat liver. *Toxicology* 62: 285-295.

Crabb, D.W., Bosron, W.F. and Li, T.-K. (1983) Steady-state kinetic properties of purified rat liver alcohol dehydrogenase: Application to predicting alcohol elimination rates *in vivo*. *Arch. Biochem. Biophys.* 224: 299-309.

Cronholm, T. (1987) Effect of ethanol on the redox state of the coenzyme bound to alcohol dehydrogenase studied in isolated hepatocytes. *Biochem. J.* 248: 567-572.

Cronholm, T., Larsén, Sjövall, J., Theorell, H. and Åkeson, Å. (1975) Steroid oxidoreductase activity of alcohol dehydrogenases from horse, rat, and human liver. *Acta Chem. Scand.* B29: 571-576.

Cronholm, T., Jones, A.W. and Skagerberg, S. (1988) Mechanism and regulation of ethanol elimination in humans: Intermolecular hydrogen transfer and oxidoreduction in vivo. *Alcoholism Clin. Exp. Res.* 12: 683-686.

Deetz, J.S., Luehr, C.A. and Vallee, B.L. (1984) Human liver alcohol dehydrogenase isozymes: Reduction of aldehydes and ketones. *Biochemistry* 23: 6822-6828.

Delmas, C., de Saint Blanquat, G., Freudenreich, C. and Biellmann, J.-F. (1983) New inhibitors of alcohol dehydrogenase: Studies in vivo and in vitro in the rat. *Alcoholism Clin. Exp. Res.* 7: 264-270.

Ehrig, T., Bohren, K.M., Wermuth, B. and von Wartburg, J.-P. (1988) Degradation of aliphatic alcohols by human liver alcohol dehydrogenase: Effect of ethanol and pharmacokinetic implications. *Alcoholism Clin. Exp. Res.* 12: 789-794.

Ehrig, T., Bosron, W.F. and Li, T.-K. (1990) Alcohol and aldehyde dehydrogenase. *Alcohol Alcoholism* 25: 105-116.

Eklund, H., Horjales, E., Vallee, B.L. and Jörnvall, H. (1987) Computer-graphics interpretations of residue exchanges between the α, β and γ subunits of human-liver alcohol dehydrogenase class I isozymes. *Eur. J. Biochem.* 167: 185-193.

Eklund, H., Müller-Wille, P., Horjales, E., Futer, O., Holmquist, B., Vallee, B.L., Höög, J.-O., Kaiser, R. and Jörnvall, H. (1990) Comparison of three classes of human liver alcohol dehydrogenase. Emphasis on different substrate binding pockets. *Eur. J. Biochem.* 193: 303-310.

Freudenreich, C., Samama, J.-P. and Biellmann, J.-F. (1984) Design of inhibitors from the three-dimensional structure of alcohol dehydrogenase. Chemical synthesis and enzymatic properties. *J. Amer. Chem. Soc.* 106: 3344-3353.

Gershman, H. and Abeles, R.H. (1973) Deuterium isotope effects in oxidation of alcohols *in vitro* and *in vivo*. *Arch. Biochem. Biophys.* 154: 659-674.

Goldberg, L. and Rydberg, U. (1969) Inhibition of ethanol metabolism *in vivo* by administration of pyrazole. *Biochem. Pharm.* 18: 1749-1762.

Greenfield, N.J. and Pietruszko, R. (1977) Two aldehyde dehydrogenases from human liver. Isolation via affinity chromatography and characterization of the isozymes. *Biochim. Biophys. Acta* 483: 35-45.

Israel, Y., Khanna, J.M. and Lin, R. (1970) Effect of 2,4-dintrophenol on the rate of ethanol elimination in the rat *in vivo*. *Biochem. J.* 120: 447-448.

Juliá, P., Farrés, J. and Parés, X. (1987) Characterization of three isoenzymes of rat alcohol dehydrogenase. Tissue distribution and physical and enzymatic properties. *Eur. J. Biochem.* 162: 179-189.

Kalant, H. (1971) Absorption, diffusion, distribution, and elimination of ethanol: Effects on biological membranes. In: Kissin, B., and Begleiter, H. (eds.): *The Biology of Alcoholism*, Vol. I, Plenum Press, New York, pp. 1-62.

Korsten, M.A., Matsuzaki, S., Feinman, L. and Lieber, C.S. (1975) High blood acetaldehyde levels after ethanol administration. Difference between alcoholic and nonalcoholic subjects. *N. Engl. J. Med.* 292: 386-389.

Li, T.-K. and Theorell, H. (1969) Human liver alcohol dehydrogenase: Inhibition by pyrazole and pyrazole analogs. Acta Chem. Scand. 23: 892-902.

Light, D.R., Dennis, M.S., Forsythe, I.J., Liu, C.-C., Green, D.W., Kratzer, D.A. and Plapp, B.V. (1992) α-Isoenzyme of alcohol dehydrogenase from monkey liver. Cloning, expression, mechanism, coenzyme and substrate specificity. *J. Biol. Chem.* 267: 12592-12599.

Lindros, K.O. (1978) Acetaldehyde - Its metabolism and role in the actions of alcohol. In: Israel, Y., Glaser, F.B., Kalant, H., Popham, R.E., Schmidt, W. and Smart, R.G. (eds.): *Research Advances in Alcohol and Drug Problems*, Plenum, pp. 111-176.

McEvily, A.J., Holmquist, B., Auld, D.S. and Vallee, B.L. (1988) 3β-Hydroxy-5β-steroid dehydrogenase activity of human liver alcohol dehydrogenase is specific to γ-subunits. *Biochemistry* 27: 4284-4288.

Mezey, E. and Potter, J.J. (1982) Effect of dihydrotestosterone on rat liver alcohol dehydrogenase activity. *Hepatology* 2: 359-365.

Mizoi, Y., Tatsuno, Y., Adachi, J., Kogame, M., Fukunaga, T., Fujiwara, S., Hishida, S. and Ijiri, I. (1983) Alcohol sensitivity related to polymorphism of alcohol-metabolizing enzymes in Japanese. *Pharmacol. Biochem. Behav.* 18 Suppl. 1: 127-133.

Moreno, A. and Parés, X. (1991) Purification and characterization of a new alcohol dehydrogenase from human stomach. *J. Biol. Chem.* 266: 1128-1133.

Murono, E. P. and Fisher-Simpson, V. (1985) Ethanol directly increases dihydrotestosterone conversion primarily to 5α-androstan-3β,17β-diol in rat Leydig cells. *Life Sci.* 36: 1381-1387.

Mårdh, G. and Vallee, B.L. (1986) Human class I alcohol dehydrogenases catalyze the interconversion of alcohols and aldehydes in the metabolism of dopamine. *Biochemistry* 25: 7279-7282.

Nuutinen, H., Lindros, K.O. and Salaspuro, M. (1983) Determinants of blood acetaldehyde level during ethanol oxidation in chronic alcoholics. *Alcoholism Clin. Exp. Res.* 7: 163-168.

Orpana, A.K., Härkönen, M. and Eriksson, C.J.P. (1990) Ethanol-induced inhibition of testosterone biosynthesis in rat Leydig cells: Role of mitochondrial substrate shuttles and citrate. *Alcohol Alcoholism* 25: 499-507.

Park, D.-H. and Plapp, B.V. (1992) Interconversion of E and S isoenzymes of horse liver alcohol dehydrogenase. Several residues contribute indirectly to catalysis. *J. Biol. Chem.* 267: 5527-5533.

Plapp, B.V. (1975) Rate-limiting steps in ethanol metabolism and approaches to changing these rates biochemically. In: Majchrowicz, E. (ed.): *Biochemical Pharmacology of Ethanol*, Plenum Press, New York, pp. 77-109.

Plapp, B.V., Leidal, K.G., Smith R.K. and Murch, B.P. (1984) Kinetics of inhibition of ethanol metabolism in rats and the rate-limiting role of alcohol dehydrogenase. *Arch. Biochem. Biophys.* 230: 30-38.

Plapp, B.V., Parsons, M., Leidal, K.G., Baggenstoss, B.A., Ferm, J.R.G. and Wear, S.S. (1987) Characterization of alcohol dehydrogenase from cultured rat hepatoma (HTC) cells. In: Weiner, H. and Flynn, T.G. (eds.): *Enzymology and Molecular Biology of Carbonyl Metabolism. Aldehyde Dehydrogenase, Aldo-Keto Reductase, and Alcohol Dehydrogenase, Prog. Clin. Biol. Res.* Vol. 232, Liss, New York, pp. 203-215.

Porter C.C., Titus, D.C. and DeFelice, M.J. (1976) Liver alcohol dehydrogenase inhibition by fatty acid amides, *N*-alkylformamides and monoalkylureas. *Life Sci.* 18: 953-960.

Sekhar, V.C. and Plapp, B.V. (1990) Rate constants for a mechanism including intermediates in the inter-conversion of ternary complexes by horse liver alcohol dehydrogenase.

Biochemistry 29: 4289-4295.

Sellin, S., Holmquist, B., Mannervik, B. and Vallee, B.L. (1991) Oxidation and reduction of 4-hydroxy-alkenals catalyzed by isozymes of human alcohol dehydrogenase. *Biochemistry* 30: 2514-2518.

Shearer, G.L., Kim, K., Lee, K.M., Wang, C.K. and Plapp, B.V. (1993) Alternative pathways and reactions of benzyl alcohol and benzaldehyde with horse liver alcohol dehydrogenase. *Biochemistry* 32, 11186-11194.

Stone, C.L., Li, T.-K. and Bosron, W.F. (1989) Stereospecific oxidation of secondary alcohols by human alcohol dehydrogenases. *J. Biol. Chem.* 264: 11112-11116.

Tank, A. W. and Weiner, H. (1979) Ethanol-induced alteration of dopamine metabolism in rat liver. *Biochem. Pharmacol.* 28: 3139-3147.

Van Doorn, E.J., Burns, B., Wood, D., Bird, C.E. and Clark, A.F. (1975) *In vivo* metabolism of ^{3}H-dihydrotestosterone and ^{3}H-androstanediol in adult male rats. *J. Steroid Biochem.* 6: 1549-1554.

Winer, A.D. and Theorell, H. (1960) Dissociation constants of ternary complexes of fatty acids and fatty acid amides with horse liver alcohol dehydrogenase-coenzyme complexes. *Acta Chem. Scand.* 14: 1729-1742.

Toward a Molecular Basis of Alcohol Use and Abuse
ed. by B. Jansson, H. Jörnvall, U. Rydberg, L. Terenius & B. L. Vallee
© 1994 Birkhäuser Verlag Basel/Switzerland

Angiotensin converting enzyme inhibitors and alcohol abuse

James F. Riordan

Center for Biochemical and Biophysical Sciences and Medicine, Harvard Medical School, 250 Longwood Avenue, Boston, MA 02115 U.S.A.

Summary

Angiotensin converting enzyme plays a key role in the regulation of blood pressure and inhibitors of the enzyme are effective antihypertensive agents. An association between hypertension and alcohol abuse has long been recognized and manipulations of the renin-angiotensin system in laboratory animals has been shown to alter their consumption of ethanol. Procedures that decrease the renin-angiotensin system increase ethanol consumption. Paradoxically, inhibitors of angiotensin converting enzyme also diminish drinking. Several possible explanations for this observation have been proposed. However, observations on the relationship between stress-induced drinking and the antidipsogenic action of a fragment of adrenocorticotropic hormone suggest another possibility: angiotensin converting enzyme may be involved in the metabolism of this peptide and thereby exert an influence on drinking behavior.

Introduction

Angiotensin converting enzyme (ACE) is a central participant in the renin angiotensin system and inhibitors of the enzyme have gained widespread acceptance as effective therapeutic agents for the control of hypertension (Gavras, 1990). Recent interest in the potential of ACE inhibitors for controlling alcohol abuse as well, stems from the long recognized relation that exists between ethanol consumption and hypertension (Klatsky, 1990). Early in this century, in a study of the factors affecting the health of a group of 150 French soldiers, Lian (1915) reported that hypertension was more frequent among those who consumed more than 2.5 liters of wine per day. After a long hiatus additional reports along this line began to appear. D'Alonzo and Pell (1968) reported that hypertension was 2.3 times more frequent among employees at DuPont who were "problem drinkers". Importantly, they also found that this hypertension was reversible. Soon thereafter, a finding from the Framingham heart study, reported by Kannel and Sorlie (1975), noted that hypertension is more common in those whose consumption of ethanol exceeds 3 liters per month. Other studies, such as one from the Kaiser Permanente group (Klatsky *et al.*, 1977) continued to find a positive, although not necessarily linear, correlation between alcohol consumption and hypertension. Indeed, as of 1990, 41 out of 43 studies that had been conducted showed a link between ethanol use and hypertension (Klatsky, 1990).

A number of underlying mechanisms have been invoked to account for the blood pressure/alcohol relationship. For one a stimulatory effect of alcohol on the renin-angiotensin system (RAS) has been noted. However, alcohol could also increase noradrenergic activity, induce chronic production of excess corticosteroids, or induce a chronic hypermetabolic state (Doyal *et al.*, 1988). Further, it could also directly affect the heart and/or the peripheral vasculature (Doyal *et al.*, 1988).

This report will focus on the possible role of the RAS in the etiology of alcohol abuse by discussing the relationship between the activity of the RAS and ethanol consumption. It will examine the paradoxical effect of ACE inhibitors on ethanol consumption in laboratory animals, review the role of stress-related molecules in drinking behavior, and propose an alternative explanation of these effects based on the structural and functional properties of ACE. Although completely speculative, this report supports the need for further study of the role of the RAS in alcohol-related behavior.

Results and Discussion

The renin-angiotensin system

Angiotensin converting enzyme (EC 3.4.15.1) is a ubiquitous mammalian dipeptidyl carboxypeptidase that plays a central role in blood pressure regulation by converting the inactive decapeptide, angiotensin I (Asp-Arg-Val-Tyr-Ile-His-Pro-Phe-His-Leu), to the potent vasopressor octapeptide angiotensin II by cleavage of the dipeptide, His-Leu, from the carboxyterminal end of the molecule. In addition ACE also abolishes the activity of bradykinin, a vasodepressor, by an analogous mechanism. The substrate, angiotensin I, is generated from a liver-derived precursor, angiotensinogen, in blood plasma by the action of another enzyme, renin, which is released from the kidney in response to a number of blood pressure-related signals (Ehlers and Riordan, 1990). The only known function of angiogenin I is to serve as a substrate for ACE. The product, angiotensin II, exerts a generalized effect on vascular smooth muscle cells to produce arteriolar vasoconstriction and thereby increase blood pressure. It also reduces blood flow to the kidney, acts directly on the kidney to alter cation reabsorption and acts on the adrenal cortex to induce the secretion of aldosterone. All three of these effects result in sodium (and water) retention which further increases blood pressure. In addition ACE is thought to participate in the thirst mechanism. The relationship between ACE and vascular pressure is reflected in the location of the enzyme. It is an ectoenzyme bound via a transmembrane segment of its carboxyterminal sequence to the luminal side of the endothelial cells that constitute the vascular bed. Hence it is exposed to all circulating peptides. The three-dimensional structure of the enzyme is unknown and, as a consequence, the basis for its substrate specificity has not been defined. Nevertheless, highly effective inhibitors of ACE have been available for over a decade and are used world-wide to treat hypertension.

Ethanol intake and the RAS in laboratory rats

When Sprague-Dawley rats are provided with a low sodium diet and treated with a diuretic (Lasix), a regimen that is known to increase RAS activity, their consumption of alcohol is diminished (Grupp *et al.*, 1984). In contrast, Dahl rats that are salt sensitive have a low RAS activity and drink more ethanol than salt-resistant Dahl rats (Grupp *et al.*, 1986). Rats that have been surgically manipulated to have an increased RAS activity drink less than sham operated controls (Grupp *et al.*, 1987). Direct subcutaneous infusion of angiotensin II also decreases voluntary alcohol intake in a dose related manner (Grupp *et al.*, 1988). Moreover, a single intracerebral administration of the peptide inhibited alcohol consumption in rats for up to two weeks (Kotov *et al.*, 1982).

These results have been interpreted to indicate that ethanol has a direct effect on the RAS and, indeed, it has been demonstrated that plasma levels of angiotensin II rise in response to administration of ethanol as does plasma renin activity (Grupp, 1987). It has been proposed that the production of low levels of angiotensin II may be rewarding to the animal and may play a role in initiating drinking behavior. As more angiotensin II is produced, however, it reaches a level where it acts as a stop signal indicating satiety. Within this context, then, any drug that would alter the RAS should have an effect on ethanol intake.

ACE inhibitors and alcohol consumption in laboratory rats

ACE is a zinc metalloenzyme and at least superficially bears some mechanistic resemblance to carboxypeptidase A. The latter catalyzes the hydrolysis of amino acids from the carboxy-terminus of oligopeptides one at a time whereas ACE removes them two at a time. Details of the structure and mechanism of carboxypeptidase have been known for many years (Auld *et al.*, 1989). Based on this analogy it was possible to design inhibitors of ACE that capitalize on its catalytically essential zinc ion and as a consequence could bind extremely tightly. The first of these, known as captopril, turned out to be a safe, orally active drug for treating hypertension. In experiments with laboratory rats it was found to have a statistically significant, dose related effect on alcohol consumption (Spinosa *et al.*, 1988). Remarkably, however, it decreased consumption contrary to what might have been expected from an agent that should lower the plasma levels of angiotensin II. Other ACE inhibitors were also examined and virtually the same results were obtained. Both normotensive and hypertensive rats responded as did the so-called P rats that have a genetically acquired preference for alcohol (Grupp, 1991). Several hypotheses were proposed to account for this seeming inconsistency. For example it was thought that ACE inhibition would increase plasma levels of angiotensin I thereby allowing it to cross the blood-brain barrier and accumulate in the CNS. Since all of the components of the RAS are known to be present in brain and since ACE inhibitors do not cross the blood-brain barrier, increased brain angiotensin I would lead to increased brain angiotensin II and perhaps produce a satiety signal. However, there is no evidence that angiotensin I is capable of crossing the blood-brain barrier. Alternatively, it was thought that ACE inhibitors diminish alcohol consumption by an aversion mechanism. Somehow the inhibitors cause disruption of organ function (liver) that lead to impaired ethanol metabolism and, subsequently a malaise that discourages drinking (Hubbell *et al.*, 1991). However, ACE inhibitors have no obvious effect on ethanol tolerance tests and therefore it seemed possible that there might be another reason.

Stress and the consumption of alcohol

One of the reasons that have been proposed to explain why animals that have been subjected to stress react by drinking ethanol is that they do so to relieve the tension induced by the stress. Another theory is that during the period of stress the hormones that are released confer a positive response - an ACTH "high" - and that once the stress is removed, the animals drink ethanol to maintain this rewarding response. To test this latter hypothesis, Nash and Maikel (1988) established the baseline level of ethanol consumption by adult, male Sprague-Dawley rats for a period of 7 days. They then subjected the rats to 14 days of stress (ranging from 0-4 hours per day) of either isolation or immobilization. Alcohol consumption did not increase during this so-called stress period. However, in the two weeks after cessation of daily stress alcohol consumption increased several fold. Hypophysectomy eliminated this post-stress alcohol response but adrenalectomy did not, indicating a role for ACTH in this behavior. Moreover, treatment with ACTH by i.v. injection mimicked the results of stress in that alcohol consumption increased once treatment was stopped.

In order to identify a possible active fragment of ACTH that might be responsible for this effect Krishnan and Maikel (1991) tested the effect of known metabolic products of ACTH (a 39-residue oligopeptide released from the anterior pituitary in response to signals from the hypothalamus induced by stress). One of these, containing residues 11-24, had no effect but another containing residues 4-10, was remarkable in that it decreased ethanol intake dramatically during the period of treatment. Moreover, there was no post-treatment increase in alcohol consumption. A non-peptide analogue of residues 4-9 known as HOE-427 (Ebiratide) was even more effective in that it completely abolished ethanol consumption during the treatment period.

Stress, the consumption of alcohol and ACE

Stress can be a common contributing factor along with age, race, education, adiposity and others for both alcohol consumption and hypertension. Indeed, as indicated above, peptide hormones are also mutual participants in both of these situations. ACE had long been viewed as an enzyme specific for the conversion of angiotensin I to angiotensin II. Some years ago an enzyme that was involved in the metabolism of bradykinin was identified as kininase II. Subsequently it was shown that kininase II is identical to ACE, thus broadening the specificity of the enzyme to two important blood pressure-related peptides. More recently ACE has been shown to act on a broad range of oligopeptides, and substrates have been assigned to three different classes based on their amino acid sequences around the scissile peptide bond. All specificity studies of ACE, which typically use short peptides, have been based on its action as a dipeptidyl carboxypeptidase. It only acts on peptides that have a free C-terminal carboxyl group. However, there has been suggestive evidence derived from sequence studies of the enzyme that suggests that it might have even broader specificity than suspected. Some background information may be helpful.

ACE is a zinc metallopeptidase and a member of the neutral protease class of enzymes. It is a transmembrane glycoprotein that is found on the outer luminal surface of primarily endothelial cells but also of renal tubular and intestinal epithelial cells. There are two forms of the enzyme in human tissues, the product of differential translation of mRNA from a single gene. One form, the somatic enzyme, is ubiquitous and is a 180 kD glycoprotein whose primary function is blood pressure regulation and sodium and fluid balance. The second form is only found in testes, specifically in developing spermatogonia, and its function is unknown. It is also a glycoprotein and its molecular weight is about half that of the somatic enzyme. Recombinant human testis ACE has a sequence of 701 amino acids (Ehlers *et al.*, 1989) and contains a characteristic short sequence, HEMGH, associated with the zinc binding site of metalloendoproteases (Vallee and Auld, 1990). The recombinant human endothelial cell ACE contains 1277 amino acids and has two such HEMGH sequences (Soubrier *et al.*, 1988). The protein is the product of a duplicated gene with two active sites. One half of the sequence, the C-terminal half, is identical to that of testicular ACE. The somatic enzyme binds 2 g.at. zinc per mol while the testicular enzyme binds 1 g.at. zinc per mol (Ehlers and Riordan, 1991). Thus, each of the two putative metal binding sequences in the somatic enzyme would seem to indeed bind zinc. This raises the question as to whether both of these sites are catalytically active and equally so.

As noted, the HEMGH sequence is typical of endoproteases, enzymes that catalyze the hydrolysis of peptide bonds in the interior of a polypeptide chain not at either terminus. Could it be that ACE has endoprotease activity in addition to its well-known dipeptidyl carboxypeptidase activity? The answer to this is yes although this was discovered empirically rather than on the basis of the above reasoning (Inokuchi and Nagamatsu, 1981). Indeed a number of oligopeptides blocked at their C-terminus by e.g. amidation, have been found to be reasonably good substrates for somatic ACE although less so for testicular ACE.

In a survey of peptides considered to be atypical substrates for ACE, relative activities toward the two forms of the enzyme were compared (Ehlers and Riordan, 1991). Luteinizing hormone-releasing hormone (LH-RH), a decapeptide released from the hypothalamus to activate the anterior pituitary to release LH, is preferentially cleaved by somatic ACE compared to testicular ACE. There is good reason to suspect that this cleavage is catalyzed by the active site located in the N-terminal domain of somatic ACE. For one, these separate domains have been expressed by recombinant technology and the N-terminal domain has been shown to be catalytically active (Wei *et al.*, 1991). Moreover, a mutant form of somatic ACE in which the

C-terminal domain has been inactivated is actually 5 times more active toward LH-RH than intact ACE and even more active than the C-terminal domain (Jaspard *et al.*, 1993).

For another, the testicular enzyme and by analogy the C-terminal domain of the somatic enzyme can be inactivated by chemical modification with fluoro-dinitrobenzene (Bünning *et al.*, 1990). This reagent specifically blocks an essential tyrosine (and a lysine) important for catalysis. Remarkably, these essential residues are only present in the C-terminal half of the somatic enzyme (Chen and Riordan, 1990). Yet the modified somatic enzyme still exhibits substantial activity toward LH-RH and other peptides including bradykinin.

These results indicate that the N-terminal domain of somatic ACE is catalytically active and may play a special role in peptide metabolism. Limited studies have examined whether or not ACE inhibitors are equally effective at both catalytic sites. Thus far, the data are insufficient to provide a definitive answer. It is clear that two moles of inhibitor can bind per mole of enzyme, at least for the inhibitors tested (Ehlers and Riordan, 1991; Wei *et al.*, 1992), but little is known about the relative specificity of the two sites for different peptides. This is particularly true for peptides of biological significance.

How can this added dimension of the specificity of ACE relate to the role of ACE inhibitors in alcohol consumption? Consider the possibility that the metabolism of ACTH (1-39) - once it reaches its target organ, the adrenal - leads to the formation of ACTH(4-10) a product known to suppress alcohol consumption in laboratory rats. Consider also that ACE may be involved in the further metabolism of ACTH(4-10). This peptide has a sequence that seems extremely susceptible to ACE-catalyzed hydrolysis although this has not yet been tested. If the product, ACTH(4-8) does not alter drinking, this would abolish the antidipsogenic action of ACTH(4-10). ACE inhibitors would stop the degradation of this peptide and perhaps allow it to accumulate to the point where it would interfere with further drinking. This view would extend the range of substrates for ACE and might provide a plausible rationale for the otherwise perplexing second active site.

Conclusion

It is perhaps appropriate to conclude by reiterating the closing comments of Hubbell *et al.* (1991). "These data lead to the suggestion of an alternative way of explaining the effects of ACE inhibitors [on drinking] than that proposed previously. These data support the idea that further explorations of the role of the renin-angiotensin system in the propensity to take alcohol beverages are needed."

References

Auld, D.S., Riordan, J.F. and Vallee, B.L. (1989) Probing the mechanism of action of carboxypeptidase A by inorganic, organic and mutagenic modifications. In: Sigel, H. (ed.): *Metal Ions in Biological Systems*, vol. 25, Marcel Dekker, Inc., Basel, pp. 359-394.

Bünning, P., Kleemann, S.G. and Riordan, J.F. (1990) Essential residues in angiotensin converting enzyme: Modifications with 1-fluoro-2,4-dinitrobenzene. *Biochemistry* 29: 10488-10492.

D'Alonzo, C.A. and Pell, S. (1968) Cardiovascular disease among problem drinkers. *J. Occup. Med.* 10: 244-250.

Doyal, L.E., Morton, W.A. and Crane, D.F. (1988) Antihypertensive drug therapy in alcohol dependence. *Psychosomatics* 29: 301-306.

Ehlers, M.R.W., Fox, E.A., Strydom, D.J. and Riordan, J.F. (1989) Molecular cloning of human testicular angiotensin converting enzyme: The testis isozyme is identical to the C-terminal half of endothelial angiotensin converting enzyme. *Proc. Natl. Acad. Sci. U.S.A.* 86: 7741-7745.

Ehlers, M.R.W. and Riordan, J.F. (1990) Angiotensin converting enzyme. Biochemistry and molecular biology. In: Laragh, J.H. and Brenner, B.M. (eds.): *Hypertension. Pathology, diagnosis and management*, Raven Press, New York, pp. 1217-1231.

Ehlers, M.R.W. and Riordan, J.F. (1991) Angiotensin-converting enzyme: Zinc- and inhibitor-binding stoichiometries of the somatic and testis isozymes. *Biochemistry* 30: 7118-7126.

Gavras, H. (1990) Angiotensin converting enzyme inhibition and its impact on cardiovascular disease. *Circulation* 81: 381-388.

Grupp, L.A. (1987) Alcohol satiety, hypertension and the renin-angiotensin system. *Med. Hypotheses* 24: 11-19.

Grupp, L.A. (1991) Effects of angiotensin II and an angiotensin converting enzyme inhibitor on alcohol intake in P and NP rats. *Pharmacol. Biochem. Behav.* 41: 105-108.

Grupp, L.A., Killian, M., Perlanski, E. and Stewart, R.B. (1988) Angiotensin II reduces voluntary alcohol intake in the rat. *Pharmacol. Biochem. Behav.* 29: 479-482.

Grupp, L.A., Perlanski, E., Wanless, I.R. and Stewart R.B. (1986) Voluntary alcohol intake in the hypertension prone Dahl rat. *Pharmacol. Biochem. Behav.* 24: 1167-1174.

Grupp, L.A., Perlanski, Leenen, F.H.H. and Stewart, R.B. (1987) Renal artery stenosis: An example of how the periphery can modulate voluntary alcohol drinking. *Life Sci.* 40: 563-570.

Grupp, L.A., Stewart, R.B. and Perlanski, E. (1984) Salt restriction and the voluntary intake of ethanol in rats. *Physiol. Psychol.* 12: 242-246.

Hubbell, C.L., Chrisbacher, G.A., Bilsky, E.J. and Reid, L.D. (1991) Manipulations of the renin-angiotensin system and intake of a sweetened alcohol beverage among rats. *Alcohol* 9: 53-61.

Inokuchi, J.-I. and Nagamatsu, A. (1981) Tripeptidyl carboxypeptidase activity of kininase II (angiotensin-converting enzyme). *Biochim. Biophys. Acta* 662: 300-307.

Jaspard, E., Wei, L. and Alhenc-Gelas, F. (1993) Differences in the properties and specificities of the two active sites of angiotensin I-converting enzyme (kininase II). *J. Biol. Chem.* 268:

9496-9503.

Kannel, W.B. and Sorlie, P. (1975) Hypertension in Framingham. In: Paul, O. (ed.): *Epidemiology and control of hypertension*, Stratton Intercontinental Medical Book Co., New York, pp. 553-592.

Klatsky, A.L. (1990) Blood pressure and alcohol intake. In: Laragh, J.H. and Brenner, B.M. (eds.): *Hypertension. Pathophysiology, diagnosis and management*, Raven Press, New York, pp. 277-294.

Klatsky, A.L., Friedman, G.D., Siegelaub, A.B. and Gerard, M.J. (1977) Alcohol consumption and blood pressure. *N. Engl. J. Med.* 296: 1194-1200.

Kotov, A.V., Kelesheva, L.F., Kuznetsov, S.L. and Pal'tsev, M.A. (1982) Role of angiotensin II in the mechanisms of ethanol consumption regulation in the rat. *Biull Eksp. Biol. Med.* 93: 68-71 (Publ. in Russian).

Krishnan, S. and Maikel, R.P. (1991) The effect of HOE-427 (an $ACTH_{4-9}$ analog) on free-choice ethanol consumption in male and female rats. *Life Sci.* 49: 2005-2011.

Lian, C. (1915) L'alcoholisme, cause d'hypertension arterielle. *Bull. Acad. Natl. Med. (Paris)* 74: 525-528.

Nash, J.F. Jr. and Maikel, R.P. (1988) The role of the hypothalamic-pituitary-adrenocortical axis in post-stress induced ethanol consumption in rats. *Prog. Neuro-Psychopharmacol. Biol. Psychiat.* 12: 653-671.

Soubrier, F., Alhenc-Gelas, F., Hubert, C., Allegrini, J., John, M., Tregear, G. and Corvol, P. (1988) Two putative active centers in human angiotensin I-converting enzyme revealed by molecular cloning. *Proc. Natl. Acad. Sci. U.S.A.* 85: 9386-9390.

Spinosa, G., Perlanski, E., Leenen, F.H.H., Stewart, R.B. and Grupp, L.A. (1988) Angiotensin converting enzyme inhibitors: Animal experiments suggest a new pharmacological treatment for alcohol abuse in humans. *Alcohol. Clin. Exp. Res.* 12: 65-70.

Vallee, B.L. and Auld, D.S. (1990) Zinc coordination, function and structure of zinc enzymes and other proteins. *Biochemistry* 29: 5647-5659.

Wei, L., Alhenc-Gelas, F., Corvol, P. and Clauser, E. (1991) Two homologous domains of human angiotensin I-converting enzyme are both catalytically active. *J. Biol. Chem.* 266: 9002-9008.

Wei, L., Clauser, E., Alhenc-Gelas, F. and Corvol, P. (1992) The two homologous domains of human angiotensin I-converting enzyme interact differently with competitive inhibitors. *J. Biol. Chem.* 267: 13398-13405.

Toward a Molecular Basis of Alcohol Use and Abuse
ed. by B. Jansson, H. Jörnvall, U. Rydberg, L. Terenius & B. L. Vallee
© 1994 Birkhäuser Verlag Basel/Switzerland

Treatment of alcoholism

Bengt Jansson

The Karolinska Institute, Dept. of Psychiatry, Huddinge Hospital, S-141 86 Huddinge, Sweden

Summary

Special problems in the treatment of alcoholism are discussed. Risk groups, for instance children of alcoholics, should be identified as early as possible so that special attention can be paid to them. Treatment motivation plays a very central role, and motivating a patient for long-term treatment often is the most difficult part of the work. For alcohol dependent alcoholics, total abstinence should be the goal of treatment. To achieve this, psychotherapy consisting of supportive, behavioral and dynamic elements is recommended. To achieve effective therapy it should be carried out for a considerable length of time.

Treatment of alcoholism presents several difficulties which are not met with in most other medical situations. Firstly, it is a question of finding those individuals who should be treated. The next step is to motivate them for a relevant treatment program and to carry them through it. If the program is successful it is important to find measures to make the result obtained last as long as possible. Prior to the first step there is also a question of prevention, directed towards a whole population or towards special risk groups.

Prevention

The question what is possible to achieve by <u>information</u> or <u>restrictions</u> has been debated for a long time. Information about the risks of alcohol in most countries has not proven particularly fruitful - the positive properties of the drug alcohol as a drug make people who do not consider themselves at risk immune to risk propaganda. There is a clear discrepancy between positive attitudes towards information among laymen and the press on the one hand and the more sceptical, resigned attitude of the medical profession on the other. One exception from the main rule should be mentioned, however: Information to pregnant women about the risks of alcohol consumption during pregnancy in Stockholm has resulted in reduced consumption and, over time, disappearance of the fetal alcohol syndrome (Larsson, 1983).

Perhaps restrictions are more powerful: there are several indications that there is a significant negative correlation between prize level and consumption (Nielsen, 1965), and further that the abolition of restrictions leads to an increased consumption and increase in complications (Nilsson and Frey, 1958). It has been shown more clearly that there is a positive relation between the total amount of alcohol consumed in the population and the number of patients with severe complications due to alcohol (de Lint and Schmidt, 1971; Bruun *et al.*, 1975).

Many students of alcohol, however, consider it more important to identify risk groups of individuals disposed to alcoholism to decrease this group's consumption. The next step is the identification of the risk groups.

Diagnosis of risk groups and/or of alcoholics

Family studies of alcoholism (Knop *et al.*, 1993) are most important in order to evaluate the relative importance of genetic or environmental influences on the development of alcoholism in children of alcoholics. Studies from this group show that children of alcoholics tend to have a 3-4 fold increased susceptibility to alcoholism, whether they are raised by their biological alcoholic parent or by non-alcoholic adoptive parents, indicating a considerable genetic component of the etiology. Such studies are not easy to design, however. Most of them investigate only the sons of alcoholic fathers (Velleman, 1992). Additional studies also reveal considerable variation in the definition of alcoholism, as well as differences concerning the time when the children are studied. The vast majority of research has been concerned with the examination of the immediate effects (if any) during childhood of having a parent with a drinking problem. Many studies show that antisocial adolescents, coming from homes with social disadvantages, develop problems with both alcohol use and sociopathy more frequently than controls. "These studies all suffer from the problem, however, that their results were derived from samples that premorbidly exhibited a disproportionate amount of antisocial behavior" (Velleman, 1992). One of the studies which has tried to overcome this bias is that of Nylander and Rydelius (1982) comparing children of alcoholic fathers from excellent versus poor social conditions. Their conclusion is that if the father is a "chronic alcoholic", social class of origin does not affect outcome.

Velleman and Orford (1990) investigated a sample of 170 offspring of problem drinkers and 80 young controls. They did not find any intermediate differences between offspring and controls in terms of current quantity of alcohol consumption. "Some support, however, was found for the detrimental effect of having two parents with drinking problems, and of having a problemdrinking parent who drinks at home."

With this background, it seems important to create programs directed towards children of alcoholics, to increase their awareness of the problem and their surroundings.

One very important question in the treatment of alcoholism of course is: how does one diagnose an alcoholic and at what stage should treatment begin? Babor (1990) quotes expert committees which recommend new initiatives in the secondary prevention of alcohol problems through physician-based interventions at the primary care level. Babor complains, however, about the difficulties of introducing new technologies, especially behavioural technologies, into medical practice. Babor is surprised about the fact that it is often very difficult to interest physicians in questioning patients regarding alcohol.

Bradley (1992) advocates early treatment in primary care to prevent relapse. This may include the phase of detoxification; a hospital stay is not necessarily the first step to rehabilitation. A randomized, controlled study of 164 patients with mild-to-moderate withdrawal symptoms but in the absence of active medical or psychological problems found outpatient detoxification safe and effective (Hayashida et al., 1989). And, inpatient detoxification is 9 to 20 times more expensive, when lost opportunity costs are included. Holt (1989) stresses the importance of early diagnosis by taking a thorough drinking history in everyday practice in conjunction with measuring some biochemical markers, such as gamma-glutamyl transpeptidase.

It is ideal situation, of course, where a primary care patient exhibits an increased gamma-glutamyl transpeptidase activity and has motivation to reduce his drinking, and is willing to make regular out-patient visits to check his enzyme levels. Experience (Kristenson et al., 1983) has shown that such a procedure in materials consisting of socially well-adjusted heavy consumers obviously reduces their drinking considerably.

The situation is different, however, when dealing with groups of definite drug-dependent alcoholics, especially if they get into social problems. In this population treatment is often initiated in acute situations and discontinued when the situation improves and the patient is not longer motivated to seek further treatment. Thus, the alcoholic may take the initiative to contact a psychiatric department when he is overwhelmed by anxiety in connection with detoxification or by an insipient delirious state. After 2-3 days he often prefers to leave the hospital. Many patients repeat this experience several times a year before they might ultimately be motivated to seek more serious treatment programs, which can be very expensive and a waste of care frustrating the staff.

Other types of patients seek help in medical departments because of liver cirrhosis, in surgical departments for pancreatitis or fractures brought about by accidents or violence in connection with a drinking bout. If the responsible doctors are ambitious they consult their psychiatric colleagues, but in most cases this patient group is not motivated to seek further treatment of their alcohol dependence.

But even if there is no further treatment at that time the first step is taken: patients such as those mentioned above are diagnosed and identified as alcoholics, and hopefully, sooner or later, in connection with a relapse they may be motivated to obtain further treatment.

Motivation for treatment

First, a short discussion about treatment goals. Is it a definite demand that the patient should be totally abstinent from alcohol for the rest of his life?

That depends on what kind of patient is being discussed. It is important to differentiate between different groups of alcoholics. For severe cases it is obviously all or nothing. On the other hand, some patients who abuse alcohol appear to be able to return to nonproblem drinking. Vaillant *et al.* (1983) found that, in the course of eight years, 6% of alcoholics studied reverted to social drinking, while only a third became dependent on alcohol. I agree with Bradley (1992) that "after firmly recommending abstinence and after a frank discussion with patients about the serious risks of continued drinking, alcohol-abusing patients who refuse to stop drinking should be advised to reduce their intake. This approach is not recommended for patients who are alcohol-dependent, because it is unlikely that they would be able to control the amount they drink."

My conclusion is that "the Kristenson model" is well adapted for socially well adjusted abusers in primary care and obviously reduces their drinking, but total abstinence should be the goal among alcohol dependent patients in a psychiatric department. The problem then is how to motivate the patient. There is no easy solution for this, but it seems very important the staff believe in the treatment programs used and shows enthusiasm in their attitude for the therapy recommended.

Treatment programs

Alcoholism is most definitely one of those disorders where no one can claim to have an unfailing therapy. It is easy to see how prevailing therapies have changed according to the theories that have dominated the psychiatric field at a given moment. Some decades ago, under the influence of psychodynamic theories, it went without saying that it was impossible to try to treat an alcoholic without knowledge of the <u>cause</u> of his abuse. This period was succeeded by a more realistic attitude when the therapist concentrated upon the more immediate problem of making the patient stop drinking, often by simpler, more behavior oriented methods, using simple reward and punishing mechanisms.

During the last decades, The Minnesota model has gained some attention although it does not present a definitely new approach, but because it has adopted many good things from different therapies and carries through its program with great consequence, not least the after-care. The importance of including the family in the treatment has also been stressed heavily in the Minnesota model and its treatment goal is abstinence; non-abstinence goals are considered unrealistic and dangerous, as pointed out by many other authors Wallace (1989).

Psychotherapy

Kaufman (1990-91) has presented a very balanced overview of the psychodynamics of substance abuse and its application to a psychotherapeutic approach. He proposes a model of psychotherapy which integrates psychodynamic theory, AA and other 12-step groups, family therapy and dual-diagnosis issues into a phase-specific method, a model consisting of three stages: dryness, sobriety and wellness. He quotes the VA-Penn study (Okpatu, 1986) which found that patients receiving supportive-expressive psychotherapy or cognitive-behavioral psychotherapy had greater gains and used fewer both prescribed and self-obtained medications than patients treated by drug counseling alone.

However, it is not easy to perform clinical studies of therapeutic effects which are free from objections as is discussed in detail by Mattson and Allen (1991). How should one match alcoholic patients to the most adequate treatment? Five approaches are highlighted: randomness, availability/convenience, patient self selection, clinical judgement and algorithm or formal rule. In 40 studies which the authors analysed, they differentiate between four classes of patient variables: demographics, alcohol-specific characteristics, psychopathology and finally social/personal characteristics. The authors conclude with three clinically important questions:

1/ will clinicians accept adjunct treatment planning guidelines that may be at variance with their own experience?

2/ many types of matches will require much greater variety of interventions than is currently provided in treatment programs. Will providers be willing to perform the necessary assessments?

3/ will third party payers be willing to absorb the extra costs that will be incurred if treatment programs develop multiple interventions?

I believe the authors are right when they consider it doubtful that clinicians and/or payers will accept such rules.

Drugs

Drugs like disulfiram which make the alcoholic unable to consume alcohol have been used extensively since about 1950. The American College of Physicians (1989) in a position paper agrees with the statement that disulfiram can be effective over the short term in reducing the frequency of alcohol consumption in the compliant patient. For some alcoholics, reducing the frequency of alcohol use may result in physical, mental, and behavioral improvement, even if total abstinence is not achieved. "At this time however, disulfiram has not been shown to alter the long-term course of alcoholism".

Other drugs, directed mainly towards reducing the component of anxiety of alcoholics, have been of comparatively little use. According to Linnoila (1989) "benzodiazepines are not indicated in the treatment of alcohol dependence, other than withdrawal". Alcohol dependence is a lifelong disorder characterized by remissions and relapses and it is important not to prescribe drugs with potential habit forming tendencies. "In fact, because of the likelihood of development of a second drug dependence, anxiolytic agents are contraindicated in the long-term management of alcoholism" (Wallace, 1989). In individuals suffering from panic attacks, secondary amine tricyclic antidepressants or serotonin reuptake inhibitors are recommended (Linnoila, 1989). He also states that medications that have strong sedative interactions with ethanol should be avoided. An antianxiety agent, Buspirone hydrochloride, that neither chemically nor pharmacologically resembles benzodiazepines is reported to have low potential to produce physical dependence, and possibly is applicable in the treatment of coexisting anxiety disorders in alcoholics (Wallace, 1989).

New Drugs

In the future, however, we may hope for drugs which may alter drinking behavior without creating a new dependence, as in some cases, serotonin re-uptake blockers have done. Janssen and collaborators (Meert *et al.*, 1991) have shown that ritanserin, a specific central serotonin 5-HT2 antagonist, in low doses rapidly reversed drug preference without changing total fluid intake in rats. Quantitatively, the reduction in drug consumption was greater for alcohol than for cocaine. Nathan (1990) believes that it is only a matter of time before receptor blocking agents that reduce or eliminate the reinforcing properties of alcohol are introduced into the therapeutic armamentarium. However, to translate this into an effective therapy a series of step-by-step behavioral procedures that will ensure the regular ingestion of these agents are necessary. Cox (1986) and Miller (1985) emphasize the central role motivation plays in treatment response. It seems clear to me that also the most adequate drugs which may be introduced in the future in the treatment of alcoholism have to be combined with psycho-social treatment programs in order to achieve lasting results.

References

American College of Physicians (1989) *Disulfiram Treatment of Alcoholism*. Annals of Internal Medicine 111: 943-945.

Babor, T.F. (1990) Brief intervention strategies for harmful drinkers: new directions for medical education. *Canad. Med. Assoc. J.* 143: 1070-1076.

Bradley, K.A. (1992) Management of Alcoholism in the Primary Care Setting. *West. J. Med.* 156: 273-277.

Bruun, K., Edwards, G., Lumio, M., Mäkelä, K., Pan, L., Popham, R.E., Room, R., Schmidt, W., Skog, O.-J., Sulkunen, P. and Österberg, E. (1975) Alcohol Control Policies in public health perspective. *The Finnish Foundation for Alcohol Studies*, vol. 25. Forssa: Aurasen Kirjapaino.

Cox, W.M. (1986) Research on the personality correlates of alcohol use. *Drugs Society* 1: 61-83.

Hayashida, M., Alterman, A.I., McLellan A.T., O'Brien, C.P., Purtill, J.J., Volpicelli, J.R., Raphaelson, A.H. and Hall, C.P. (1989) Comparative effectiveness and costs of inpatient and outpatient detoxification of patients with mild to moderate alcohol withdrawal syndrome. *N. Engl. J. Med.* 320: 358-365.

Holt, S. (1989) Identification and intervention for alcohol abuse. *J. S. C. Med. Assoc.* 85: 554-

340

559.

Kaufman, E. (1990-91) Critical aspects of the psychodynamics of substance abuse and the evaluation of their application to a psychotherapeutic approach. *Int. J. Addict.* 25: 97-116.

Knop, J., Goodwin, D.W., Jensen, P., Penick, E., Pollock, V., Gabrielli, W., Teasdale, T.W. and Mednick, S.A. (1993) A 30-year follow-up study of the sons of alcoholic men. *Acta Psychiatr. Scand.* Suppl 370:48-53.

Kristenson, H., Öhlin, H., Hultén-Nosslin, M.-B., Trell, E. and Hood, B. (1983) Identification and intervention of heavy drinking in middle-aged men: Results and follow-up of 24-60 months of long-term study with randomized controls. *Alcoholism: Clin. Exp. Res.* 7: 203-209.

Larsson, G. (1983) Prevention of fetal alcohol effects. An antenatal program for early detection of pregnancies at risk. *Acta. Obstet. Gynecol. Scand.* 62: 171-178.

de Lint, J. and Schmidt, W. (1971) Consumption averages and alcoholism prevalence. A brief review of epidemiological investigations. *Br. J. Addict.* 66: 97-107.

Linnoila, M.I. (1989) Anxiety and alcoholism. *J. Clin. Psychiatry* 50: 26-29.

Mattson, M.E. and Allen, J.P. (1991) Research on matching alcoholic patients to treatments: findings, issues, and implications. *J. Addict. Dis.* 11: 33-49.

Meert, T.F., Awouters, F., Niemegeers, C.J., Schellekens, K.H. and Janssen, P.A. (1991) Ritanserin reduces abuse of alcohol, cocaine, and fentanyl in rats. *Pharmacopsychiatry* 24: 159-163.

Miller, W.R. (1985) Motivation for treatment: a review with special emphasis on alcoholism. *Psychol. Bull.* 98: 84-107.

Nathan, P.E. (1990) Integration of biological and psychosocial research on alcoholism. *Alcohol Clin. Exp. Res.* 14: 368-374.

Nielsen, J. (1965) Delirium tremens in Copenhagen. *Acta Psychiatr. Scand.* 41: 187.

Nilsson, L. and Frey, T. S:son (1958) Iakttagelser angående vissa alkoholpsykoser. *Läkartidn.* 55: 2997.

Nylander, I. and Rydelius, P. (1982) A comparison between children of alcoholic fathers from excellent versus poor social conditions. *Acta Paediatr. Scand.* 71: 809-813.

Okpatu, S.O. (1986) Psychoanalytically oriented psychotherapy of substance abuse. *Adv. Alcohol Subst. Abuse* 6: 17-33.

Vaillant, G.E., Clark, W., Cyrus, C., Milofsky, E.S., Kopp, J., Wells Wulsin, V. and Mogielnicki, N.P. (1983) Prospective study of alcoholism treatment: Eight-year follow-up. *Am. J. Med.* 75: 455-463.

Wallace, J. (1989) Treatments for alcoholism. *Occup. Med.* 4: 275-287.

Velleman, R. (1992) Intergenerational effects - a review of environmentally oriented studies concerning the relationship between parental alcohol problems and family disharmony in the genesis of alcohol and other problems. I: The intergenerational effects of alcohol problems. *Int. J. Addict.* 27: 253-280.

Velleman, R. and Orford, J. (1990) Young adult offspring of parents with drinking problems: Recollections of parents' drinking and its immediate effects. *Br. J. Clin. Psychol.* 29: 297-317.

Toward a Molecular Basis of Alcohol Use and Abuse
ed. by B. Jansson, H. Jörnvall, U. Rydberg, L. Terenius & B. L. Vallee
© 1994 Birkhäuser Verlag Basel/Switzerland

Alcohol sensitivity and dependence

Marc A. Schuckit

The San Diego Veterans Affairs Medical Center and University of California, San Diego, School of Medicine, Department of Psychiatry (116A), 3350 La Jolla Village Drive, San Diego, CA 92161, U.S.A.

Summary

The relationship between the intensity of response to alcohol and the future development of alcohol abuse or dependence was evaluated in 454 18- to 25-year-old drinking but not alcohol dependent subjects. This sample consisted of sons of alcohol dependent fathers and family history negative controls matched on demography as well as alcohol and substance use histories. In response to an alcohol challenge at approximately age 20, 40% of the sons of alcoholics but less than 10% of the controls demonstrated less alcohol-related changes in subjective feelings, physiological measures, and motor performance. An average of 9.3 years later, a successful follow-up of 100% of the first 223 individuals revealed that the diminished response to alcohol was a potent predictor of future alcohol abuse or dependence. Neither the family history of alcoholism nor the intensity of response to alcohol at age 20 predicted drug dependence or major psychiatric disorders during the follow-up period.

Introduction

The development of severe and repetitive alcohol-related life problems, also called alcoholism or alcohol dependence, is genetically influenced (Schuckit, 1994a). For example, sons and daughters of alcoholics are 3 to 4 times more likely to develop this disorder than controls, and identical twins of alcoholics have a significantly higher risk than fraternal twins of alcohol dependent individuals. The heightened vulnerability for this disorder is also observed in adopted away children of alcoholics raised without knowledge of their biological parents' problems (Cotton, 1979; Goodwin *et al.*, 1974; Schuckit, 1992). However, the genetic influences operating in alcoholism are complex, interacting with environment and contributing to the risk through either polygenic and/or dominant mechanisms with incomplete penetrance (Chakravarti and Lander, 1990; Devor and Cloninger, 1989). Studies in this area are further complicated by the need to focus on the clearest cases of alcoholism, where the severe life problems are not likely to be the result of additional psychiatric disorders or severe early onset personality disturbances such as the antisocial personality disorder (Schuckit, 1994a).

As a result of the importance of environmental influences, probable genetic heterogeneity, and the absence of powerful Mendelian genetic mechanisms, the search for specific genes involved in the predisposition is complicated (Schuckit, 1994a). Further difficulties are added by the probability that a substantial number of individuals who inherit the genetic mechanisms that increase the risk might never demonstrate the disorder. Thus, several research groups are searching for phenotypic characteristics of vulnerable individuals that can be used to identify men and women actually at high risk, even if they do not become impaired. Once found, these characteristics can be used as markers within pedigrees to aid in the search for specific contributory genes.

Most studies of such phenotypic characteristics begin with children of alcoholics. These individuals can be identified and evaluated relatively early in life, before severe alcohol dependence is apparent. Comparisons of family history positive (FHP), high risk samples with appropriate family history negative (FHN) controls can identify differences that might be related to the future risk. Subsequent follow-ups of both FHP and FHN subsamples can then help determine which, if any, of the characteristics observed earlier in life most powerfully predict the future development of alcohol dependence.

This paper reviews a study initiated in 1975 that evaluates matched FHP and FHN subjects both at baseline and after an alcohol challenge. Both FHPs and FHNs are being followed up 8 to 12 years after initial testing, and results on the first 223 evaluations are presented.

Material and Methods

Following several years of pilot study, a mailed questionnaire and subsequent structured personal interview were used each year between 1978 and 1988 to gather a sample of 18- to 25-year-old healthy, drinking, but not yet alcohol dependent sons of primary alcoholic fathers (Schuckit and Gold, 1988). Whenever possible, subjects with multiple alcoholic relatives were selected. For each FHP individual, a FHN control was identified from the same pool of initial questionnaire respondents through careful matching with a FHP on demography, quantity and frequency of alcohol use, drug use histories, smoking histories, education, and height-to-weight ratio. The total sample of FHPs and FHN controls eventually included 454 subjects.

Using a series of related paradigms over the years, subjects were brought to the laboratory on an average of three occasions where they received placebo and two different doses of alcohol (0.75 and 1.1 ml/kg of ethanol). Following the placebo and active dose challenges, at 15 to 30 minute intervals over the 3 hours of testing, subjects were assessed for their subjective feelings of intoxication, levels of alcohol-induced increases in body sway or static ataxia, alcohol-related changes in prolactin, cortisol, and adrenocorticotropin hormone (ACTH), as well as on background cortical electroencephalogram (EEG) and event-related potential (ERP) changes.

Between 1988 and 1993 all subjects (100%) were located and asked to participate in a follow-up evaluation carried out by interviewers blind to the initial family history status and the response to alcohol at age 20. The structured interview used was based on the Alcohol Research Center interview and the Schedule of Affective Disorders and Schizophrenia, and developed by focusing on alcohol and substance use disorders, the major affective disorders, major anxiety disorders, and other major psychiatric disorders described in the Third Revised Diagnostic and Statistical Manual of the American Psychiatric Association (American Psychiatric Association, 1987; Schuckit et al., 1988c; Spitzer and Endicott, 1977). In addition, for purposes of corroborating the information provided by the subject, a separate follow-up interview was carried out with an additional informant (such as the subject's spouse) who helped describe the characteristics of the subject since the time of initial testing. Finally, blood and urine samples were taken from subjects to test for state markers of heavy drinking and urine drug levels (Irwin et al., 1988).

Results

A series of publications has described the differences between FHP and FHN subjects when initially tested at approximately age 20 (Schuckit, 1980, 1990; Schuckit and Gold, 1988). Consistent with the care with which the two groups were matched on their recent drinking histories, FHPs and FHNs were almost identical on their baseline levels of state markers of heavy drinking such as gamma-glutamyltransferase (GGT). The two groups also demonstrated virtually identical blood alcohol concentrations (BACs) over time, including identical levels of disappearance of ethanol from the bloodstream after the challenge.

Despite these similarities in drinking history and BACs, and despite the lack of any differences between FHPs and FHNs following the placebo challenge, approximately 40% of the FHPs but less than 10% of the FHNs demonstrated low levels of responses to alcohol. This reduced responsivity was apparent on subjective feelings throughout the 3 hours of testing, alcohol-induced increases in body sway, changes in 3 hormones, alterations in the stability of the background cortical EEG, and alcohol-induced increases in the amplitude of the P300 wave of the ERP (Ehlers and Schuckit, 1990; Schuckit, 1980, 1988, 1994b; Schuckit and Gold, 1988; Schuckit *et al.*, 1988a, 1988b).

The follow-up evaluations carried out about 10 years later on the first 223 subjects revealed that the diminished response to alcohol at age 20 was a potent predictor of the future development of alcohol abuse or dependence (Schuckit, 1994b). During the follow-up interval, approximately one-third of the FHPs but only 13% of the controls developed alcohol abuse or dependence, including figures for dependence of 26% for FHPs and 9% for FHNs. The two family history groups did not differ on the prevalence of any major psychiatric disorder such as major depression or any major anxiety disorder, although there was a nonsignificant trend for FHPs to demonstrate an increased prevalence of abuse or dependence on cannabinols (14% vs. 5%, p=.10) and stimulants (13% vs. 8%, p=.36).

At the time of follow-up, 55 men (including 42 FHPs) had developed alcohol abuse or dependence. The initial intensity of reaction to alcohol of these men (i.e., at age 20) was compared to that of the 168 men who did not go on to develop alcohol abuse or dependence by the time of follow-up. As predicted, the 55 men who became alcoholic demonstrated significantly lower levels of subjective feelings of intoxication and body sway. Among the FHPs who demonstrated the most obvious diminished response, the risk for alcohol abuse or dependence by age 30 was 56%, while for those FHPs who fell into the highest 20% on sensitivity to alcohol, the risk was only 14% (p<.003). While results are more difficult to interpret because of the smaller sample sizes, evaluations were also carried out of the relationship between the intensity of reaction to alcohol at age 20 and the future risk for developing alcoholism separately within FHNs. Once again, a low level of response to alcohol was associated with an enhanced future alcoholism risk for the 13 alcoholic individuals among the 98 FHNs. One individual was dropped from the analyses when he discovered he was adopted.

Finally, a structured series of regressions, a path analytical model, was preliminarily evaluated on the approximately 90 individuals who were most clearly FHP, and for whom the outcome was clearly alcohol dependent or clearly free of severe alcohol problems (Schuckit, 1993). These series of analyses revealed that the reduced responsivity to alcohol at age 20 was responsible for the majority of the relationship between family history and an alcoholic outcome.

Furthermore, these path analyses revealed that the relationship between the reduced responsivity to alcohol at age 20 and the future development of alcoholism remained robust even when additional factors such as the quantity, frequency, or quantity times frequency of alcohol use at age 20 were entered into the equation.

Discussion

The predisposition toward severe and repetitive alcohol-related life problems involves the complex interaction between genetic and environmental influences. Some individuals are likely to develop psychological and/or physical dependence on alcohol as a result of a severe personality disturbance such as the antisocial personality disorder. Others could develop their alcohol-related life problems as a consequence of the poor judgment inherent in major psychiatric syndromes such as schizophrenia or manic depressive disorder. However, the average alcoholic man and woman does not demonstrate these major psychiatric syndromes, developing his or her alcohol dependence independent of major preexisting psychopathology (Brown and Schuckit, 1988; Brown et al., 1991; Schuckit, 1985, 1994a). For these primary alcoholic men and women, it is possible that any biological or environmental factor that increases the probability that they will drink heavily, regularly, increases the probability of alcohol-related life problems and, subsequently, a diagnosis of alcoholism. In a heavy-drinking society it is possible that the need for higher levels of alcohol to achieve even a mild level of intoxication might be associated with either more frequent and/or more intense drinking. The delay between the oral consumption of alcohol and the maximal brain effect could result in some individuals with lower levels of sensitivity becoming more intoxicated than initially intended. Through these or other mechanisms, the subsequent greater intensity of drinking might then contribute to affiliation with other heavier drinkers which would further increase the probability of heavy consumption of alcohol. Regular intake of high doses of alcohol would also contribute to the development of greater levels of tolerance, with subsequent higher and higher doses of the drug required to achieve the desired effect. Thus, a reduced response to alcohol operating either through a genetically influenced decreased initial sensitivity to the drug and/or an enhanced rate of development of acute or chronic tolerance could interact with environmental forces to contribute to a probability of higher levels of drinking and associated problems.

The human studies of reduced responsivity cited in the Introduction do not in themselves show that the phenomenon of a reduced responsivity to alcohol is genetically influenced. However, there are studies of twins carried out in Australia where comparisons of identical and fraternal twin pairs following an alcohol challenge demonstrated a high level of heritability for the post-ethanol increases in body sway, with a probability of a similar genetic loading for levels of subjective intoxication (Martin *et al.*, 1985). Further support for the probable heritability of the initial sensitivity to alcohol and/or the development of tolerance comes from animal studies. Li and colleagues have selectively bred alcohol preferring and nonpreferring strains of rats, with the former regularly consuming enough alcohol to produce blood concentrations of 60 mg/dl or higher (Li *et al.*, 1993; Stewart *et al.*, 1993). One prominent characteristic of this alcohol preferring strain is the genetically controlled decreased sensitivity to alcohol and/or higher level of acute tolerance to this drug demonstrated by these alcohol imbibing animals.

Conclusion

This paper has attempted to present an overview of a series of human studies comparing sons of alcoholics and controls. The data consistently indicate that a reduced responsivity to alcohol is a reliable characteristic of the FHP subjects, and is associated with a heightened risk for alcohol abuse or dependence almost a decade later. However, such studies of responses to an alcohol challenge are complex, and require large samples of subjects and highly standardized research paradigms. A recent meta-analysis of alcohol challenge results corroborates the relationship between a diminished subjective response to alcohol and the family history status (Pollock, 1992). However, studies using relatively small samples, lower doses of alcohol than those reported above, alcohol challenge protocols that involve repeated alcohol dosing over an hour or more, and those that incorporate additional stressors such as electric shock do not uniformly replicate the diminished response to alcohol in FHPs reported above (Schuckit, 1994a). It is also important to remember the relevance of genetic heterogeneity as well as the probable difficulties inherent in any of the diagnostic approaches for alcoholism. These caveats emphasize the need for continuing these types of studies in an attempt to replicate the findings reported above and to identify environmental and additional genetic influences likely to contribute to the alcoholism risk.

Despite these caveats, the data demonstrate a potentially important lead in studies attempting to further knowledge about risk factors in alcoholism. These results can help identify a phenotype that can be applied to pedigrees of alcoholics as an aid in linkage analyses. The data can also be useful to clinicians who are attempting to work with young children of alcoholics, presenting them with data demonstrating their very high level of risk, and underscoring reasons why total abstinence, or at a minimum very rigidly controlled drinking practices, would be appropriate. These results might also be useful in prospective studies of children of alcoholics attempting to identify environmental and interpersonal attributes that might be useful in developing an alcoholism prevention protocol.

Acknowledgements

This work was supported by the Veterans Affairs Research Service as well as the National Institute on Alcohol Abuse and Alcoholism Grants #85226, #08401, and #08403.

References

American Psychiatric Association. (1987) *Diagnostic and Statistical Manual of Mental Disorders*, Third Edition, Revised, American Psychiatric Press, Washington, D.C.

Brown, S.A. and Schuckit, M.A. (1988) Changes in depression among abstinent alcoholics. *J. Stud. Alcohol* 49: 412-417.

Brown, S.A., Irwin, M. and Schuckit, M.A. (1991) Changes in anxiety among abstinent male alcoholics. *J. Stud. Alcohol* 52:55-61.

Chakravarti, A. and Lander, E.S. (1990) Genetic approaches to the dissection of complex diseases. *Banbury Report* 33: 307-315.

Cotton, N.S. (1979) The familial incidence of alcoholism. *J. Stud. Alcohol* 40: 89-116.

Devor, E.J. and Cloninger, C.R. (1989) Genetics of alcoholism. *Annu. Rev. Genet.* 23: 19-36.

Ehlers, C.L. and Schuckit, M.A. (1990) EEG fast frequency activity in the sons of alcoholics. *Biol. Psychiatry* 27: 631-641.

Goodwin, D.W., Schulsinger, F., Moller, N., Hermansen, L., Winokur, G. and Guze, S.B. (1974) Drinking problems in adopted and nonadopted sons of alcoholics. *Arch. Gen. Psychiatry* 31: 164-169.

Irwin, M., Baird, S., Smith, T.L. and Schuckit, M.A. (1988) Use of laboratory tests to monitor heavy drinking by alcoholic men discharged from a treatment program. *Am. J. Psychiatry* 145: 595-599.

Li, T.K., Lumeng, L. and Doolittle, D.P. (1993) Selective breeding for alcohol preference and associated responses. *Behav. Genet.* 23: 163-170.

Martin, N.G., Oakeshott, J.G., Gibson, J.B., Starmer, G.A., Perl, J. and Wilks, A.V. (1985) A twin study of psychomotor and physiological responses to an acute dose of alcohol. *Behav. Genet.* 15: 305-347.

Pollock, V.E. (1992) Meta-analysis of subjective sensitivity to alcohol in sons of alcoholics. *Am. J. Psychiatry* 149: 1534-1538.

Schuckit, M.A. (1980) Self-rating alcohol intoxication by young men with and without family histories of alcoholism. *J. Stud. Alcohol* 41: 242-249.

Schuckit, M.A. (1985) The clinical implications of primary diagnostic groups among alcoholics. *Arch. Gen. Psychiatry* 42: 1043-1049.

Schuckit, M.A. (1988) Reactions to alcohol in sons of alcoholics and controls. *Alcohol Clin. Exp. Res.* 12: 465-470.

Schuckit, M.A. (1990) A prospective study of children of alcoholics. *Banbury Report* 33: 183-194.

Schuckit, M.A. (1992) Advances in understanding the vulnerability to alcoholism. In: O'Brien, C.P. and Jaffe, J.H. (eds.): *Addictive States*, Raven Press, Ltd., New York, pp. 93-108.

Schuckit, M.A. (1993) *The reaction to alcohol as a predictor of alcoholism.* The Distinguished Research Awardee Plenary Lecture. The Research Society on Alcoholism Annual Meeting, San Antonio, Texas, June 12, 1993.

Schuckit, M.A. (1994a) A clinical model of genetic influences in alcohol dependence. *J. Stud. Alcohol.* 55: 5-17.

Schuckit, M.A. (1994b) Low level of response to alcohol as a predictor of future alcoholism. *Am. J. Psychiatry.* 151: 184-189.

Schuckit, M.A. and Gold, E.O. (1988) A simultaneous evaluation of multiple markers of ethanol/placebo challenges in sons of alcoholics and controls. *Arch. Gen. Psychiatry* 45: 211-216.

Schuckit, M.A., Risch, S.G. and Gold, E.O. (1988a) Alcohol consumption, ACTH level, and family history of alcoholism. *Am. J. Psychiatry* 145: 1391-1395.

Schuckit, M.A., Gold, E.O., Croot, K., Finn, P. and Polich, J. (1988b) P300 latency after ethanol ingestion in sons of alcoholics and in controls. *Biol. Psychiatry* 24: 310-315.

Schuckit, M.A., Irwin I., Howard, T. and Smith T. (1988c) A structured diagnostic interview for identification of primary alcoholism: A preliminary evaluation. *J. Stud. Alcohol* 49: 93-99.

Spitzer, R.L. and Endicott, J. (1977) *Schedule for Affective Disorders and Schizophrenia*, New York Psychiatric Institute, New York.

Stewart, R.B., Gatto, G.J., Lumeng, L., Li, T.K. and Murphy, J.M. (1993) Comparison of alcohol-preferring (P) and nonpreferring (NP) rats on tests of anxiety and for the anxiolytic effects of ethanol. *Alcohol* 10: 1-10.

Toward a Molecular Basis of Alcohol Use and Abuse
ed. by B. Jansson, H. Jörnvall, U. Rydberg, L. Terenius & B. L. Vallee
© 1994 Birkhäuser Verlag Basel/Switzerland

Treatment of alcoholism as a chronic disorder

Charles P. O'Brien

University of Pennsylvania/VA Medical Center, 3900 Chestnut Street, Philadelphia, PA 19104-6178 USA

Summary

Alcoholism is a common disorder that tends to be chronic and relapsing. Although there is clear evidence that treatment can be expected to induce a period of remission or at least decreased symptoms, treatment of alcoholism is generally regarded as unsuccessful. Alcoholism should be approached as a chronic medical disorder such as diabetes or arthritis. Complete abstinence is the preferred goal, but "cures" or permanent abstinence from alcohol are rare. In this model, treatment benefits may be measured by length of remission, reduction in alcohol use, improvement in health and enhancement of social functioning. Treatment continues over a period of years, mainly on an outpatient basis with increasing intensity if symptoms recur. Medications that reduce craving for alcohol or diminish the euphoric effects of alcohol would be very helpful in the management of this chronic disorder. Pre-clinical studies have produced evidence for involvement of the endogenous opioid system in the reinforcing effects of alcohol. Recent controlled clinical trials of the opiate receptor antagonist naltrexone suggest that medications of this type may improve the results of treatment for alcoholism.

Introduction

Dependence on alcohol or other drugs is so common that almost everyone has known a relative, neighbor or co-worker who suffers from the problem. This person most probably had some treatment, stopped the addicting substance for a while, perhaps a long while and then returned to use of the substance. Thus the judgment of the general community is that the treatment was ineffective. In contrast, most also know someone suffering from diabetes, heart disease or arthritis, disorders that also persistently return. Yet the community perception is not that treatment for these medical disorders is a waste of time or ineffective. In assessing the treatment of alcoholism, researchers in the past have tacitly supported the total abstinence or failure concept of alcoholism by measuring success mainly by the proportion of patients who remain completely abstinent after treatment for a set period of time (Nathan, 1986). If return of any symptoms of alcohol drinking denotes treatment failure, then most treatment for heart disease, arthritis or diabetes would also have to be classified as ineffective. Studies of treatment effectiveness for alcoholism that measure changes in quantity of drinking after treatment and not simply abstinence are more likely to be a true measure of the effects of treatment (McKay and

Maisto, 1993). Using these improved outcome measures, there have been numerous controlled studies that demonstrated efficacy for specific treatment techniques (Miller, 1992; McKay and Maisto, 1993).

Medical treatment of alcoholics

Until recently, medical approaches to the treatment of alcoholism focused on detoxification while long-term rehabilitation was the province of drug-free, non-medical programs based on the 12 Steps/Alcoholics Anonymous philosophy. Many patients have been helped greatly by this approach and some have achieved long term abstinence, but the overall rate of return to symptomatic drinking is substantial even in the best drug-free programs. Thus, although there is clear evidence of efficacy for many of the existing treatments for alcoholism, there is ample room for improvement. Therapists in alcohol programs have previously opposed including medications in the treatment program because all drugs are seen as a "crutch." But during the past decade there has been increasing interest in the potential for a new approach that includes a medication that decreases the tendency to resume alcohol drinking combined with a psychotherapy program or other behavioral intervention.

Patients could be maintained indefinitely on medication if it was effective and frequency of treatment visits would depend on the needs of the patient over time. Frequent attendance at self-help groups such as Alcoholics Anonymous would be encouraged. In this model, therapists would strongly advise the patient against return to any drinking, but some "lapses" or even periods of relapse to excessive drinking would not necessarily be considered treatment failure. Return of this symptom is merely a sign that another period of intense treatment is required.

A major problem with a combined medication/psychotherapy approach has been the unavailability of a medication that is effective in the long term management of alcoholism. Medications currently are available for treating accompanying psychiatric disorders such as affective or anxiety disorders, but medications for alcoholism itself have been elusive. Disulfiram, a medication that blocks the metabolism of alcohol, has been available for a long time. This medication reliably produces noxious consequences when patients drink any alcohol while taking the medication. There is evidence for some clinical benefit when disulfiram is combined with a behavioral program to ensure that the patient ingests the medication regularly.

Unfortunately, large controlled studies have failed to show that disulfiram is more effective than placebo in preventing relapse in clinics where there is variable medication compliance (Fuller et al 1986). Because it has clear pharmacological effectiveness, disulfiram may still have a role with specific alcoholic patients who can be protected from impulsive drinking by this medication, but it is not helpful for general use with alcoholics.

Medication development

In order to develop a new medication that might aid in a long term alcoholism treatment program, psychopharmacologists have developed animal models of excessive alcohol drinking. Studies using these rodent and non-human primate models have produced clues to the neurochemical factors involved in the reinforcement produced by alcohol. Using these models, medications that may block or diminish alcohol reinforcement have been tested. The implicit hypothesis has been that if a medication, acceptable to patients, could be found that reduces the urges to resume alcohol drinking or diminishes the rewarding effects of alcohol, relapse would be delayed and patients would remain accessible to counseling thus improving overall treatment results.

Endogenous opioid hypothesis

One line of experimental evidence links the endogenous opioid system to reinforcement of alcohol drinking. Alcohol preferring strains of mice and rats have increased basal β-endorphin levels in the pituitary gland and increased metenkephalin and β-endorphin in some brain areas relative to alcohol non-preferring rodents (Gianoulakis, 1990). Drugs that block opiate receptors such as the μ/δ opiate receptor antagonist naltrexone have been found to reduce alcohol drinking in monkeys (Altschuler et al., 1980; Myers et al., 1986) and in strains of rodents selected for alcohol preference (Froelich et al., 1990; Samson and Doyle, 1985; Marfaing-Jallet et al., 1983). Naltrexone also blocks the post-stress drinking of alcohol observed in rats (Volpicelli et al., 1986). Drugs that stimulate opiate receptors have effects depending on the dose. Low doses of morphine have been reported to stimulate alcohol drinking in rats (Hubbell et al., 1988) while higher doses suppress the drinking of alcohol (Sinclair et al., 1973).

A few studies have been conducted in human subjects. Gianoulakis and colleagues (1990), studied non-alcoholic volunteers with a strong family history of alcoholism (High Risk) and compared them to volunteers with no family history of alcoholism (Low Risk). Baseline levels of plasma β-endorphin were lower in the High Risk subjects, but a 0.5 mg/kg test dose of alcohol produced a significantly greater rise in plasma β-endorphin in the High Risk group. In another study of plasma β-endorphin, alcoholics were found to have depressed levels shortly after cessation of drinking, but the β-endorphin levels returned to normal range after six weeks of abstinence (Vescovi et al., 1992). Of course plasma β-endorphin levels reflect pituitary rather than brain activity, but the sum of the animal and human studies support the concept that alcohol drinking involves the endogenous opioid system which may be more sensitive to alcohol in those at high risk of becoming alcoholics.

These data are consistent with the hypothesis that alcohol ingestion stimulates the release of endogenous opioids which increases some of the rewarding effects of alcohol through opioid mechanisms. A small amount of morphine may stimulate alcohol drinking in rodents by priming this system, but larger doses of an opioid drug would produce rewarding effects itself thus replacing alcohol and reducing the drive to drink alcohol. Naltrexone, by blocking opiate receptors, would block this mechanism of opioid reinforcement and thus reduce or eliminate alcohol preference. This hypothesis was generated from animal data and it is supported by evidence from clinical trials of naltrexone in alcoholics reviewed below. Since opioid reinforcement has also been linked to limbic dopamine activation (Widdowson and Holman, 1992), the opioid effects on alcohol drinking behavior are consistent with the data implicating dopamine in the alcohol reinforcement mechanism (Koob, this volume). Serotonergic mechanisms have also been implicated in alcohol drinking in animals. Studies of serotonin mechanisms and subsequent clinical trials are reviewed elsewhere in this volume by Naranjo and by Janssen. Clearly there are multiple mechanisms involved in the effects of alcohol on behavior, but the endogenous opioid systems appears to have an important role.

Clinical trials

The clinical course of alcoholism is quite variable so in any test of a medication or a new treatment approach, a randomly assigned control group is essential. We cannot assess any new treatment without knowing what would have occurred in the absence of that treatment.

Evaluation prior to beginning treatment should be comprehensive. Simply recording the daily intake of alcohol is not adequate. Important variables that may influence outcome include age, history of prior treatment, use of other drugs, presence of another psychiatric disorder, and the degree of problems in the family, social, occupational, legal or medical areas. The sample size should be large enough so that there is general equality between treatment and control groups across all of these important variables. Pre-, during and post-treatment evaluations should be conducted by investigators who are not aware of the patient's group assignment.

Compliance issues

Compliance with medication is an important variable in clinical trials of all chronic disorders including alcoholism. Those who are sufficiently motivated to comply with the medication regimen may be more likely to remain abstinent irrespective of medication. It is particularly misleading to measure compliance in the medication group without also measuring it in the placebo group. With most medications, one can measure compliance directly by monitoring plasma levels of the compound, but measures of placebo compliance are problematic. One technique is to add riboflavin to both experimental drug and placebo. Riboflavin ingestion results in urine that fluoresces under U/V light and this can be tested on each clinic visit. Errors can occur because of dietary riboflavin and false negatives occur when medication is ingested within one hour or more than 12 hours before the clinic visit. Another method is to count the number of pills remaining in the medication container at each visit. Of course this method too can be misleading if a patient is determined to deceive the investigator while not dropping out of the study. A comparison of the riboflavin method and pill counts indicated comparable accuracy (Sullivan et al., 1989). Pill counts can be enhanced by giving research patients a container that has a different number of pills each week and determining how many were ingested by counting the remainder. Yet another method involves special pill bottles containing a micro-chip in the lid that records the dates and times that the lid is removed. Some method of checking compliance should be included in all clinical trials of new medications.

Another variable that should be quantitated in clinical trials is the level of psycho-social intervention. It is unrealistic to expect any medication to have significant effects on alcoholism without the support of some program to help patients cope with the many psychological and social complications of alcoholism. As reviewed above, there is significant efficacy in existing treatment programs without medication and therefore inequality in the amount of behavioral intervention could skew the results of a medication trial. The type of psychosocial treatment should be specified and equal for all research patients. In order to insure this equality and to assess the psychosocial component in all patients, a treatment assessment instrument (McLellan et al., 1992) has been developed that is patterned on the Addiction Severity Index (ASI). Data are obtained directly from the patient as the number of minutes spent each week with a helping person or a self-help group working on problems in each of the seven areas of the ASI.

Any discussion of clinical trials methodology should include a consideration of the population that volunteers for a treatment requiring random assignment. Studies (Strohmetz et al., 1990) of the alcoholics who refuse to volunteer after the study is explained to them indicate that in general, the volunteer study subjects are more severely ill than the refusers. Thus double-blind trials may under-estimate medication efficacy because of the nature of the patients being treated.

Trials of opioid antagonists

As reviewed above, there are several animal models of alcohol drinking that suggest a role for endogenous opioids in the reinforcement produced by alcohol. It was these reports from animal studies that caused Volpicelli and colleagues (Volpicelli et al., 1992) to conduct a double-blind trial of naltrexone versus placebo in chronic alcoholics. The patients in the study were applying for treatment at the Philadelphia Veterans Affairs Medical Center. The 70 male subjects were on average 43 years old with a 20 year history of heavy drinking and all met at least five of the nine DSM-III-R criteria for alcohol dependence including physical signs of withdrawal sufficient to require medication. After completion of detoxification, the patients began an outpatient rehabilitation program that consisted of one month of daily participation in a day hospital program tapering to weekly counseling sessions during months two and three. At the time of entering the rehabilitation program, the patients were randomly assigned to 50 mg naltrexone daily or to naltrexone placebo. During the three months of the study, the naltrexone-treated patients reported significantly less craving for alcohol and they reported significantly fewer days of alcohol use. The overall rate of relapse for the two groups during the three-month study is shown in Fig. 1. As is typical of chronic alcoholics in a rehabilitation program, 54% of the placebo patients met criteria for relapse but only 23% of those in the naltrexone group met these criteria. Among the patients randomized to placebo, having a "slip" and drinking any alcohol almost invariably (95%) led to uncontrolled drinking that met the criteria for a relapse. In contrast, those receiving naltrexone who used any alcohol had only a 50% relapse rate. These

RELAPSE RATES - (From Volpicelli, 1992)

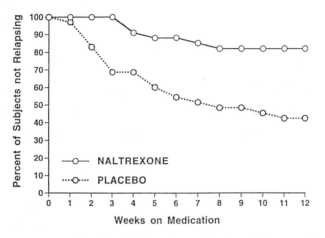

Fig. 1. Taken from Volpicelli et al., 1992, this shows the rate of patients during outpatient treatment meeting criteria for relapse. Note the significantly slower rate for those patients randomly assigned to naltrexone medication.

data support the notion that use of even small amounts of alcohol after detoxification usually leads to loss of control and relapse. In contrast, the chances of relapsing were much less if the patient who drank alcohol happened to be on naltrexone. In other words, in spite of the slip, the patient could remain in treatment and continue the rehabilitation program. Relapse while not implying treatment failure does mean that a new course of treatment has to be instituted usually beginning again with detoxification.

All patients were followed for three months including early drop outs. Thus two naltrexone patients who dropped out due to nausea were included in the outcome analysis. There was no evidence that naltrexone produced hepatotoxicity. To the contrary, there were more improvements in liver enzyme data in the naltrexone group probably reflecting their greater abstinence, but the differences did not reach statistical significance.

Positive results in a single study of a new treatment for alcoholism would not be remarkable because of the variability in the course of this disorder. For a treatment to be regarded as effective, it should be replicated by several groups of investigators. Thus it is particularly important that an attempt was made to repeat this study very rapidly. O'Malley and colleagues (1992) heard a preliminary report by Volpicelli et al. and decided to initiate a similar study among 97 alcohol dependent subjects in New Haven, Connecticut. Instead of a day hospital, the rehabilitation program consisted of coping skills/relapse prevention or supportive psychotherapy. The results of this independent study involving a predominantly white male population were quite similar to those of the Philadelphia group. Naltrexone was found to be clearly superior to placebo irrespective of the type of psychosocial intervention to which the patients were assigned. Those randomly assigned to naltrexone drank on half as many days and consumed one third the number of standard drinks during the trial as did subjects who received placebo. Craving for alcohol was significantly decreased in those naltrexone subjects who completed the study. Relapse rates were significantly lower in the naltrexone-treated patients (Fig. 2). As in the Volpicelli et al study, relapse rate differences were particularly high among patients who drank some alcohol. Among the patients who had at least one drink, those randomized to naltrexone and coping skills therapy had one fourth the risk of relapse compared with subjects taking placebo who received coping skills treatment. Medication compliance was measured by the addition of riboflavin to the study medication and the testing of urines weekly with UV light. This method indicated high compliance for both groups, but significantly higher for the naltrexone treated patients. As in the previous study, there was no evidence of naltrexone induced-hepatotoxicity. Aspartate aminotransferase levels were significantly lower in the naltrexone-treated patients at endpoint and a similar trend was noted for alanine aminotransferase. O'Malley and colleagues did a follow up evaluation of these patients six months after the end of the study and termination of the naltrexone treatment. At follow up, the patients who had been randomly assigned to naltrexone at the beginning of the study continued to show significantly less return of symptomatic drinking than those randomized to placebo.

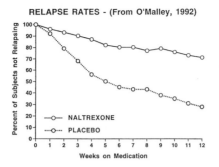

Fig. 2. This shows the relapse rates for the O'Malley et al.study. Note the similarity to the curves generated from the study by Volpicelli and colleagues.

Possible mechanism for naltrexone efficacy

The patients in the Volpicelli et al study who had a "lapse" and drank some alcohol were asked about the effects of the alcohol. Those receiving placebo reported that the effects were no different from those experienced prior to the study. Those receiving naltrexone reported significantly less euphoria from their drinking than they expected based on prior experience. This finding which suggests an effect of naltrexone on the immediate subjective effects of alcohol may give a clue as to the mechanism of action by which naltrexone might decrease relapse rates in alcoholics. Double-blind studies in human subjects of the acute effects of alcohol and alcohol placebo in the presence of naltrexone or naltrexone placebo are needed. These studies should be conducted in normal volunteers, in alcoholics and in non-alcoholics with a family history of alcoholism.

These two studies of naltrexone in middle-aged alcoholics involved in good outpatient rehabilitation programs and receiving adjunctive medication consisting of either naltrexone or placebo had strikingly similar results. More studies are required, but the available clinical evidence is completely consistent with evidence from rodent and non-human primate models. It suggests that naltrexone by blocking μ and δ opiate receptors is able to reduce the reinforcement produced by alcohol drinking. Both studies had total abstinence as the stated goal of treatment and there was no intention of using naltrexone to teach controlled drinking. Of course, naltrexone clearly did not directly prevent alcoholics from relapsing. Patients can drink while taking naltrexone with no noxious consequences. They can also stop the medication at any time with no withdrawal symptoms. Naltrexone may, however, support the effects of the rehabilitation program by reducing the immediate reward produced by alcohol thus enabling the patient to remain in treatment longer and receive the benefits of long term behavior change.

Acknowledgements

Figs. 1 and 2 are reprinted with permission from Archives of General Psychiatry. This chapter was supported in part by the Medical Research Service of the US Dept of Veterans Affairs and by USPHS Grant No. DA 05186.

References

Altschuler, H.L., Phillips, P.E. and Feinhandler, D.A. (1980) Alteration of ethanol self-administration by naltrexone. *Life Sci.* 26: 679-688.

Gianoulakis, C., Angelogianni, P., Meaney, M., Thavundayil, J. and Tawar, V. (1990) Endorphins in individuals with high and low risk for development of alcoholism. In: Reid, L.D., (ed.9. *Opioids, Bulimia and Alcohol Abuse and Alcoholism.* New York: Springer-Verlag, pp. 229-246.

Froehlich, J.C., Harts, J., Lumeng, L. and Li, T.K. (1990) Naloxone attenuates voluntary ethanol intake in rats selectively bred for high ethanol preference. *Pharmacol Biochem Behav.* 35: 385-390.

Fuller, R.K. and Roth, H.P. (1979) Disulfiram for the treatment of alcoholism: an evaluation in 128 men. *Ann. Intern. Med.* 90: 901-904.

Hubbell, C.L., Abelson, M.L., Burkhardt, C.A., Herlands, S.E. and Reid, L.D. (1985) Constant infusions of morphine and intakes of sweetened ethanol solution among rats. *Alcohol* 5: 409-415.

Marfaing-Jallet, P., Miceli, D. and Le Magnen, J. (1983) Decrease in ethanol consumption by naloxone in naive and dependent rats. *Pharmacol Biochem Behav.* 18: 537-539.

McKay, J. R. and Maisto, S.A. (1993) An overview and critique of advances in the treatment of alcohol use disorders. *Drugs and Society* (in press).

McLellan, A.T., Alterman, A.I., Cacciola, J., Metzger, D and O'Brien, C.P. (1992) A new measure of substance abuse treatment: initial studies of the treatment services review. *J Nervous and Mental Disease* 180: 101-110.

McLellan, A.T., O'Brien, C.P., Metzger, D., Alterman, A.I., Cornish, J. and Urschel, H. (1992) How effectice is substance abuse treatment - compared to what? In: C.P. O'Brien and J.H. Jaffe (eds.): *Addictive States*, pp. 231-252.

Miller, W.R. (1992) The effectiveness of treatment for substance abuse. *J. of Substance Abuse Treatment* 9: 93-102.

Myers, R.D., Borg, S. and Mossberg, R. (1986) Antagonism by naltrexone of voluntary alcohol selection in the chronically drinking macaque monkey. *Alcohol* 3: 383-388.

Nathan, P.E. (1986) Outcomes of treatment for alcoholism: current data. *Annals Behav. Medicine* 8: 40-46.

O'Malley, S.S., Jaffe, A.J., Chang, G, Schottenfeld, R.S., Meyer, R.E. and Rounsaville, B. (1992) Naltrexone and coping skills therapy for alcohol dependence. *Arch. Gen. Psychiatry* 49: 881-887.

Samson, H.H. and Doyle, T.F. (1985) Oral ethanol self-administration in the rat: effect of naloxone. *Pharmacol Biochem Behav.* 22: 91-99.

Sinclair, J.D., Adkins, J. and Walker, S. (1973) Morphine-induced supression of voluntary alcohol drinking in rats. *Nature* 246: 425-427.

Strohmetz, D.B., Alterman, A.I. and Walter, D. (1990) Selection factors in alcoholics volunteering for a treatment study. *Alcoholism: Clin & Exp Res.* 14: 736-738.

Sullivan, J.T., Naranjo, C.A and Sellers, E.M. (1989) Compliance among heavy alcohol users in clinical drug trials. *J Sub Abuse* 1: 184-194.

Vescovi, P.P., Coiro, V., Volpi, R., Giannini, A. and Passeri, M. (1992) Plasma beta-endorphin but not met-enkephalin levels are abnormal in chronic alcoholics. *Alcohol & Alcoholism* 27(5): 471-175.

Volpicelli, J.R, Alterman, A.I., Hayashida, M. and O'Brien, C,P. (1992) Naltrexone in the treatment of alcohol dependence. *Arch Gen Psychiatry* 49: 876-880.

Volpicelli, J.R., Davis, M.A. and Olin, J.E. (1986) Naltrexone blocks the post-shock increase of ethanol consumption. *Life Sci.* 38: 841-847.

Widdowson, P.S. and Holman, R.B. (1992) Ethanol-induced increase in endogenous dopamine release may involve endogenous opiates. *J Neurochem.* 59: 157-162.

Toward a Molecular Basis of Alcohol Use and Abuse
ed. by B. Jansson, H. Jörnvall, U. Rydberg, L. Terenius & B. L. Vallee
© 1994 Birkhäuser Verlag Basel/Switzerland

Addiction and the potential for therapeutic drug development

Paul A. J. Janssen

Janssen Research Foundation, Turnhoutseweg 30, B-2340 Beerse, Belgium

Summary

Therapeutic drug development in alcoholism could be targeted at any of the following: direct antagonism, substitution, treatment of abstinence, enhancement of aversion, modification of biodisposition, or craving. Ritanserin is a potent, centrally acting, highly selective $5\text{-HT}_{1C/2}$ antagonist which, in addition to having a sleep-regulating and anti-depression/anti-anxiety effect, displays a unique pharmacological action in several animal paradigms of substance abuse which assess drug-craving. In fact, the latter pharmacological action was demonstrated after initial clinical observations suggested an effect of ritanserin in the chronic withdrawal phase after detoxification from alcohol in patients.

   The results of a recent double-blind, placebo-controlled, trial indicated that ritanserin did not induce aversion to drink alcohol in normal volunteers who display social drinking, but are not suffering alcohol dependence.

   Currently, a full clinical development program of ritanserin in cocaine and alcohol abuse is ongoing. Three major double-blind, placebo-controlled trials in alcohol dependent patients are in progress. Patients of different severity levels, ranging from mild to very severe, are studied. The dosages of ritanserin tested (2.5 mg, 5 mg, and 10 mg o.d.) are known to be well tolerated and safe. Two trials aim for relapse prevention - clinically defined in one, biochemically defined in the other-, an one trial has improved (reduced) drinking behaviour as a therapeutic goal. This program, which involves close to 900 alcohol-dependent patients, is well under way, and is still picking up momentum.

Introduction

Systematic approaches to the development of pharmacological treatments for alcoholism have to concentrate not only on the short but also on the long-term reduction of alcohol use and on the problems associated with alcohol withdrawal. Alcohol withdrawal results, after acute physical symptoms, in withdrawal anxiety, depressed mood and sleep disturbances with a decrease in deep sleep and an increase in the number of stage changes. These sleep disturbances may persist for several years. It is suggested that certain features of sleep disturbances may increase the risk of renewed drug abuse or lead to dependence on other drugs such as minor tranquillizers. These physiological and mood shifts are integrated in the concept of craving.

   Theoretically, various approaches can be used to treat alcohol abuse and alcohol addiction. Firstly, one can try to develop an *antagonist for the effects* of alcohol. Such an antagonist can be used as an antidote during alcohol intoxication and the agent might reduce the positive reinforcing properties of alcohol, resulting in an extinction. An example of this kind of approach is the partial benzodiazepine inverse agonist Ro 15-4513.

Secondly, there is the possibility of *substitution*. Replacing alcohol by a comparable agent can reduce problems of withdrawal and craving. However, this approach includes the risk of cross-tol erance. The use of methadone in heroin addicts is an excellent example of such strategy.

Thirdly, one can focus on the immediate and/or protracted *treatment of the abstinence effects*. Doing so, one reduces the negative reinforcement, facilitates detoxification and prevents direct relapse. Typical for alcohol withdrawal treatment is the use of the benzodiazepine loading technique and the administration of calcium entry blockers and anticonvulsants.

Fourthly, it is possible to *enhance the aversive effects* of alcohol which theoretically might lead to a reduction in drinking. The use of disulfiram is based on this concept.

In addition, it is possible to influence alcohol abuse by *modifications of the biodisposition* of alcohol through changes in metabolism processes such as alterations in the onset and facilitation of elimination. However, a reduction of toxicity associated with the long term use of alcohol and the treatment of underlying psychopathology, such as anxiety, depression, mood shifts, obsessive compulsive disorders and sleep abnormalities, is possible. Within this last therapeutic avenue, various serotonergic agents have been tested.

A final approach is to directly interact with the fundamental processes of *craving*. Doing so, it is possible to treat a variety of substance abuses because one is not limited to a given substance of abuse. The employment of ritanserin in the treatment of addiction reflects this idea.

Ritanserin in substance abuse: Early observations

Ritanserin is a potent long acting $5-HT_{2/1C}$ antagonist. In humans, ritanserin was demonstrated to increase deep slow wave sleep, to improve mood and to promote liveliness in a variety of psychiatric disorders (e.g. dysthymia, depressive states, insomnia, agitation and anxiety) and to facilitate participation in behaviour therapy. During clinical trials, in which ritanserin was well tolerated, unexpected observations indicated that ritanserin is of value in treating drug addicts. Furthermore, initial clinical observations confirmed the efficacy of ritanserin in the chronic withdrawal phase after detoxification from alcohol (Monti and Alterwain, 1991). There are indications that ritanserin, as in dysthymic patients, may support alcohol addicts to cope better with their problems of daily life.

Because until now, mainly drugs that enhance serotonergic transmission (5-HT agonists and 5-HT-reuptake inhibitors) have been studied in addiction, animal experiments were carried out to support and evaluate further a possible role of ritanserin in drug addiction. In different tests, ritanserin was compared with various other serotonin antagonists, including 5-HT$_{1A}$, 5-HT$_2$, 5-HT$_{1C/2}$ and 5-HT$_3$ antagonists. Many structural analogues of ritanserin were also tested. These comparative studies underlined the unique activity profile of ritanserin. A review of some available preclinical data of ritanserin in these animal models of drug abuse are presented below.

Ritanserin in substance abuse: Pharmacological data

Important for a new compound in drug addiction is the lack of any abuse potential and of toxic effects after combination of the test compound with drugs of abuse. There are various arguments indicating that ritanserin lacks an abuse potential. In a drug discrimination test procedure in rats, ritanserin appeared devoid of discriminative stimulus properties. Rats could not be trained to discriminate between the presence and absence of various doses of ritanserin (Meert and Janssen, 1989). Furthermore, ritanserin does not produce any stimulus generalization to various drugs with an abuse potential including alcohol, nor does it directly interfere with the generalization gradients of these drugs (Meert and Janssen, 1991, 1992a, 1992b). In a conditioned place preference test, rats did not develop any preference for the ritanserin treatment associated side, indicating that ritanserin lacks reinforcing properties. Also in an oral drinking model in rats, in which the animals were given the choice between a ritanserin solution and water after a period of forced ritanserin drinking, no ritanserin preference was measured. Various additional behavioural observations, including LD50 toxicity studies, excluded any direct interactions between ritanserin and various drugs of abuse.

In order to evaluate the effects of ritanserin on alcohol intake and alcohol preference, oral drinking models were used in rats (Meert and Janssen, 1991; Panocka *et al.*, 1992; Rammsayer and Vogel, 1991; Meert, 1993). In these models, ritanserin was demonstrated to reduce alcohol intake and alcohol preference in a dose-related manner without interfering with consumatory processes. The activity of ritanserin varied to some degree with the test and treatment procedure as well as with the rat strain used. Long term efficacy and gradual increases in alcohol drinking after the cessation of ritanserin treatment were also demonstrated. Additional experiments excluded that the effects of ritanserin could be simply attributed to an interaction with the discriminative stimulus or any other properties of alcohol or a direct induction of alcohol aversion.

Intracerebral injections with ritanserin further illustrated that the nucleus accumbens, as part of the dopaminergic rewarding system often implied in drug abuse, might be a primary site of action (Panocka *et al.*, 1993). Biochemical observations confirmed this idea by demonstrating that ritanserin was able to change the extracellular concentrations of DA and 5-HT in the nucleus accumbens (Devaud *et al.*, 1992). Also from electrophysiological studies, interactions between ritanserin and the dopaminergic system were reported (Udego *et al.*, 1989)

With regard to alcohol withdrawal, ritanserin was demonstrated to reduce the alcohol withdrawal-induced inhibition of exploratory behaviour in rats chronically exposed to a liquid diet containing 10 % alcohol. Ritanserin had only a limited activity against the alcohol withdrawal-induced supersensitivity to tremorogenic agents within the same animals (Meert, 1993).

Based on these experimental confirmations of the efficacy of ritanserin in drug abuse, a clinical development plan was started.

Ritanserin in substance abuse: Ritanserin in alcoholism

To explore further the effects of ritanserin, a controlled trial was performed in normal volunteers who display social drinking, in a specific experimental set-up. Heavy social drinkers (≥ 28 drinks/ week), but without alcohol-related problems, and under no incentive to change their habit, were studied. They participated in a seven day placebo baseline period, and were then randomised into one of three treatment groups (ritanserin 10 mg o.d., ritanserin 5 mg o.d., or placebo) for 14 days. An experimental bar session, during which the subjects were offered standard drinks, was conducted at the end of the baseline, and at the end of the treatment period. Treatment with ritanserin did not induce aversive effects towards alcohol in these normal volunteers, neither during the outpatient phase, nor during the experimental bar sessions. It should be emphasized that craving which is in part caused by the need to alleviate unwanted withdrawal phenomena was not present in these social drinkers without alcohol-related problems.

Because of these demonstrated effects of ritanserin, a large scale development program of ritanserin in substance abuse is currently ongoing. This program consists of several trials in both cocaine dependence and alcohol dependence. Clearly, there are many methodological problems inherent to clinical development in these indications. First, different levels of severity are associated with substance abuse, and the preclinical data do not allow predicition of which degree of severity is most likely to benefit from ritanserin treatment. Second, the legal, social, and economic backgrounds against which human substance abuse occurs must be taken into account and controlled when setting up clinical trials. Third, compliance with the drug regimen is expected to be lower, and drop-out rates are expected to be higher than average in these clinical trials, because of the particular patient population. Fourth, which methods are to be used to assess efficacy ? Several psychometric instruments exist, for diagnostic purposes, as well as for severity assessments and for estimation of consumption, but their validity in a clinical trial setting has not always been documented extensively. Last, in the case of alcohol, there is not always consensus on the therapeutic goal to strive for: total abstinence, or reduction of consumption.

For these reasons, we are currently doing three different large clinical trials to assess the effect of ritanserin in alcoholism. All three are double-blind, placebo-controlled, parallel group studies, but they differ in other aspects of design and methodology. Different dosages, known to be safe and well tolerated from clinical trials in other indications, are tested. This will enable us to ascertain the effect of treatment in different patient populations, in different circumstances and with different therapeutic goals.

The first trial is a relapse-prevention trial in detoxified alcohol addicts. The trial includes patients with alcohol dependence (DSM-III-R 303.90 criteria) who have been successfully detoxified (withdrawal symptoms have disappeared after seven days) while being hospitalised in the specialised clinic in which this single-centre study takes place. These patients are severely affected. Within one year, the overwhelming majority of them usually have a relapse necessitating re-admission to the hospital.

The patients are randomised to receive ritanserin 10 mg o.d. or placebo. In this study, the primary interest is the difference in time to clinical relapse since discharge from the hospital. Clinical relapse is defined as drinking behaviour to such an extent that a readmission is necessary according to the trial nursing team or investigator. To ascertain this, the patients are interviewed weekly at their home. During these weekly visits, the approximate weekly alcohol consumption is estimated, and trial medication for the next visit is dispensed. Every fourth visit, biochemical and hematological tests are performed, to evaluate the somatic repercussions of alcohol consumption. On a three-monthly basis, a symptom check-list-90 (SCL-90) (Arindell and Ettema, 1986) is completed, and a patient's self-evaluation of craving and alcohol-related problems, a clinician's global impression, and the patient's global impression are performed.

Group-size estimations were based on the assumption that, under standard conditions at this centre, which includes the weekly visits, 20% of the patients will complete one year without clinical relapse (E. Mostert, personal communication). Assuming this percentage to be 45% in the ritanserin group, and 20% in the placebo group, 52 patients per group are required to detect a statistically significant difference with a power of 80% at a two-tailed 5% probability level. Therefore, a total of 120 patients will be included in this trial. Currently, approximately two thirds of the patients have been enrolled.

The second trial is also a relapse prevention study in abstinent chronic alcoholics in the post-detoxification period. Alcohol dependent patients (DSM-III-R 303.90) with a dependence of moderate to severe can participate in the study, provided they have been successfully detoxified. The latter is ascertained by means of the Withdrawal Scale for Alcohol and related psychoactive drugs (WSA) (Bech et al., 1989), on which the patient must obtain a weighted score <2, two to six weeks after detoxification, before being eligible for the trial. In addition, the patient must be completely abstinent from alcohol, and have the intention to remain abstinent, before enrolment.

The patients are randomly assigned to treatment with ritanserin 2.5 mg, ritanserin 5 mg, ritanserin 10 mg or placebo for a period of 24 weeks.

In contrast to the first study, where relapse is defined by re-hospitalization, a methodology based upon three liver function tests outlined above is used to define relapse. Recovering alcoholics who remain abstinent can be distinguished from those who resume drinking, with a sensitivity of over 95% and a specificity of over 80%, when using an increase of $\geq 20\%$ in gamma-GT, or 20% in ALAT, or 40% in ASAT over baseline values as criteria (Irwin *et al.*, 1988). To decrease the likelihood of false positive responses, the threshold values used in this trial were increased to $\geq 40\%$ for gamma-GT, $\geq 60\%$ for ALAT, or $\geq 60\%$ for ASAT. This raises the specificity of the method to over 96%. In addition to these biochemical variables, the patient's self-report on quantity and frequency of consumed drinks, and a Clinical Global Impression (CGI) (Guy, 1976) given by the investigator on the severity of the patient's illness and the occurrence of relapse are registered as primary criteria. Also included in this study as secondary criteria are assessments of alcohol craving by visual analog scale and questions, the Hamilton Depression Rating Scale (HDRS) (Hamilton, 1960), assessment of sleep quality and morning vigilance (by questions), and the Social Functioning Questionnaire (SFQ) (Tyrer, 1990).

The number of subjects needed in this trial has been calculated by assuming an expected relapse rate in the placebo treatment group of 70 % during the 24-week trial period. This figure is arrived at by averaging the widely different data in the literature. A reduction of the relapse rate under pharmacotherapy to 50% was considered to be clinically relevant. Based on these assumptions, and expecting a 20% drop-out rate, a minimum of 116 patients per treatment group are to be recruited, to detect such a difference with a power of 80% at a 5% significance level (two-tailed). Therefore, this study aims to enrol a minimum of 464 patients. Currently, well over half have been enrolled.

The third large trial targets yet another type of patient. To be enrolled, the subjects must have a diagnosis of alcohol dependence (303.90), but with a mild to moderate dependence, according to DSM-III-R criteria. They also must have a Short Michigan Alcoholism Screening Test (MAST) (Selzer *et al.*, 1976) score of ≥ 3. However, subjects with severe alcohol dependence, or subjects where medical supervision would be necessary for alcohol withdrawal, are excluded.

Another distinguishing feature of this third study is that the patients who enter the trial must, besides expressing a desire to stop or to reduce alcohol consumption, be willing to enrol in a standard psychosocial treatment program. This program consists of individual sessions (one per week) in Cognitive-Behavioral Coping Skills (Kadden *et al.*, 1992).

After having completed a single-blind placebo phase of one week, the subjects are randomly assigned to treatment with ritanserin 2.5 mg, ritanserin 5 mg, or placebo for 11 weeks.

The following primary efficacy measurements are used in this study. First, daily alcohol consumption for the week prior to each study visit is assessed using the Time-Line Follow-Back (TLFB) technique. Drinking is recorded as the number of standard drinks per day. Second, alcohol craving is measured at each visit using a 100 mm visual analog scale. Third, a Clinical Global Impression (CGI) (Guy, 1976) is rated relative to change from initial assessment (1- to 7-point scale). Other efficacy assessments include the Addiction Severity Index (ASI) (McLellan *et al.*, 1980), the Profile of Mood States (POMS) (McNair *et al.*, 1971), and a Social Functioning Questionnaire (SFQ) (Tyrer, 1990). In addition, estimations of blood alcohol concentration using breath analysis (breathalyzer) will be made at each study visit and before each counselling session.

Sample size has been calculated assuming a mean improvement of the ritanserin group over placebo at endpoint of 10 mm on the 100 mm alcohol craving scale with a power of 80% at a 5% significance level (two-tailed), and assuming a 30% drop-out rate. Per treatment group, at least 100 patients will be enrolled (a minimum of 300 patients in total). This study has recently started.

Conclusion

The clinical development of ritanserin in treating substance abuse is based on the unique property of this highly specific and potent centrally acting 5-HT$_{2/1C}$ antagonist, namely its propensity to reduce both craving, and also, at least in the case of alcohol, to limit the incentive and extent of social drinking.

Preclinical data have shown that ritanserin reduces drug preference, without reducing total fluid intake. Most importantly, pharmacological experiments have shown that ritanserin does not have an abuse potential itself, that it does not act via a drug substitution effect, or via an aversive mechanism. Early clinical observations have indicated that ritanserin exerts a beneficial effect in addicted patients, by reducing craving for the substance of abuse. It was demonstrated that ritanserin did not cause aversion to alcohol in normal socially drinking volunteers.

A full-scale development of ritanserin in the treatment of alcohol and cocaine dependence is currently ongoing, and still gaining momentum. Different therapeutic goals are explored in three large, placebo-controlled trials involving close to 900 patients with alcohol dependence of various levels of severity.

References

Arindell, W.A. and Ettema, J.H.M. (1986) *Handleiding bij een multi-dimensionele psychopathologie-indicator*. Swets & Zeitlinger B.V., Lisse.

Bech, P., Rasmussen, S., Dahl, A. and Lauritsen, B. (1989) The Withdrawal Syndrome Scale for alcohol and related psychoactive drugs: interobserver reliablility and construct validity. *Nordisk Psykiatrisk Tidsskrift* 43 (4): 291-294.

Devaud L.L., Hollingsworth, E.B. and Cooper B.R. (1992) Alterations in extracellular and tissue levels of biogenic amines in rat brain induced by the serotonin₂ receptor antagonist, ritanserin. *J. Neurochem* 59: 1459-1466.

Guy, W. and Ed. (1976) ECDEU Assessment manual for psychopharmacology, *revised DHEW Publication* No. (ADM) 76-338, Rockville, Md., National Institute of Mental Health.

Hamilton, M. (1960) A rating scale for depression. *J. Neurol. Neurosurg. Psychiatry* 23: 56-62.

Irwin, M., Baird, S., Smith, T.L. and Schuckit, M. (1988) Use of laboratory tests to monitor heavy drinking by alcoholic men discharged from a treatment program. *Am. J. Psychiatry* 145 (5): 595-599.

Kadden, R., Carroll, K., Donovan, D., Cooney, N., Monti, P., Abrams, D., Litts, M. and Hester, R. (1992) Cognitive-Behavioral Coping Skills Therapy Manual, National Institute on Alcohol Abuse and Alcoholism Project MATCH Monograph Series, Volume 3, *DHHS Publication* No. (ADM) 92-1895.

McLellan, T., Luborsky, L., Woody, G. and O'Brien, C. (1980) An improved diagnositc evaluation instrument for substance abuse patients: the addiction severity index. *J. Nerv. Ment.* Dis. 168: 26-33.

McNair, D.M., Lorr, M. and Droppleman, L.F. (1971) EITS Manual for the Profile of Mood States. *Educational and Industrial Testing Service, San Diego, Ca.*

Meert T.F. (1993) Effects of various serotonergic agents on alcohol intake and alcohol preference in Wistar rats selected at two different levels of alcohol preference. *Alcohol & Alcoholism* 28: 157-170.

Meert T.F. (1993) Pharmacological evaluation of alcohol withdrawal-induced inhibition of exploratory behaviour and supersensitivity to harmine-induced tremor. *Alcohol & Alcoholism*, press.

Meert T.F. and Janssen P.A.J. (1989) The psychopharmacology of ritanserin: Comparison with chlordiazepoxide. *Drug. Dev. Res.* 18: 119-144.

Meert T.F. and Janssen P.A.J. (1991) Ritanserin, a new therapeutic approach for drug abuse. Part 1: Effects on alcohol. *Drug. Dev. Res.* 24: 235-249.

Meert T.F. and Janssen P.A.J. (1992a) Ritanserin, a new therapeutic approach for drug abuse. Part 2: Effects on cocaine. *Drug. Dev. Res.* 25: 39-53.

Meert T.F. and Janssen P.A.J.(1992b) Ritanserin, a new therapeutic approach for drug abuse. Part 3: Effects on fentanyl and sucrose. *Drug. Dev. Res.* 25: 55-66.

Monti J.M. and Alterwain P. (1991) Ritanserin decreases alcohol intake in chronic alcoholics. *The Lancet* 337: 60.

Nomikos G.G. and Spyraki C. (1988) Effects of ritanserin on the rewarding properties of d-amphetamine, morphine and diazepam revealed by conditioned place preference in rats. *Pharmacol. Biochem. Behav.* 30: 853-858.

Panocka, I., Massi., M. and Pozzi F. (1992) Long lasting suppression of alcohol preference in ats following serotonin receptor blockade by ritanserin. *Brain. Res. Bulletin* 28 (3): 493-496.

Panocka, I., Ciccocioppo, R., Polidori, C. and Massi, M. (1993) The nucleus accumbens is a site of action for the inhibitory effects of ritanserin on alcohol intake in rats. *Pharmacol. Biochem. Behav.*, in press.

Rammsayer, T. and Vogel, W.H. (1991) Differential effects of a 5-HT2 receptor blocker on alcohol intake in rats selectively bred for high and low catecholamine responses to stress. *Integrative Physiol. Behav. Science* 26: 189-199.

Selzer, M.L., Vinokur, A. and Vam, R.L. (1976) A self administered short Michigan alcoholism screening test (SMAST). *J. Stud. Alcohol.* 36: 124.

Tyrer, P. (1990) Personality disorder and social functioning. In: Peck, D.F. and Shapiro, C.M. (eds): *Measuring Human Problems*, A Practical Guide. Revised DHWE Pub. (ADM) National Institue of Mental Health, Rockville, MD, pp. 218-222.

Udego, L., Grenhoff, J. and Svensson, T.H. (1989) Ritanserin, a 5-HT$_2$ receptor antagonist, activates midbrain dopamine neurons by blocking serotonergic inhibition. *Psychopharmacology* 98: 45-50.

Toward a Molecular Basis of Alcohol Use and Abuse
ed. by B. Jansson, H. Jörnvall, U. Rydberg, L. Terenius & B. L. Vallee
© 1994 Birkhäuser Verlag Basel/Switzerland

Therapeutic lessons from traditional Oriental medicine to contemporary Occidental pharmacology

Wing-Ming Keung and Bert L. Vallee

Center for Biochemical and Biophysical Sciences and Medicine, Harvard Medical School, 250 Longwood Avenue, Boston, MA 02115.

Summary

An extract of *Radix Puerariae* (RP), an herb long used in traditional Chinese medicine for alcohol addiction and intoxication, was shown to suppress the free-choice ethanol intake of ethanol-preferring Syrian golden hamsters. Two isoflavones, daidzein (4',7-dihydroxyisoflavone) and daidzin (7-glucoside of daidzein), isolated from the extract were shown to account for this effect. Daidzin administered intraperitoneally at 150 mg/kg/day suppressed free-choice ethanol intake by $\geq$ 50%. Such effect has been confirmed in a total of 79 consecutive hamsters studied over a period of more than a year. Daidzein was less potent and a higher dose (230 mg/kg/day) was required to produce similar effect. RP-, daidzin-, and daidzein-treated hamsters appeared to remain healthy and exhibited no significant change in body weight and water or food intake. *In vitro,* daidzin and daidzein inhibited human mitochondrial aldehyde dehydrogenase (ALDH-2) and $\gamma\gamma$-alcohol dehydrogenase ($\gamma\gamma$-ADH), respectively. However, at doses that suppressed ethanol intake, daidzin and daidzein had no effect on overall acetaldehyde and ethanol metabolism in hamsters. These findings clearly distinguish the action(s) of daidzin and daidzein from those of the classic, broad acting inhibitors of ALDH (e.g. disulfiram) and class I ADH isozymes (e.g. 4-methylpyrazole), and identify them as a new class of compounds that offer promise as safe and effective therapeutic agents for alcohol abuse.

Introduction

Excessive uncontrolled ethanol intake is a serious behavioral problem for some humans that has major social, economic and medical consequences. Recognition of suitable pharmacological products and their development into effective therapeutic agents for this disorder has been a major objective of alcohol research. However, the lack of a well defined biochemical and etiological basis for this disease has hampered the formulation of a rational scientific approach to the discovery of such agents. A number of traditional Chinese medicines for the empirical treatment of alcohol addiction and abuse were described in ancient Chinese pharmacopoeias. Among them, *Radix Puerariae* (RP)- and *Flos Puerariae* (FP)-based medications are still used in China by traditional medical therapists (often referred to as herbalists) for the treatment of alcohol abuse. A search among these traditional medicines for leads/remedies for a disease whose etiology is understood poorly may prove to be a fruitful approach.

Monitoring the search for new therapeutic agents for alcohol abuse requires a laboratory animal that consumes ethanol voluntarily, preferably, in large quantities. Toward this end, we have found the Syrian golden hamster (*Mesocricetus auratus*) to be exceptionally suitable because, by nature, this animal species when given a choice prefers and consumes large and remarkably consistent quantities of ethanol on a daily basis (Arvola and Forsander, 1961). Furthermore, this species exhibits excellent "predictive validity": agents that have been shown to attenuate ethanol consumption in alcohol-dependent humans also suppress ethanol intake in the golden hamster. We have used this animal model to confirm the putative antidipsotropic effect in humans given RP in China and show that two major constituents of the herb, daidzin and daidzein, primarily account for this action.

Materials and Methods

Crude RP extract, daidzin and daidzein were isolated (Keung and Vallee, 1993) or synthesized (Farkas and Varady, 1959) by published procedures. Plasma acetaldehyde was analyzed by the method of Peterson and Polizzi (1987). Plasma ethanol was measured enzymatically using a Sigma assay kit.

Animal drinking experiment

Adult male Syrian golden hamsters (Sasco, Inc., Omaha, NE, 131-135 g) were housed (4-5 per cage) in a room maintained at 23°C, 35-45% humidity, with a 12/12 hr light/dark cycle (light on 0600-1800 hr), *ad libitum* access to Purina Chow (5001) and a 15% v/v solution of ethanol in filtered (MLQ, Millipore) tap water. After a week, each hamster was transferred to an individual stainless steel metabolic cage (26x18x17.5 cm). Two 50 ml drinking bottles, one containing filtered water and the other a 15% ethanol solution, were provided continuously. The bottles were placed in holders equipped with tilted platforms which direct spillage to tubes placed outside of the cages. The positions of the two drinking bottles on each cage were alternated daily to prevent development of positional preference. Fluid intake was measured at 0900 hr each day. Hamsters that drank significant (>8 ml/day) and consistent (daily variance $< \pm 20\%$) amounts of ethanol solution were selected for drug testing. To establish baseline ethanol and water intake, 1 ml sterile saline was administered daily to each hamster between 1500-1600 hr for 6 days (saline control period, Day -5 to Day 0). A daily dose of test

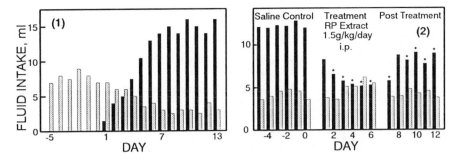

Fig. 1. Free-choice ethanol (solid bars) and water (shaded bars) intake by an adult hamster (130g). Day -5 to Day 0: only water was provided, Day 1 to Day 13: water and a 15% ethanol solution were provided.
Fig. 2. Effect of crude RP extract on free-choice ethanol (solid bars) and water (shaded bars) intake by hamsters. Values are the mean for 8 hamsters. *p<0.05.

compound was then administered for 6 days (treatment period, Day 1 to Day 6). After the last dose of test compound, ethanol and water intake were usually monitored for another 6 days (post treatment period, Day 7 to Day 12). All test compounds were dissolved or suspended in 1 ml sterile saline and administered intraperitoneally (i.p.).

Results

Free-choice water and ethanol intake

Ethanol-naive hamsters ($\sim$ 130 g) given free access to food and water consumed $\sim$8 ml of fluid each day. With free choice between water and a 15% ethanol solution, water intake declined steadily while consumption of ethanol increased concomitantly. After $\sim$ 1 week, the daily water and ethanol intake stabilized at relatively constant levels, and total fluid intake increased 2- to 3-fold. Fig. 1 shows the result obtained with one of the hamsters studied. Although the ethanol intake of different hamsters varied from 7 to 19 g/kg/day, the daily ethanol intake of each individual hamster is remarkably constant.

Effects of RP extract, daidzin, daidzein and puerarin on free-choice ethanol intake

RP extract has been used and claimed to be effective in treating alcohol addicts in China (Keung, 1991). However, to our knowledge, its efficacy has never been examined critically by Western standards. We therefore prepared a RP extract and tested its effect on free-choice ethanol intake in a group of 8 hamsters. During the saline control period, this group of hamsters consumed

374

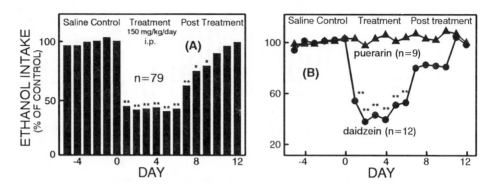

Fig. 3. Effect of (A) daidzin, and (B) daidzein and puerarin on free-choice ethanol intake by hamsters. Saline (1ml/day), daidzin (150 mg/kg/day), daidzein (230 mg/kg/day) and puerarin (150mg/kg/day) were injected i.p. The mean of ethanol intake measured during saline control period is taken as 100%. Values are the means of measurements obtained from n hamsters. **$p < 0.001$, *$p < 0.01$.

an average of ~ 12 ml of a 15% ethanol solution per day. This was suppressed significantly ($p < 0.05$) by the injection of the crude extract (Fig. 2). RP extract administered i.p. at 1.5 g/kg/day decreased ethanol intake by $\geq 50\%$. The ethanol intake remained suppressed throughout the treatment period but was partially reversed after treatment was terminated. RP extract had no significant effect on the water intake of these hamsters.

To identify the active principle(s) in RP that suppresses ethanol intake, three major constituents, daidzin, daidzein and puerarin (Fang, 1980), were examined for their effect on free-choice ethanol intake. Daidzin injected at 150 mg/kg/day suppressed ethanol intake by $\geq 50\%$ (Fig. 3A). The effect of daidzin was reversible: ethanol intake gradually returned to that of the saline control after treatment was terminated. We have now studied a total of 79 consecutive hamsters in a period of more than a year and in each instance the ethanol intake was significantly suppressed (31% to 79%) by daidzin. Suppression data obtained from this group of hamsters approximated a normal distribution with a mean $\pm$ s.d. = 57 $\pm$ 12%. Daidzein also suppressed ethanol intake in hamsters but appeared to be less potent. Thus a higher dose of daidzein (230 mg/kg/day) is necessary to produce ~ 50% suppression (Fig. 3B). Daidzin and daidzein had no effect on the water intake of the hamsters (data not shown). Puerarin, the most abundant isoflavone in RP had no effect on the free-choice ethanol intake of hamsters (Fig. 3B). Hamsters receiving RP extract, daidzin, daidzein or puerarin treatment appeared to remain healthy and did not evidence any significant change in food intake or body weight throughout

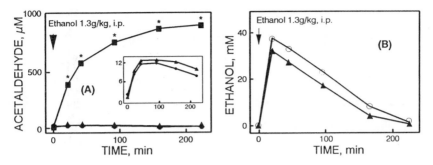

Fig. 4. Effects of (A) daidzin and disulfiram on acetaldehyde metabolism, and (B) daidzein on ethanol metabolism. Hamsters were treated with 6 doses of daidzin (150mg/kg/day, •), disulfiram (150mg/kg/day, ■), daidzein (230mg/kg/day, ○), or saline (1ml/day, ▲). A test dose of ethanol was injected 12 hr after the last dose of agents were administered. Values are means for 6-9 hamsters. *p<0.01.

the experiment (data not shown).

Both daidzin and daidzein are only slightly soluble in aqueous solution and, hence, had to be administered as suspensions in sterile saline. To rule out that their effect on ethanol intake was due to stress induced by injection of crystalline suspensions of these isoflavones into the peritoneum, two flavones (chrysin and 7,8-dihydroxyflavone) which like daidzin and daidzein are virtually insoluble in water, were studied as controls. Both compounds were administered as a suspension in 1 ml sterile saline at the same dose as daidzin (150 mg/kg/day). Neither of them suppressed ethanol intake.

Acetaldehyde metabolism in daidzin- and disulfiram-treated hamsters
Daidzin has been shown recently to inhibit human mitochondrial ALDH selectively (Keung and Vallee, 1993). Since ALDH inhibitors, such as disulfiram are thought to suppress ethanol intake by inhibiting acetaldehyde metabolism (Johansson *et al.*, 1991), we examined the effect of daidzin on acetaldehyde metabolism in hamsters. Disulfiram- and saline-treated hamsters were also studied for comparison. After receiving a test dose of ethanol, disulfiram-treated hamsters registered very high levels of plasma acetaldehyde which reached ~900 μM after 4 hr and showed no sign of decline (Fig. 4A). In contrast, the plasma acetaldehyde levels of daidzin-treated hamsters challenged with the same dose of ethanol stayed low (<12 μM) and were not significantly different (p>0.05) from those observed for the saline-treated controls (Fig. 4A, insert).

Ethanol elimination in daidzein-treated hamsters

Daidzein does not inhibit human ALDH isozymes but selectively and reversibly inhibits human γγ-ADH (Keung 1993). To determine whether or not daidzein suppresses ethanol intake by inhibiting ethanol metabolism we examined its effect on ethanol elimination in hamsters. Ethanol tolerance curves obtained for the saline- and daidzein-treated hamsters are shown in Fig. 4B. No significant difference ($p > 0.05$) was found in the peak plasma ethanol level attained and in the ethanol elimination rates estimated from the two ethanol tolerance curves (0.18 mM/min).

Discussion

RP, an herbal medicine prepared from the root of the *Leguminosae Pueraria lobata* (commonly known as kudzu), was first described in *Shen-nung Pen-t'sao Ching* (ca. 200BC) as a medication safe for human consumption with antipyretic, antidiarrheic, diaphoretic and antiemetic properties. The use of RP and FP (the flower of kudzu) as "anti-drunkenness" agents was described ca. 600 A.D. in *Beiji-Qianjin-Yaofeng*. Treatment of alcohol addiction with a FP-based medication was documented about 1000 years later in *Lan-tai-kuei-feng*. Today, FP- and RP-based medications are still used in China both for this and other purposes and are described as safe and effective by traditional physicians (Keung, 1991). The present study is the first to confirm the "putative" antidipsotropic effect of RP in an ethanol-preferring laboratory animal.

Syrian golden hamsters, in contrast to most outbred strains of laboratory rodents, display a high preference for and consume large quantities of ethanol on a simple two-bottle free-choice regimen (Arvola and Forsander, 1961), as confirmed in this study. However, relatively few investigators have used hamster for alcohol research largely because it lacks "face validity": after prolonged consumption of large quantities ($\sim$ 12-20 g/kg/day) of ethanol, golden hamsters do not develop behavioral patterns, symptoms and signs (Fitts and Dennis, 1981) that have been postulated as criteria for a "good" animal model of "alcoholism" (Keane and Leonard, 1989). The term "alcoholism" is meant to include both the behavioral disorder of excessive, uncontrolled intake of ethanol, and the clinical abnormalities that are the consequences of alcohol abuse. Excessive, uncontrolled ethanol intake may or may not necessarily lead to medical or neurologic illnesses that require medical treatment but it may nevertheless interfere seriously with normal marital, social and economic life. Thus, an agent that abolishes or significantly suppresses the desire to drink ethanol without producing undesirable side effects would seem to be most desirable for the treatment of "alcoholism". On this basis, the key feature of an animal species to monitor the search for such agent(s) would not be its "face validity" but rather its capacity to predict whether or not an agent will effectively suppress ethanol consumption in alcohol-dependent humans, i.e., its "predictive validity" (Sinclair, 1987). The finding that RP, a "putative" antidipsotropic agent that has been used safely and effectively for hundreds of years in humans who abuse alcohol also suppresses the free-choice ethanol intake of hamsters indicates that this animal system can be used effectively in the search for and identification of novel pharmacological agent(s) for alcohol abuse. Toward this end, we also examined the effects on free-choice ethanol intake in hamsters of pharmacological agents that have been shown to reduce craving for alcohol in alcohol-dependent humans (Litten and Allen, 1991). The results (Table I) indicate that suppression of ethanol intake in hamsters is consistent with the beneficial effects of these agents observed in alcohol-dependent humans and thus confirm the "predictive validity" of the hamster model.

Two major constituents of RP, daidzin and daidzein, have been shown to suppress the free-choice ethanol intake of hamsters (Fig. 3A and 3B). Daidzin appears to be the more potent with a ED_{50} value of ≤ 150 mg/kg (0.36 meq/kg), 2- to 3-times lower than that of ~ 230 mg/kg (0.9 meq/kg) for daidzein. The antidipsotropic effect of daidzin has been confirmed in a total of 79 consecutive alcohol preferring hamsters studied over a period of more than a year. Changes in seasons and differences in batches of animals appeared to have no significant effect on ethanol intake and its suppression by daidzin.

Daidzin and daidzein are isoflavones, a group of natural products found mostly in leguminous plants (Harborne, 1986). The pharmacological effects of isoflavones on animals have been studied since the 1950s when a weak estrogenic activity was first detected in most isoflavone aglycones including daidzein (Cheng et al., 1955). Since then, isoflavones have been claimed to have antifebrile, antispasmodic, antihypertensive and antidysrhythmic activities (Nakamoto et al., 1977; Fan et al.,1985). However, an effect of isoflavones on ethanol drinking behavior has never been reported. Biochemically, isoflavones in general, and genistein in particular, inhibit a number of enzymes including DNA topoisomerase II (Okura et al., 1988) and tyrosine- (Akiyama et al., 1987) and histidine- (Huang et al., 1992) specific protein kinases. Genistein

Table I. Effect of some CNS agents on ethanol intake in golden hamsters

Agents	Dose (mg/kg/day)[a]	Suppression, %[b] (n)
Bromocryptine	5	42 ± 11 (4)
buspirone	20	41 ± 12 (4)
Lithium Carbonate	38	56 ± 4 (4)
Zimelidine	38	40 ± 5 (4)

[a] Doses are 10-20 times of those tested in human. [b] % suppression = $(V_c - V_t) \times 100/V_c$, where V_c = the mean of ethanol intake measured during saline control period, and V_t = the mean of ethanol intake measured during treatment period. Values are mean ± S.E.M for n hamsters.

also inhibits endothelial cell proliferation and *in vitro* angiogenesis (Fotsis *et al.*, 1993). Daidzin and daidzein affect some of these activities, but only at high, > 50μM, concentrations.

We have shown recently that daidzin is a potent (K_i = 40 nM) and selective inhibitor of human mitochondrial ALDH (Keung and Vallee, 1993). Alcohol abuse is rare among the ~50% of Asians who have inherited an inactive variant form of this isozyme, and it is widely accepted that the high level of blood acetaldehyde that accumulates when these individuals ingest ethanol acts as a deterrent to further ethanol intake (Goedde and Agarwal, 1990). Similarly, alcohol-sensitizing agents like disulfiram and other inhibitors of ALDH are thought to suppress ethanol intake by inhibiting the metabolism of acetaldehyde which leads to its accumulation to aversive levels after consuming ethanol (Johansson *et al.*, 1991). However, we have found that daidzin has no effect on overall acetaldehyde metabolism in hamsters and, hence, its action must be different from that suggested for the classic, broad-acting ALDH inhibitors, such as disulfiram (Fig. 4). This is consistent with what we have learned from traditional Chinese physicians in China: In their experience, RP treatment does not induce a "hypersensitive" or flushing response to alcohol (Keung, 1991).

Daidzein does not inhibit human ALDH isozymes. Instead, it inhibits the class I isozymes of human ADH, preferentially, the γ-type ADH isozymes (Keung, 1993). Another ADH inhibitor, 4-methylpyrazole which inhibits all human class I ADH isozymes, has been shown to suppress ethanol intake in P rats by inhibiting the metabolic elimination of ingested ethanol (Waller *et al.*, 1982). Daidzein, however, at a dose that significantly suppresses ethanol intake has no effect on the overall ethanol metabolism in hamsters. Thus daidzein must suppress ethanol intake via a mechanism that is different from that of 4-methylpyrazole. At present, the mode of action of daidzin and daidzein is not known.

The physical, chemical and enzymatic properties of human and other mammalian ADH and ALDH isozymes have been explored thoroughly. Importantly, ADH and ALDH isozymes exhibit different specificities toward a wide spectrum of alcohols and aldehydes other than ethanol and acetaldehyde. Among these, many have significant physiological roles and it is possible that some may be involved in the mediation of ethanol drinking behavior. Indeed it may be that interference by daidzin and daidzein with these normal metabolic processes causes the decrease of ethanol consumption in hamsters. In this context, it should be noted that the structural analogs puerarin, chrysin and 7,8-dihydroxyflavone which inhibit neither ADH (Keung 1993) nor ALDH (Keung and Vallee, 1993), do not suppress ethanol intake. This finding does not prove but is consistent with the notion that a physiological pathway catalyzed by the γ-type ADH and mitochondrial ALDH may play a critical role in the regulation of ethanol intake.

The metabolic fates of daidzin and daidzein in hamsters are as yet unknown. Also it is not known whether daidzin and daidzein themselves are the pharmacologically active molecules that suppress ethanol intake directly or whether they serve as pro-drugs. Thus, to uncover the site(s) and mechanism(s) of action of daidzin and daidzein it will be important to learn the metabolic fates and identify the active species of these isoflavones. Results from these studies will not only allow a better understanding of the biochemical basis for the action of these drugs on free-choice ethanol intake in hamsters and thereby shed light on the mechanistic basis underlying alcohol abuse in humans, but will also provide a rationale for the design of much needed, safe therapeutic agents for alcohol addiction.

References

Akiyama, T., Ishida, J., Nakagawa, S., Ogawara, H., Watanabe, B., Itoh, N., Shibuya, M. and Fukami, Y. (1987) Genistein, a specific inhibitor of tyrosine-specific protein kinase. *J. Biol. Chem.* 262: 5592-5595.

Arvola, A. and Forsander, O. (1961) Hamsters in experiments of free choice between alcohol and water. *Nature* 191: 819-820.

Cheng, E.W., Yoder, L., Story, C.D. and Burroughs, W. (1955) Oestrogenic activity of some naturally occurring isoflavones. *Ann. N Y Acad. Sci.* 61: 637-736.

Fan, L.L., Zhao, D.H., Zhao, M.Q. and Zeng, G.Y. (1985) The antidysrhythmic effect of

380

puerariae isoflavones. *Acta Pharma. Sin.* 20(9): 647-651.

Fang, Q. (1980) Some current study and research approaches relating to the use of plants in the traditional Chinese medicine. *J. Ethnopharmacol.* 2: 57-63.

Farkas, L., and Varady, J. (1959) Synthese des daidzins und des ononins. *Ber. Dtsch. Chem. Ges.* 92: 819-821.

Fitts, D.A., and Dennis, C. (1981) Ethanol and dextrose preferences in hamsters. *J. Stud. Alcohol* 42: 901-907.

Fotsis, T., Pepper, M., Adlercreutz, H., Fleischmann, G., Hase, T., Montesano, R. and Schweigerer, L. (1993) Genistein, a dietary-derived inhibitor of in vitro angioesis. *Proc. Natl. Acad. Sci. U.S.A.* 90: 2690-2694.

Goedde, H.W., and Agarwal, D.P. (1990) Pharmacogenetics of aldehyde dehydrogenase (ALDH). *Pharmac. Ther.* 45: 345-371.

Harborne, J.B. (1986) The role of phytoalexins in natural plant resistance. In: Green, M.B. and Hedin, P.A. (eds.): *Natural Resistance of Plants to Pests.* American Chemical Society, p. 22.

Huang, J., Nasr, M., Kim, Y. and Matthews, H.R. (1992) Genistein inhibits protein histidine kinase. *J. Biol. Chem.* 267: 15511-15515.

Johansson, B., Angelo, H.R., Christensen, J.K., Moller, I.W. and Ronsted, P. (1991) Dose-effect relationship of disulfiram in human volunteers. *Pharmacol. Toxicol.* 68: 166-170.

Keane, B., and Leonard, B.E. (1989) Rodent models of alcoholism: a review. *Alcohol Alcohol.* 24: 299-309.

Keung, W.M. (1991) It is not a general practice of the traditional Chinese physicians (herbalists) to publish or otherwise publicize their modes of therapy. As a consequence, written clinical records on the identity of drugs, their efficacy, dosage and side effects are not readily available. To collect more information on the use of RP as remedial agent for alcohol abuse, WMK visited China and interviewed a number of herbalists and research scientists. The herbalists interviewed collectively recalled that of the thousands of patients treated with RP- or FP-based medications, ~ 300 cases were related to alcohol abuse. In all of these cases, the medications were considered effective both in controlling and suppressing appetite for alcohol and improving the general health of the patients. Significant improvement is usually said to be observed within a week of treatment. After 2-4 weeks, most patients (~ 80%) no longer experienced a craving for alcohol. Alleviation of some of the alcohol induced damage of some vital organs usually required longer treatment (4-6 months). No adverse side effects associated with the use of RP or FP have ever been reported. Moreover, patients receiving RP treatment for other purposes do not develop flushing in response to alcohol. It is of interest to note that RP or FP are not prescribed for any purposes to individuals who exhibit a flushing response to alcohol. Although the herbalists recognize "flushers" as a group, they are unaware of the biochemical basis for this well-known clinical state.

Keung, W.M., and Vallee, B.L. (1993) Daidzin: a potent, selective inhibitor of human mitochondrial aldehyde dehydrogenase. *Proc. Natl. Acad. Sci. U.S.A.* 90: 1247-1251.

Keung, W.M. (1993) Biochemical studies of a new class of alcohol dehydrogenase inhibitors from *Radix puerariae. Alcohol. Clin. Exp. Res.* in press.

Litten, R.Z., and Allen, J.P. (1991) Pharmacotherapies for alcoholism: promising agents and clinical issues. *Alcohol. Clin. Exp. Res.* 15: 620-633.

Nakamoto, H., Iwasaki, Y. and Kizu, H. (1977) The study of aqueous extract of *Puerariae*

Radix. IV. The isolation of daidzin from the active extract (MTF-101) and its antifebrile and spasmolytic effect. *Yakugaku Zasshi* 97(1): 103-105.

Okura, A., Arakawa, H., Oka, H., Yoshinari, T. and Monden, Y. (1988) Effect of genistein on topoisomerase activity and on the growth of [Val 12]Ha-ras-transformed NIH 3T3 cells. *Biochem. Biophys. Res. Commun.* 157: 183-189.

Peterson, C.M., and Polizzi, C.M. (1987) Improved method for acetaldehyde in plasma and hemoglobin-associated acetaldehyde: results in teetotalers and alcoholics reporting for treatment. *Alcohol* 4: 477-480.

Sinclair, J.D. (1987) The feasibility of effective psychopharmacological treatments for alcoholism. *British J. Addict.* 82: 1213-1223.

Waller, M.B., McBride, W.J., Lumeng, L. and Li. T.-K. (1982) Effects of intravenous ethanol and of 4-methylpyrazole on alcohol drinking in alcohol-preferring rats. *Pharmacol. Biochem. Behav.* 17: 763-768.

Toward a Molecular Basis of Alcohol Use and Abuse
ed. by B. Jansson, H. Jörnvall, U. Rydberg, L. Terenius & B. L. Vallee
© 1994 Birkhäuser Verlag Basel/Switzerland

Potential gene therapy for alcoholism

Roscoe O. Brady

Developmental and Metabolic Neurology Branch, National Institute of Neurological Disorders and Stroke, National Institutes of Health, Bethesda, Maryland 20892, USA

Summary

Genes that have an actual or a potential relationship to alcoholism may be useful targets for therapy. Candidate genes are considered in relationship to family studies, differences in alcohol preferences in various rodent strains, biochemical reactions, physiologic response mechanisms, and alterations in brain pharmacology. Suggestions are made concerning the identification of candidate genes, design of gene antisense constructs, and techniques for their organ-specific delivery. The complexities surrounding alcoholism in humans make it likely that several simultaneous approaches may be required for effective therapy for alcoholism.

Introduction

That there are genetic influences involved in alcoholism appears indisputable. Convincing evidence has been forthcoming from studies of alcohol ingestion by identical and fraternal male twins raised in disparate environments. Various strains of rats and mice clearly exhibit differences in the preference or avoidance of alcohol. Moreover, such differences may not only be inherent in the biochemical or pharmacological events involved in consumption of alcohol, but perhaps in the acquisition (learning) of these traits as well (Gauvin *et al.*, 1993). If the latter concept is valid, and if it extends to humans, it raises the level of sophistication required to penetrate the etiology of high levels of alcohol ingestion and the development of alcoholism.

Conceptual background

Alcohol consumption by various strains of rodents

Evidence of genetic influences regarding the quantity of alcohol consumed by mammals became apparent during the identification and development of specific strains of rats and mice that exhibit strong preference to, or the avoidance of, alcohol such as the University of Chile UChA/UChB lines (Mardones *et al.*, 1983), the Alko (Finland) AA preferring and ANA non-preferring lines (Eriksson, 1969), the Indiana University P preferring and NP non-preferring lines, the HAD/LAD lines (Li *et al.*, 1981, 1991), the Sardinian sP and sNP lines (Fadda, *et al.*, 1989), and alcohol preferring fawn hooded rats (Rezvani *et al.*, 1990). This deduction is enhanced by human studies of male twins raised in separate environments where an hereditary predisposition to high alcohol consumption appears to have been demonstrated (Goodwin *et al.*, 1973; Cloninger *et al.*, 1981).

Possible role of brain opioids on alcohol consumption

In certain situations, genetic component(s) involved in alcohol preference by rodents may be subserved by reinforcing pharmacological changes following chronic ingestion of high levels of alcohol. For example, Sprague-Dawley rats that did not consume alcohol when it was offered to them in a free choice paradigm where either water or a 20% (v/v) solution of ethanol were freely available from measured drinking tubes, could be induced to consume large quantities of alcohol if placed on a diet deficient in the B-vitamin complex (Mardones and Onfray, 1942). Addition of the missing vitamins temporarily lowered the high alcohol consumption. However, on supplementation with full vitamin B complex (Fig. 1, Arrow 1), rats previously conditioned to high levels of alcohol intake gradually resumed ingesting large quantities of alcohol (Brady and Westerfeld, 1947). After adding further nutritional supplements at arrow 2, or if given normal rat chow diet, alcohol consumption by the rats did not completely return to the baseline value, and it rapidly returned to a high level. The fact that the rats temporarily ceased drinking large quantities of alcohol suggests that impairment of lipid or carbohydrate metabolism in the liver or other organs probably was not a dominant factor in the resumption of alcohol ingestion. The recurrence of large amounts of alcohol consumption and the failure to restore alcohol intake to the low initial levels of ingestion suggest that this amount of alcohol may have activated some reward mechanism in the brain.

Following the publication of these observations, a considerable amount of evidence has accrued leading to the deduction that increases of endogenous brain opioid levels may potentiate alcohol-seeking behavior. This reasoning is supported by the finding that ingestion of alcohol can increase brain enkephalin (Schultz *et al.*, 1980; Seizinger *et al.*, 1983) as well as β-endorphin (Cheng and Tseng, 1982). Furthermore, naltrexone suppressed the intravenous self-administration of alcohol in monkeys (Altschuler *et al.*, 1980) and naloxone reduced alcohol drinking by alcohol-preferring rats (Froehlich *et al.*, 1990). This concept was further

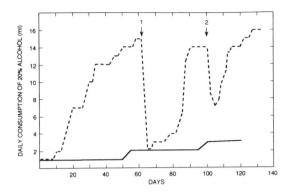

Fig. 1. Voluntary consumption of alcohol by Sprague-Dawley rats. The solid line is the average daily consumption of a 20% solution of alcohol by rats on basal rat chow diet. The dashed line is the average consumption of 20% alcohol by rats on a vitamin B-deficient diet. Principal components of the Vitamin B group were added at arrow 1. Additional defined supplements plus Wilson's Liver Fraction L were added at arrow 2. (Redrawn from Figs. 1 and 2, Brady and Westerfeld, 1947).

strengthened by the observation that ICI 174864, a highly selective inhibitor of the delta subtype of opioid receptors, decreased alcohol ingestion in an auto-selection paradigm (Froehlich *et al.*, 1991). Additional evidence for the involvement of endogenous opioids was adduced from a corollary experiment where the catabolism of enkephalins was delayed by the administration of thiorphan, an inhibitor of enkephalinase activity (Roques *et al.*, 1980). Administration of thiorphan led to a temporary increase in the voluntary alcohol consumption by rats (Froehlich *et al.*, 1991) implicating an elevation of brain opioids on alcohol-seeking behavior. This deduction seems consistent with the findings illustrated in Fig. 1.

Mutations in mitochondrial aldehyde dehydrogenase gene and alcohol consumption

It has been known for a number of years that certain individuals, particularly those of Oriental extraction, experience flushing of the face, neck and chest following the consumption of even small quantities of alcoholic beverages. These signs may be accompanied by tachycardia, headache, nausea, hypotension, drowsiness and even urticaria and asthma (Thomasson *et al.*, 1993). These signs were shown to follow slow intravenous infusions of acetaldehyde (Asmussen *et al.*, 1948). A probable role of abnormal increases of blood acetaldehyde level was supported by investigations by Mizoi *et al.* (1979). A causative genetic influence in certain humans was strongly indicated by studies of alcohol consumption by Orientals. Oxidation of acetaldehyde to acetate is considered to be primarily catalyzed by an aldehyde dehydrogenase in mitochondria that has a low K_m for this substrate. This isozyme is known as ALDH2. The gene for ALDH2 is located on human chromosome 12. Individuals with a mutation in the gene for mitochondrial acetaldehyde dehydrogenase (ALDH2) that reduces the catalytic activity of this enzyme experience the untoward effects of increased levels of blood acetaldehyde (flushing, etc). This reaction is thought to exert a strong deterring influence on the ingestion of alcohol by individuals with the mutated gene. The mutation in the gene for ALDH2 resulted in substitution of the amino acid lysine for glutamate at position 487 (^{487}Glu→Lys) of the 500 amino acid mature polypeptide chain of ALDH2 (Yoshida *et al.*, 1984). The mutation occurs in exon 12 of the 13 exons of ALDH2. This alteration appears to be a dominant trait (Schwitters *et al.*, 1982; Crabb *et al.*, 1989), and the mutated allele is designated as ALDH2*2. The aversion to alcohol is even more predominant in persons who are homozygous for the ALDH2*2 allele. As compelling as this evidence appears to be, it should be noted that other factors over and above the contribution of the ALDH2*2 allele may contribute to the ingestion of or aversion to alcohol (Thomasson *et al.*, 1993). One of these, another genetic alteration, may involve various alleles of alcohol dehydrogenase.

Further support for a role of ALDH2 in controlling alcohol ingestion is found in the observation that low-alcohol consuming rodents have higher acetaldehyde levels than the strains that ingest larger quantities of alcohol (summarized by Sinclair and Lindros, 1981). A possible inverse aspect of such a genetic influence on alcohol consumption has been uncovered in studies in alcohol-preferring rats. A mutation in murine ALDH2 has been reported in alcohol preferring (P) rat lines at amino acid position 67 where glutamine has largely replaced arginine in the alcohol non-preferring (NP) rats (^{67}Arg→Gln) (Carr *et al.*, 1991). Glutamine is normally the amino acid at this site in Sprague-Dawley rats. However, arginine is found in the human ALDH2 sequence at amino acid position 67. The significance of this polymorphism in the P and NP rats remains to be determined. It has been speculated that <u>acceleration</u> of acetaldehyde oxidation by the P rats might "reduce aversive effects of acetaldehyde" (Carr *et al.*, 1991).

Technological developments

Subtractive hybridization

An important emerging technology that may prove useful for identifying differences in the genetic background or even the actual genes involved in alcohol preference is the use of subtractive hybridization (Listisyn *et al.*, 1993; Myers, 1993). Genetic distinction(s) between the UChA and UChB, AA and ANA, P and NP, HAD/LAD, and the Sardinian sP and sNP lines may be discerned by this technique. This technology might initially be easier to use with inbred mouse lines, but progress is being made in its application to rat strains. DNA differences in specific organs such as the brain or liver should be examined. Once a difference is noted, it should be fairly straightforward to identify the gene and the gene product. Previously unsuspected genes may play a role in limiting excessive consumption of alcohol. For example, altered function of a chaperonin may impair the import of ALDH2 into mitochondria leading to a physiological reduction of aldehyde dehydrogenase activity. If animals are as sensitive as humans to increased levels of acetaldehyde, reduction of ALDH2 activity for any of several reasons may exert a restraint on alcohol ingestion. Important and eventually therapeutically useful leads may be obtained in this fashion. DNA subtraction studies to elucidate the genetics of alcohol preference by rodents provide for the exploration of several genetic alternatives and appear to have considerable potential.

Antisense DNA and RNA

One of the important emerging possibilities of applied molecular biology is the use of antisense molecules to block the expression of specific genes. Therapeutic applications of this approach are under development for cancer and AIDS as well as a number of other conditions (Murray, 1992; Crooke and Lebleu, 1993). It seems likely that appropriate use of antisense technology may have an important role in the treatment of alcoholism. This deduction stems from the aversion to alcohol that is produced by inhibitors of acetaldehyde dehydrogenase such as disulfiram, calcium cyanamide and other agents (Sinclair and Lindros, 1981). Oligonucleotide phosphorothioates are frequently used to block gene expression. A recent study has indicated that sequence-specific antisense oligomers of 28 nucleotides in length directed to the *gag* and *rev* mRNA's of HIV-1 inhibited virus replication at a concentration of $1\mu M$ over a period of 3 months (Lisziewicz *et al.*, 1993). Moreover, after the initial dose at this level, the inhibition of virus production continued at doses of $0.1\mu mM$ and $0.01\mu M$ of the antisense agents. Of particular interest was the claim that these compounds were not cytotoxic.

Ribozymes

Another molecular biology strategy that is receiving much attention at the present time is the design and use of ribozymes to cleave specific intracellular RNA transcripts (Chrisey *et al.*, 1991). Viral vectors and non-viral delivery systems such as cationic liposomes are under investigation for the intracellular delivery of these potentially important agents.

Applications

Site-directed antisense oligonucleotides to ALDH2 mRNA

Since the cDNA for human ALDH2 has been cloned and extensively characterized (Hsu *et al.*, 1985) and the genomic structure is known (Hsu *et al.*, 1988), it should be possible to select an appropriate target site to which an antisense oligonucleotide might be directed. In particular, nucleotide sequences that code for the suggested coenzyme-binding fold from the amino acid alanine at +194 to isoleucine at +232 may be an appropriate target (Hempel *et al.*, 1985). Alternatively, antisense oligomers may be directed to nucleotides at the catalytically active site of the enzyme which is considered to be a hydrophobic cleft that involves the cysteine at +302 and another cysteine residue (Hempel *et al.*, 1985). Delivery of the antisense oligonucleotide(s) may require applications of glycobiology similar to that which was developed for the successful treatment of patients with Gaucher's disease (Brady and Furbish, 1982; Barton *et al.*, 1991). Here the exogenous glucocerebrosidase had to be targeted to macrophages such as the Kupffer cells in the liver. In treating alcoholism, one would expect that hepatocytes would be the primary target. In this case, taking advantage of the high-affinity galactose lectin on the surface of hepatocytes would seem advisable. The antisense oligomer could be complexed with polylysine coupled to asialo-orosomucoid (Wu and Wu, 1987) or to a galactose-terminal neoglycoprotein such as galactosylated human serum albumin. The delivery to hepatocytes may be improved by incorporating binding-incompetent adenovirus into the conjugate (Michael *et al.*, 1993). These investigators used a monoclonal anti-fiber antibody to block adenovirus binding. The adenovirus-antibody-ligand complex prevents lysosomal degradation of internalized complexes in endosomes (Fig. 2). Encapsulation in small liposomes may be another way to deliver the antisense oligomer to hepatocytes (Aliño *et al.*, 1993). Once the target site has been identified and the strategy for the delivery of appropriate antisense oligomers to hepatocytes has been developed, one could determine whether hepatic ALDH2 activity is reduced by the administration of the oligomer to rodents. If it is, the effect of this molecular strategy could be readily examined in alcohol-preferring rats.

Ribozymes

The use of ribozymes to reduce the activity of ALDH2 by inactivation of ALDH2 mRNA seems to be a more distant prospect. If suitable delivery and targeting strategies can be developed, it is likely that this technology will become more widely applied. The principal reason for this assumption is that only small amounts of the appropriate ribozyme would have to be delivered

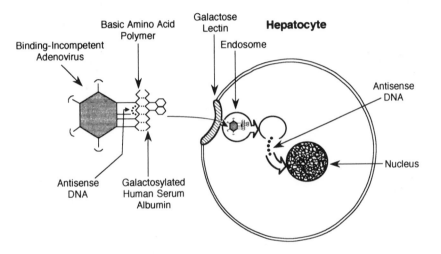

Fig. 2. Proposed delivery of ALDH2 DNA antisense oligomer to hepatocytes. The antisense component is targeted to hepatocytes by a galactosylated human serum albumin neoglycoprotein (Gal-HSA) via the galactose lectin on the plasma membrane of hepatocytes. Adenovirus may be included in the polybasic amino acid-Gal-HSA-antisense oligomer complex to disrupt endosomes. Disruption of endosomes should permit delivery of the antisense oligomer to the cytoplasm without lysosomal degradation and eventually to the nucleus of hepatocytes.

to the target cell (hepatocytes) since ribozymes act catalytically. If liposomes were to be used, one certainly could conceive of targeting them to hepatocytes by incorporation into, or by covalent attachment of, galactose-terminal oligosaccharides to the outer lipid leaflet (Das et al., 1985).

Discussion

The suggestion for the use of antisense nucleotides to the ALDH2 mRNA to reduce alcohol consumption is based on four premises. The first is that reduction of ALDH2 activity will elicit an aversion to alcohol. The second is that the requisite antisense molecules are non-toxic. The third is that the oligomer can be effectively delivered to hepatocytes, the major site of alcohol metabolism. The fourth is the assumption that knock-out of the ALDH2 gene may not necessarily have to be complete or permanent. Once an effective reduction of ALDH2 activity is obtained, progressively smaller doses of the antisense oligonucleotide may be used until pathologic drinking is brought under control. Whether the recipient will escape and resume high levels of alcohol consumption and require resumption of the treatment will probably require tailoring to particular individuals. An additional strategy for an attack on the mRNA for ALDH2 may be the construction and delivery of an appropriate ribozyme. The cellular delivery of ribozymes is currently less well developed than that for antisense DNA molecules. However, this approach should be kept in mind because delivery of very small quantities of ribozymes may be all that is necessary. Ribozymes may therefore eventually be used to reduce ALDH2 activity, or that of another appropriate enzyme, or even to alter a protein component of δ-opioid receptors on neurons. Finally, either antisense DNA or ribozyme RNA may eventually be produced by introducing genes that code for these agents. Gene therapy of this sort may well preclude recidivism that occurs with considerable frequency when the treatment of alcoholics is attempted with pharmacological agents such as Antabuse$^{®}$

It seems likely that along with antisense or ribozyme therapy a number of additional supportive measures should be provided. These ancillary elements might including psychological counseling, restoration of caloric and nutritional status, and comprehensive B-vitamin supplementation. Should these measures not be sufficient, one should consider the use of δ-opioid receptor inhibitors such as naltrexone (Trexan^{TM}) or naloxone (Narcan$^{®}$). The latter agents would seem to assist in overcoming the pleasure-like response(s) that have been engendered during periods of high alcohol consumption. Inhibitors of δ-opioid receptors may be particularly necessary during the period of decay of endogenous ALDH2 gene expression before a sufficient reduction of mitochondrial acetaldehyde dehydrogenase activity has occurred to discourage alcohol abuse.

Conclusions

A combination of molecular biological and pharmacological strategies is proposed for the treatment of alcoholism. Such a strategy will only be beneficial if it can be shown to be non-injurious. To this end, successful application of the antisense or ribozyme arms of this plan in animal models such as alcohol-preferring rats or mice would seem to be a reasonable prerequisite before attempting this treatment in humans. In any event, the present proposal can serve as a point of departure for gene therapy for alcoholism. It seems likely that additional candidate genes will appear with time. The delivery and targeting strategies suggested here may have considerable utility for other efforts that may be applied in the future for gene therapy for alcoholism.

References

Altshuler, H.I., Phillips, P.E. and Feinhandler, D.A. (1980) Alteration of ethanol self-administration by naltrexone. *Life Sci.* 26: 679-688.

Aliño, S.F., Bobadilla, M., Garci-Sanz, M., Lejarreta, M., Unda, F. and Hilario, E. (1993) *In vivo* delivery of human α 1-antitrypsin gene to mouse hepatocytes by liposomes. *Biochem. Biophys. Res. Commun.* 192: 174-181.

Asmussen, E., Hald, J. and Larsen, V. (1948) The pharmacological action of acetaldehyde on human organism. *Acta Pharmacol. Toxicol.* 4: 311-320.

Barton, N.W., Brady, R.O., Dambrosia, J.M., DiBisceglie, A.M., Doppelt, S.H., Hill, S.C., Mankin, H.J., Murray, G.J., Parker, R.I., Argoff, C.E., Grewal, R.P. and Yu, K.-T. (1991) Replacement therapy for inherited enzyme deficiency - macrophage-targeted glucocerebrosidase for Gaucher's disease. *N. Engl. J. Med.* 324: 1464-1470.

Brady, R.O. and Furbish, F.S. (1982) Enzyme replacement therapy: specific targeting of exogenous enzymes to storage cells. In: Martonosi, A.N. (ed.): *Membranes and Transport*, Vol. 2, Plenum Publishing Corp., New York, pp. 587-592.

Brady, R.O. and Westerfeld, W.W. (1947) The effect of B-complex vitamins on the voluntary consumption of alcohol by rats. *Quart. J. Stud. Alcohol* 7: 499-505.

Carr, L.G., Mellencamp, R.J., Crabb, D.W., Weiner, H., Lumeng, L. and Li, T.-K. (1991) Polymorphism of the rat liver mitochondrial dehydrogenase cDNA. *Alcohol. Clin. Exp. Res.* 15: 753-756.

Cheng, S.T. and Tseng, L.F. (1982) Chronic administration of ethanol on pituitary and hypo thalamic β-endorphin in rats and golden hamsters. *Pharmacol. Res. Commun.* 14: 1001-1008.

Chrisey, L., Rossi, J. and Sarver, N. (1991) Ribozymes: progress and prospects of catalytic RNA as therapeutic agents. *Antisense Res. Develop.* 1: 57-63.

Cloninger, C.R., Bohman, M. and Sigvardsson, S. (1981) Inheritance of alcohol abuse. *Arch. Gen. Psychiatry* 38: 861-868.

Crabb, D.W., Edenberg, H.J., Bosron, W.F. and Li, T.-K. (1989) Genotypes for aldehyde dehydrogenase deficiency and alcohol sensitivity. The inactive ALDH2^2 allele is dominant.

J. Clin. Invest. 83: 314-316.

Crooke, S.T. and Lebleu, B. (1993) *Antisense Research and Applications.* CRC, Boca Raton.

Das, P.K., Murray, G.J., Zirzow, G.C., Brady, R.O. and Barranger, J.A. (1985) Lectin-specific targeting of β-glucocerebrosidase to different liver cells via glycosylated liposomes. *Biochem. Med.* 33: 124-131.

Eriksson, K. (1969) The estimation of heritability for the self-selection of alcohol in the albino rat. *Ann. Med. Exp. Biol. Fenn.* 47: 172-174.

Fadda, F., Mosca, E., Colombo, G. and Gessa, G.L. (1989) Effect of spontaneous ingestion of ethanol on brain dopamine metabolism. *Life Sci.* 44: 281-287.

Froehlich, J.C. and Li, T.-K. (1990) Enkephalinergic involvement in voluntary alcohol drinking. In: Reid, I. (ed.): *Opioids, bulemia, alcohol abuse and alcoholism.* Springer-Verlag, New York, pp. 217-238.

Froehlich, J.C., Zweifel, M., Harts, J., Lumeng, L. and Li, T.-K. (1991) Importance of delta opioid receptors in maintaining high alcohol drinking. *Psychopharmacol.* 103: 467-472.

Gauvin, D.V., Moore, K.R. and Holloway, F.A. (1993) Do rat strain differences in ethanol consumption reflect differences in ethanol sensitivity of the preparedness to learn? *Alcohol* 10: 37-43.

Goodwin, D.W., Schulsinger, F., Hermansen, L., Guze, S.B. and Winokur, G. (1973) Alcohol problems in adoptees raised apart from alcoholic biological parents. *Arch. Gen. Psychiatry* 28: 238-243.

Hempel, J., von Bahr-Lindström, H. and Jörnvall, H. (1985) Mitochondrial aldehyde dehydrogenase from human liver. Primary structure, differences in relation to the cytosolic enzyme, and functional correlations. *Eur. J. Biochem.* 153: 13-28.

Hsu, L.C., Tani, K., Fujiyoshi, T., Kurachi, K. and Yoshida, A. (1985) Cloning of cDNA's for human aldehyde dehydrogenases 1 and 2. *Proc. Natl. Acad. Sci. USA* 82: 3771-3775.

Hsu, L.C., Bendel, R.E. and Yoshida, A. (1988) Genomic structure of the human mitochondrial aldehyde dehydrogenase gene. *Genomics* 2: 57-65.

Li, T.-K., Lumeng, L., McBride, W.J. and Waller, M.B. (1981) Indiana selection studies on alcohol-related behavior. In: McClearn, G.E., Dietrich, R.A. and Erwin, V.G. (eds.): *Development of Animal Models as Pharmacogenetic Tools. NIAAA Research Monograph No. 6.* U. S. Department of Health and Human Services, Rockville, MD, pp. 171-191.

Li, T.-K., Lumeng, L., Doolittle, D.P. and Carr, L.G. (1991) Molecular associations of alcohol-seeking behavior in rat lines selectively bred for high and low voluntary ethanol drinking. *Alcohol Alcohol. Suppl.* 1: 121-124.

Lisitsyn, N., Lisitsyn, N. and Wigler, M. (1993) Cloning the differences between two complex genomes. *Science* 259: 946-951.

Lisziewicz, J., Sun, D., Metelev, V., Zamecnik P., Gallo, R.C. and Agarwal, S. (1993) Long-term treatment of human immunodeficiency virus-infected cells with antisense oligonucleotide phosphorothioates. *Proc. Natl. Acad. Sci. USA* 90: 3860-3964.

Mardones, R.J. and Onfray, B.E. (1942) Influencia de una substancia de la levadura (elemonto del complejo vitamínico B?) sobre el consumo de alcohol en ratas en experimentos de autoselección. *Rev. Med. Aliment. Chile* (1942) 5: 148-149.

Mardones, J. and Segovia-Riquelme, N. (1983) Thirty-two years of selection of rats for ethanol preference: UChA and UChB strains. *Neurobehav. Toxicol. Teratol.* 5: 171-178.

Michael, S.I., Huand, C.-H, Remer, M.U., Wagner, E., Huii, P.-C. and Curiel, D.T. (1993) Binding-incompetent adenovirus facilitates molecular conjugate-mediated gene transfer by the receptor-mediated endocytosis pathway. *J. Biol. Chem.* 268: 6866-6869.

Mizoi, Y., Ijiri, I., Tatsuno, Y., Kijima, T., Fujiwara, S. and Adachi, J. (1979) Relationship between facial flushing and blood acetaldehyde levels aafter alcohol intake. *Pharmacol. Biochem. Behav.* 10: 303-311.

Murray, J.A.H. (1992) *Antisense RNA and DNA*. A.R. Liss, New York.

Myers, R.M. (1993) The pluses of subtraction. *Science* 259: 942-943.

Rezvani, A.H., Overstreet, D.H. and Janowsky, D.S. (1990) Genetic serotonin deficiency and alcohol preference in the fawn hooded rats. *Alcohol Alcohol.* 25: 573-575.

Schultz, R., Wuster, M., Duka, T. and Herz, A. (1980) Acute and chronic ethanol treatment changes endrophin levels in brain and pituitary. *Psychopharmacol.* 68: 221-227.

Schwitters, S.Y., Johnson, R.C., Johnson, S.B. and Ahern, F.M. (1982) Familial resemblances in flushing following alcohol use. *Behav. Genet.* 12: 349-352.

Seizinger, B.R., Bovermann, K., Maysinger, D., Hollt, V. and Herz, A. (1983) Differential effects of acute and chronic ethanol treatment on particular opioid peptide systems in discrete regions of rat brain and pituitary. *Pharmacol. Biochem. Behav.* 18: 361-369.

Sinclair, J.D. and Lindros, K.O. (1981) Suppression of alcohol drinking with brain aldehyde dehydrogenase inhibition. *Pharmacol. Biochem. Behav.* 14: 377-383.

Thomasson, H.R., Crabb, D.W., Edenberg, H.J. and Li, T.-K. (1993) Alcohol and aldehyde dehydrogenase polymorphisms and alcoholism. *Behav. Genet.* 23: 131-136.

Wu, G.T. and Wu, C.H. (1987) Receptor-mediated *in vitro* gene transformation by a soluble DNA carrier system. *J. Biol. Chem.* 262: 4429-4432.

Yoshida, A., Huang, I.-Y. and Ikawa, M. (1984) Molecular abnormality of an inactive aldehyde dehydrogenase variant commonly found in Orientals. *Proc. Natl. Acad. Sci. USA* 81: 258-261.

Toward a Molecular Basis of Alcohol Use and Abuse
ed. by B. Jansson, H. Jörnvall, U. Rydberg, L. Terenius & B. L. Vallee
© 1994 Birkhäuser Verlag Basel/Switzerland

Outlook: Prospects for alcoholism treatment

Enoch Gordis

National Institute on Alcohol Abuse and Alcoholism, 5600 Fishers Lane, Room 16-105, Rockville, MD 20857 USA

Summary

Treating alcoholism reduces many alcohol-related social, economic, and medical problems. A historical lack of support for alcoholism treatment and for alcoholism treatment research is disappearing, and significant progress is being made toward developing new and improved treatments. Current research findings are the basis of this "Outlook" which is divided into three levels of confidence: clear extrapolations from present research, extrapolations that are reasonable but uncertain, and predictions that attempt to guess the unguessable. It concludes that the future looks promising for new and improved alcoholism treatments.

Introduction

The topic of this paper — the outlook for alcoholism treatment — is an important one. Untreated alcoholism results a variety of social, economic, and medical consequences. For example, a study done at Johns Hopkins University Hospital (Moore *et al.*, 1989) found that 20-40 percent of the beds at many of the medical services were occupied by people who were being treated for the complications of drinking. Patients such as these will make repeated visits to the hospital for the complications of their drinking if their alcoholism remains untreated.

To provide an "Outlook" for alcoholism treatment is a kind of prophecy. A mundane kind of prophecy simply extrapolates what is known from what is likely to happen in the near future. But even the best prophet cannot always predict the surprises. If the "Think Tanks" fashionable in the United States 20 years ago had made their "50-year" predictions in the 1920's, they would have missed a few "minor details" like the discovery of penicillin, the fission of uranium, and the identification of DNA. Because there is no reason to think that nature has yet run out of surprises, the "Outlook" presented in this paper admits that the unexpected is likely.

The discussion is organized as follows. First is a general discussion of alcoholism and alcoholism treatment. Second is my "Outlook" divided according to three levels of confidence: predictions that are clear extrapolations from present knowledge; predictions that are reasonable but not certain; and predictions that attempt to guess the unguessable. A concluding statement follows. Although the discussion is generally applicable to alcohol research and treatment worldwide, much of it draws on our experiences in the U.S.

Alcoholism and alcoholism treatment

Alcoholism treatment has been influenced by its development outside of science and medicine and by public attitudes toward alcoholism. Alcoholism treatment in the U.S. (and elsewhere) grew principally from self-help groups such as Alcoholics Anonymous (AA). These nonmedical "roots" have led to problems. First, because treatment was provided largely by a cadre of recovered individuals many of whom had recovered through AA, AA-like treatment was seen as the only way to treat alcoholism. Second, because everything was supposedly known about how to treat alcoholism, alcoholism research was considered unnecessary. Public attitudes about alcoholism, which have vacillated between viewing alcoholism as either a sin or as an illness, have resulted in a curious dichotomy—public acknowledgement of alcoholism as a major public health problem but only limited public support for alcoholism research, prevention and treatment.

Many of these problems are disappearing. For example, there is an increasing willingness to examine the effectiveness of alcoholism treatment. A striking example of this change comes from AA itself which 3 years ago published the results of a survey showing, among other things, attrition in attendance at AA meetings after initial contact (Alcoholics Anonymous 1990). AA is not a formal research group, and 10 or 20 years ago such a self-analysis would have been unthinkable. Health care financing issues also have served to increase support for alcoholism research. "Managed care" and other aspects of today's health care policy formulation demand good data; we all must now prove that what is being paid for works. Research has already played an important role in the U.S. health care policy debate; without the cost-benefit evidence provided by contemporary clinical research, it is unlikely that treatment for alcoholism would be covered under a proposed new Federal health care financing system.

Alcoholism treatment research seeks to understand why some individuals progress to alcoholism and to develop effective therapies for those who do. The progression of individuals from drinking without problems to alcoholism is commonly referred to as the "natural history of alcoholism." Different symptoms occur at each stage and different interventions are necessary to bring about positive results. The first stage is a period of development. Intervention at this stage involves understanding a number of social and regulatory issues dealing with individual exposure to alcohol, environmental pressures that influence drinking, and vulnerability to dependence, particularly how genes and the environment interact to increase vulnerability. Movement to the second stage—early dependence—happens when an individual experiences psychological dependence with habituation. Individuals in this stage place an increasing priority on drinking. From this population emerges a smaller fraction of people who show symptoms of dependence: withdrawal, tolerance, impaired control (inability to control drinking on any given occasion), and craving during the abstinent state, the most subtle and elusive of the dependence symptoms.

Alcoholism treatment has three main elements: identifying alcoholic patients, safely conducting patients through withdrawal, and managing long-term patient recovery including modifying patient drinking behavior and supporting continued sobriety. Identifiying patients with alcohol problems has been aided by a growing body of scientifically validated screening, assessment, and diagnostic instruments. Safe conduct through withdrawal is not difficult. However, there are some important research findings in this area that will be addressed below. In principle, there are two types of therapies that could be used in the long-term management of alcoholism; verbal therapy, a combination of teaching, persuasion, and coercion, and pharmacological therapy to reduce craving. Most of our current therapies are verbal. They include "twelve-step" approaches (based on the AA model) and behavior modification, which teaches patients how to avoid situations that precipitate drinking. New pharmacological therapies in the form of medications to weaken craving show promise. (Medications such as disulfiram [antabuse] also are used in alcoholism treatment. However, they belong more appropriately to the verbal therapies category because they work by threatening an adverse reaction to drinking rather than by direct pharmacologic action to reduce craving). Verbal therapies have a precise target that we wish to modify—drinking—but they are not efficient. Pharmacological therapies are powerful, especially in psychiatric treatment, but they are not precise because they affect multiple body systems and may produce unintended side effects. The study of receptor subtypes and their varied distributions in the brain may help us to develop medications that act on precise targets in the brain for treating craving and other alcohol dependence symptoms. But whether the study of individual receptors and the ligands that bind to them will be sufficient to achieve this goal is still a matter of conjecture.

Outlook: Clear extrapolations from present research

Clinical trials using controls, blinding, and randomization, while long the standard in other areas of medicine, are relatively new in alcoholism treatment. Excellent multi-site studies of disulfiram (Fuller *et al.*, 1986) and lithium (Dorus *et al.*, 1989) have been cited as models for randomized trials of alcoholism treatments. Currently, The U.S. National Institute on Alcohol Abuse and Alcoholism (NIAAA) is conducting a major multi-site clinical trial, "Project MATCH," a controlled study of 1700 patients randomized to three categories of verbal therapy. The purpose of this study is to determine whether matching specific patients to specific therapies improves treatment outcome. It is expected that the results of "Project MATCH" will help alcohol treatment providers match patients to treatments based on specific patient characteristics rather than the current practice of offering patients a smorgasbord of treatment approaches. Other advances in alcoholism treatment research tools and techniques include new standardized assessment tools, such as the Clinical Institute Withdrawal Assessment (CIWA) (Sullivan *et al.*, 1989) which permits the comparison of withdrawal from one setting to another and from one patient to another, and well-designed studies in the literature comparing treatment settings. For example, an excellent study by Walsh and Hingson (1991) of treatment alternatives offered by an American industry indicated that for a certain class of people who are socially well-supported and with their jobs at stake, inpatient treatment prevented relapse to a greater extent than did outpatient treatment or AA alone. In this study, the costs of inpatient treatment, although higher initially than the cost of outpatient treatment, were largely offset by less frequent hospital readmissions among the inpatient group compared with those who were treated as outpatients from the start.

Early intervention is going to become more important as emphasis, driven in part by economic considerations, is placed on identifying and treating early stage alcohol problem. Another question that may be settled is the long-standing controversy about the relative merits of inpatient treatment compared with outpatient treatment. There has been a general reduction in the importance of inpatient therapy in the U.S., partly in reaction to the abuse of higher reimbursements available for inpatient services. There is a danger however, that in our over-zealous attempts to redress the abuses of inpatient therapy, patients who truly need it will be denied it. Evidence from studies such as the Walsh and Hingson study mentioned above should allow us to better determine which patients are likely to fare better in which settings.

Near-term progress in the development of markers of alcohol consumption should be rapid.

What is needed is a marker in the blood that integrates the value of blood alcohol over a period of several weeks even if on the day of sampling the patient is abstinent. Carbohydrate deficient transferrin (CDT) and 5-hydroxytryptophol are among several substances that have been examined as possible markers (Carlsson *et al.*, 1993). CDT has thus far shown the most promise, with commercial CDT measurement kits under development for clinical use. The CDT test is quite specific but only moderately sensitive. Thus, CDT may be more useful as a marker in followup than in diagnosis, but this matter should be settled soon.

Finally, there is the promise of pharmacotherapy. Two recent studies (Volpicelli *et al.*, 1992, O'Malley *et al.*, 1992) have shown that the use of the opiate antagonist naltrexone in combination with verbal therapy prevented relapse more than standard verbal therapy alone. This coupling of verbal with pharmacological approaches is a new direction in the alcohol field and one that may produce profound changes in the way we view and treat alcoholism. Other pharmacotherapies under investigation include buspirone, an anti-anxiety partial serotonin 5-HT$_{1a}$ agonist (Kranzler *et al.*, 1989; Bruno, 1989; Tollefson *et al.*, 1992; Malcolm *et al.*, 1992) and the opiate antagonist nalmefene (Yeomans *et al.*, 1990).

Outlook: Reasonable but uncertain

Now we move from the "just down the road" predictions to those that are somewhat less certain. One of these concerns the practice implications of research on treating withdrawal. In the U.S. there has been a movement toward treating withdrawal without medication in nonmedical facilities to save money. In most cases, any individual episode of withdrawal can be safely seen through in a nonmedical setting. However, there is evidence showing that repeated mild withdrawals produce a "kindling effect" which can eventually lead to a severe withdrawal episode (Ballenger and Post, 1978). Repeated withdrawal also appears to threaten the integrity of the hippocampus due to secretion of cortisol during withdrawal (Sapolsky *et al.*, 1986) and possibly excess activity at the N-methyl-D-aspartate (NMDA) receptor. Based on this evidence, the use of medications to manage withdrawal appears indicated. Whether these findings will outweigh economic considerations and shift the balance toward pharmacological (possibly inpatient) treatment of withdrawal is not yet clear.

The appetitive aspects of alcoholism also are receiving renewed attention. Although much of the current discussion in alcoholism has to do with the affective side of the disease, there are a variety of issues that stem from the fact that alcohol is a food as well as a drug. For example, are there common appetitive mechanisms involved in ordinary eating and alcohol ingestion? Most people are able to regulate their food intake, and most are able to regulate their alcohol intake. We do not yet understand why some individuials become compulsive eaters or alcoholics and whether the appetitive mechanism involved in regulating food intake is the same for alcohol. Insights in this area of alcohol research might render moot the question of whether abstinence is the only appropriate treatment outcome; if treatment can relocate alcohol in a person's priorities to a more normal position, the focus on abstinence as a treatment goal may no longer be needed.

Outlook: Guessing the unguessable

Finally, the most difficult "Outlook"—guessing the unguessable—will touch on three main topics: the impact of genetics on alcoholism treatment; reconciliation through understanding of disparate viewpoints; and new typologies based on new understanding.

The alcohol field is moving from population genetics into molecular biology. Under a major NIAAA genetics study entitled the "Consortium on the Genetics of Alcoholism," (COGA) six centers are involved in a variety of family genetics and molecular biology studies to determine those genes that account for that portion of the vulnerability to alcoholism that is inherited. This study, which involves the use of candidate genes and studies of the whole genome, is in its fourth year and will continue for another six or seven years. We do not know which genes will be determined to play a role in alcoholism or what role(s) they play. They may, for example, code for receptor subtypes or enzymes that synthesize their ligands. Perhaps, these will be new kinds of genes that control higher functions in the brain and neural networks. This area is a good candidate for nature's bag of surprises, and future discoveries could be quite remarkable.

The alcohol field is one of the leaders in animal genetics. Animals have been bred over many generations for a variety of alcohol-related characteristics including withdrawal seizures, sensitivity to hypothermia, and alcohol preference. The question of how close an animal model has to be to human alcoholism before one can learn something valuable from its genetic structure is unresolved. It is hoped that genetic material from these animals will help us to better understand the physiological aspects of the disease and pave the way for new treatments such as gene therapy or the use of antisense oligonucleotides. This has been called "genomic pharmacology." Even the brain is a potential future target. For example, in one study (Wahlestedt et al., 1993), an antisense oligonucleotide to the NMDA receptor injected directly into the brain of rats reduced excitotoxicity and infarct size after cerebral artery occlusion. Obviously, even if research suggests appropriate targets for antisense oligonucleotides in alcoholism, there are many difficult questions that must be answered concerning the site of administration, safety, and efficacy of any such potential therapy.

Future alcohol research progress depends on the reconciliation of disparate viewpoints. There are four issues that I believe will require attention in this regard: reductionism as the sole way of looking at science, neural processing of multiple characteristics of alcoholism, craving, and verbal-pharmacological therapy interaction.

Reductionism is the view that appears to be shared by some that all human functions will be predictable once the human genome is completely known. This viewpoint is certainly arguable. I do not believe that neuroscience and physiology are temporary sciences that will become obsolete as soon as we know the whole genome.

With regard to neural processing, recall that many different mechanisms may be at work in alcoholism: affect and reward; taste, calories, and energy balance; cognition and learning; and tolerance and craving. These mechanisms may be processed in a manner similar to how the brain processes visual information—simultaneous processing by the brain of separate attributes of observed objects (e.g. color, shape) and eventual integration of these attributes into a complete picture. Alcoholism might be the result of similar processing—possibly a disorder not of one neurotransmitter but of simultaneous networks that do not operate in proper balance. The future study of different mechanisms and how they might individually and collectively relate to alcoholism will help shed light on this area.

The study of craving is difficult. The literature suggests at least three possible kinds of craving. The first is unadorned hunger, a desire to eat or to drink not based on cues and conditioning (except possible internal cues learned during infancy). An example from obesity is that of a 300-pound person who reduced to 185 pounds compared to a person whose has weighed 185 pounds his entire life. The formerly obese person, even without external cues, is continuously hungry whereas the always 185-pound person is not. Another kind of craving relates to cues and conditioning (Marlatt and Gordon, 1980). In this concept, the environment can trigger the discomfort of withdrawal and resumption of drinking behavior. Last is a theory by Steven Tiffany (1990) which suggests that drinking behavior is a totally automated process like driving home but remembering nothing of the trip. Tiffany believes that drinking itself is automated, but the control of the drinking is not; it is the cognitive effort to resist alcohol that is felt as craving. These three views of craving are different but might occur in the same person. Future research may be able to resolve if these are three aspects of craving, and if so, who is at risk for which type or types.

Verbal and pharmacological therapies might work together in three ways. One way is that one therapy might continue to function if the other failed. This resembles treatments using two antibiotics where one medication can serve as backup if resistance to another develops. A second way is that each therapy might increase the efficacy of the other. For example, verbal therapy may enhance compliance with pharmacological therapy which in turn reduces craving, allowing the patient's more complete attention to the verbal therapy. A third way illustrates the possibility that verbal and pharmacological therapy are not as radically different as they seem; they might act on the same neural circuits. For example, a recent study of obsessive-compulsive disorder (Baxter *et al.*, 1992) compared positron emission tomography (PET) scans of patients before and after behavioral treatment with PET scans of patients before and after pharmacological treatment. In this study, changes in glucose utilization in the head of the caudate nucleus in patients who had responded to the behavioral treatment resembled changes seen in patients who had responded to the pharmacological treatment. Even though this study has limitations, these observations are striking because they indicate that successful verbal therapies and pharmacological therapies may work on the same system.

Lastly, I believe that new typologies of alcoholism will be developed based on understanding alcohol's psycho-physiological actions, the different kinds of craving and ways to prevent it, appetitive control, the neural networks that are involved in the development of alcoholism, and the findings of genetics. Finally, a caution. When we understand essential issues in alcoholism, the disease might look more unified than is now the case. Sometimes the appearance of heterogeneity reflects incomplete understanding. A good example of this is tuberculosis where the fact that the disease presents different clinical pictures became secondary to the fact that one bacillus was responsible for all of them.

Conclusion

The alcohol field has matured splendidly in the last 15 years. Alcoholism research has the potential to remove stigma, find new treatments, and develop for alcoholism treatment the economy and precision necessary for a modern health care system.

Acknowledgements

The author wishes to acknowledge the following persons for their help in providing material for this paper: Dr. Joseph Volpicelli, Dr. Stephanie O'Malley, Dr. Sam Zakari, Dr. Raye Litton, and Dr. Markku Linnoila. The author also wishes to acknowledge Ms. Brenda G. Hewitt for her invaluable assistance in preparing this manuscript.

References

Alcoholics Anonymous (1990) 1989 Membership Survey, A.A. World Services, New York, New York, Appendix C, p.12.

Ballenger, J.C. and Post, R.M. (1978) Kindling as a model for alcohol withdrawal syndromes. *Br. J. Addict.* 133: 1-14.

Baxter, L.R., Schwartz, J.M., Bergman, K.S., Szuba, M.P., Guze, B.H., Mazziotta, J.C., Alazarki, A. Selin, C.E., Ferng, H.-K., Munford, P. and Phelps, M.E. (1992) Caudate glucose metabolic rate changes with both drug and behavior therapy for obsessive-compulsive disorder. *Arch. Gen. Psychiatry* 49: 681-689.

Bruno, F. (1989) Buspirone in the treatment of alcoholic patients. *Psychopathology* 22: 49-59.

Carlsson, A.V., Hiltunen, A.J., Beck, O., Stibler, H. and Borg, S. (1993) Detection of relapses in alcohol-dependent patients: Comparison of carbohydrate-deficient transferrin in serum, 5-hydroxytryptophon in urine, and self-reports. *Alcoholism: Clin. Exp. Res.* 17: 703-708.

Dorus, W., Ostrow, D.G., Anton, R., Cushman, P., Collins, J.F., Schaefer, M., Charles, H.L., Desai, P., Hayashidam, M., Malkerneker, U., Willenbring, M., Fiscella, R. and Sather, M.R. (1989) Lithium treatment of depressed and nondepressed alcoholics. *J. Am. Med. Assoc.* 262: 1646-1652.

Fuller, R.K., Branchey, L, Brightwell, D.R., Derman, R.M., Emrick, D.D., Iber, F.L., James, K.E., Lacoursiere, R.B., Lee, K.K., Lowenstam, I., Maany, I., Neiderhiser, D., Nocks, J.J. and Shaw, S. (1986) Disulfiram treatment of alcoholism: A Veterans Administration Cooperative Study. *J. Am. Med. Assoc.* 256: 1449-1455.

Kranzler, H.R. and Meyer, R.E. (1989) An open trial of buspirone in alcoholics. *J. Clin. Psychopharmacol.* 9: 379-380.

Malcolm, R. Anton, F.R., Randall, C.L., Johnston, A. Brady, K. and Thevos, A. (1992) A placebo-controlled trial of buspirone in anxious inpatient alcoholics. *Alcoholism: Clin. Exp. Res.* 16: 1007-1013.

Marlatt, G.A. and Gordon, J.R. (1980) Determinants of relapse: Implications of the maintenance of behavior change. In: Davidson, P. and Davidson, S.M. (eds.): *Behavioral Medicine: Changing Health Lifestyles.* New York, Brunner/Mazel, pp. 410-452.

Moore, R.D., Bone, L.R., Geller, G., Mamon, J.A., Stokes, E.J. and Levine, D.M. (1989) Prevalence, detection, and treatment of alcoholism in hospitalized patients. *J. Am. Med. Assoc.* 261: 403-407.

O'Malley, S.S., Jaffe, A.J., Chang, G., Schottenfeld, R.S., Meyer, R.E. and Rounsaville, B. (1992) Naltrexone and coping skills therapy for alcohol dependence: A controlled study. *Arch. Gen. Psychiatry* 49: 881-887.

Sapolsky, R.M, Krey, L.C. and McEwen, B.S. (1986) The neuroendocrinology of stress and aging: The glucocorticoid cascade hypothesis. *Endocrine Rev.* 7: 284-301.

Sullivan, J.T., Sykora, K., Schneiderman, J., Naranjo, C.A. and Sellers, E.M. (1989) Assessment of alcohol withdrawal: The revised clinical institute withdrawal assessment for alcohol scale (CIWA-AR). *Br. J. Addic.* 84: 1353-1357.

Tiffany, S.T. (1990) A cognitive model of drug urges and drug-use behavior: Role of automatic and nonautomatic processes. *Psychol. Rev.* 97: 147-168.

Tollefson, G.D., Montague-Clouse, J. and Tollefson, S.L. (1992) Treatment of comorbid generalized anxiety in a recently detoxified alcoholic population with a selective serotonergic drug (buspirone). *J. Clin. Psychopharm.* 12: 19-26.

Volpicelli, J.R., Alterman, A.I., Hayashida, M. and O'Brien, C.P. (1992) Naltrexone in the treatment of alcoholdependence. *Arch. Gen. Psychiatry* 49: 876-880.

Wahlestedt, C., Golanov, E., Yamamoto, S., Yee, F., Ericson, H., Yoo, H., Inturrisi, C.E. and Reis, D.J. (1993) Antisense oligodeoxynucleotides to NMDA-R1 receptor channel protect cortical neurons from excitotoxicity and reduce focal ischaemic infarctions. *Nature* 363: 260-263.

Walsh, D.C., Hingson, R.W., Merrigan, D.M., Levenson, S.M., Cupples, L.A., Heeren, T., Coffman, G.A., Becker, C.A., Barker, T.A., Hamilton, S.K., McGuire, T.G. and Kelly, C.A. (1991) A randomized trail of treatment options for alcohol-abusing workers. *N. Engl. J. Med.* 325: 775-782.

Yeomans, M.R., Wrigth, P., Macleod, H.A. and Critchley, J.H. (1990) Effects of nalmefene on feeding in humans. *Psychopharmacology* 100: 426-432.

Subject index

(The page numbers refer to the first page of the article in which the keyword occurs)

D.M. Lovinger, *Vanderbilt University, Nashville, TN, USA*
T.V. Dunwiddie, *University of Colorado Health Science Center, Denver, CO, USA*

Presynaptic Receptors
in the
Mammalian Brain

1994. Approx. 272 pages. Approx. 25 figs. Hardcover • ISBN 3-7643-3651-X

Neuroscientists and researchers interested in neurotransmission and brain function will find this volume fascinating reading. Unlike other volumes in this field have dealt primarily with the peripheral nervous system, this book focuses on synaptic modulation in the central nervous system. In addition, this volume emphasizes studies using functional (primarily electrophysiological, rather than biochemical) measures of transmitter release to study the role of release-regulating receptors in the control of normal physiological function at synapses. This gives unique insights into the importance of release modulation in neural function.

The electrophysiological approach has numerous advantages over the biochemical approach. For example, the involvement of presynaptic K^+ and Ca^{++} channels in modulation can be examined more directly, and in some cases the specific channel type that is affected by a neuromodulator can be identified by its physiological characteristics. Quantal analysis of transmitter release at single synapses also has the potential of providing entirely different kinds of insights into modulation. In addition, the role played by presynaptic receptors in physiological processes such as long-term potentiation has only been explored using electrophysiological techniques.

Please order through your bookseller or directly from:
Birkhäuser Verlag AG
P.O. Box 133
CH-4010 Basel / Switzerland
Fax ++41 / 61 721 79 50
Orders from the USA or Canada should be sent to:
Birkhäuser Boston
44 Hartz Way, Secaucus, NJ 07096-2491 / USA
Call Toll-Free 1-800-777-4643

For more information on recent and forthcoming books and journals you can order the Birkhäuser Life Sciences Bulletin, published twice a year and free of charge.

"The assembly in one volume of the contemporaneous views of the most active investigators in this field will probably come to be seen in the future as a landmark record..."

—from the Introductory Remarks by Herman N. Eisen

Cytotoxic Cells:
Recognition, Effector Function,
Generation, and Methods

Edited by
M. Sitkovsky and **P. Henkart**
National Institutes of Health, Bethesda, Maryland, USA

1993. 544 pages. Hardcover. ISBN 3-7643-3608-0

This collection of papers from the most important researches of cytoxic cells will make an excellent introduction to the study of cytotoxic T lymphocytes and other cytotoxic cells, including CTL, NK, LAK, TIL, ADCC, macrophages, mast cells, and platelets.

These topics are covered comprehensiviely, including generation, recognition, effector functions, and important methodologies. It will provide a state-of-the-art review of this important field. Special chapters cover the mechanisms of lethal hit delivery and immunopharmacological manipulations of cytotoxic cells which will be of interest to pharmacological researchers as well as cancer specialists. Adoptive immunotherapy is covered by experts in each field. The book is divided into brief, very readable chapters by experts in each field.

Leading the way as the first comprehensive work available in the field of Cytotoxic Cells and Cytotoxicity assays, this outstanding collection of papers by internationally renowned immunologists will serve as the reference source for immunology and cell biology laboratories, as well as allied fields of research.

Cytotoxic Cells provides an essential collection of methodologies, which are invaluable to every immunologist and cell biologist studying cellular regulation. The historical, molecular, cell biological, and clinical aspects of cell-mediated cytotoxicity are thoroughly covered by the leading researchers in their respective fields.

Over 50 chapters cover the following topics: • Introduction and Overview • Target Cell Recognition • Generation of Cytotoxic Cells • Molecular Mechanisms of Cellular Cytotoxicity • Granule Proteases • Alternate Mechanisms of Cytolysis • Biochemical and Immunopharmaceutical Manipulation of Cytotoxic Cells • Functions of Cytotoxic Cells In Vivo • Macrophage-Mediated Cytotoxicity • Methods

This is an invaluable introduction to the field for students, as well as a reference tool for practicing researchers, and a must for every laboratory bookshelf.

Please order through your bookseller or directly from:
Birkhäuser Verlag AG
P.O. Box 133
CH-4010 Basel / Switzerland
Fax ++41 / 61 721 79 50
Orders from the USA or Canada should be sent to:
Birkhäuser Boston
44 Hartz Way, Secaucus, NJ 07096-2491 / USA
Call Toll-Free 1-800-777-4643

For more information on recent and forthcoming books and journals you can order the Birkhäuser Life Sciences Bulletin, published twice a year and free of charge.